Frontiers of Astrobiology

Astrobiology is an exciting interdisciplinary field that seeks to answer one of the most important and profound questions: Are we alone?

In this volume, leading international experts explore the frontiers of astrobiology, investigating the latest research questions that will fascinate a wide interdisciplinary audience at all levels. What is the earliest evidence for life on Earth? Where are the most likely sites for life in the Solar System? Could life have evolved elsewhere in the Galaxy? What are the best strategies for detecting intelligent extraterrestrial life? How many habitable or Earth-like exoplanets are there?

Progress in astrobiology over the past decade has been rapid and, with evidence accumulating that Mars once hosted standing bodies of liquid water, the discovery of over 700 exoplanets, and new insights into how life began on Earth, the scientific search for our origins and place in the cosmos continues. The book is based on a meeting at the Pontifical Academy of Sciences, which gathered leading researchers to present state-of-the-art reviews on their research and address topics at the forefront of astrobiology.

CHRIS IMPEY is a University Distinguished Professor and Deputy Head of the Department of Astronomy at the University of Arizona, Tucson. His research interests are observational cosmology, gravitational lensing, and the evolution and structure of galaxies. He is a past Vice President of the American Astronomical Society, a Fellow of the American Association for the Advancement of Science, and has been NSF Distinguished Teaching Scholar, Phi Beta Kappa Visiting Scholar, and the Carnegie Council on Teaching's Arizona Professor of the Year.

JONATHAN LUNINE is the David C. Duncan Professor in the Physical Sciences in the Department of Astronomy at Cornell University. His research interests center broadly on planetary origin and evolution, in our Solar System and around other stars. He is an interdisciplinary scientist on the Cassini mission to Saturn, and on the James Webb Telescope, as well as on the Juno mission to Jupiter. He is the author of over 250 papers and is a member of the US National Academy of Sciences, as well as a Fellow of the American Association for the Advancement of Science.

JOSÉ FUNES is a Jesuit priest, Director of the Vatican Observatory, and member of the Pontifical Academy of Sciences. His field of research includes kinematics and dynamics of disk galaxies and star formation in the local universe. He proposed the Study Week on Astrobiology that was organized by the Pontifical Academy of Science and held in Casina Pio IV of the Vatican City State in November 2009, which led to this volume.

Frontiers of Astrobiology

Edited by

Chris Impey
University of Arizona

Jonathan Lunine
Cornell University

José Funes
Vatican Observatory, Vatican City

CAMBRIDGE
UNIVERSITY PRESS

University Printing House, Cambridge CB2 8BS, United Kingdom

One Liberty Plaza, 20th Floor, New York, NY 10006, USA

477 Williamstown Road, Port Melbourne, VIC 3207, Australia

314-321, 3rd Floor, Plot 3, Splendor Forum, Jasola District Centre, New Delhi - 110025, India

79 Anson Road, #06-04/06, Singapore 079906

Cambridge University Press is part of the University of Cambridge.

It furthers the University's mission by disseminating knowledge in the pursuit of education, learning and research at the highest international levels of excellence.

www.cambridge.org
Information on this title: www.cambridge.org/9781107006416

First published 2012

A catalogue record for this publication is available from the British Library

Library of Congress Cataloging in Publication data
Frontiers of astrobiology / edited by Chris Impey, Jonathan Lunine, José Funes.
  pages cm
Includes bibliographical references and index.
ISBN 978-1-107-00641-6
1. Exobiology. I. Impey, Chris. II. Lunine, Jonathan Irving. III. Funes, José G.
QH326.F76   2012
576.8′39 – dc23   2012012599

ISBN 978-1-107-00641-6 Hardback

Contents

List of contributors   *page* vii

Part I Introduction   1

Inauguration of the Study Week on Astrobiology   3
CARDINAL GIOVANNI LAJOLO

1 Astrobiology – A New Synthesis   5
JOHN BAROSS AND CHRIS IMPEY

Part II Origin of Planets and Life   23

2 Towards a Theory of Life   25
STEVEN BENNER AND PAUL DAVIES

3 Terran Metabolism: The First Billion Years   48
SHELLEY COPLEY AND ROGER SUMMONS

4 Planet Formation   73
SEAN N. RAYMOND AND WILLY BENZ

Part III History of Life on Earth   87

5 The Early Earth   89
FRANCES WESTALL

6 Evolution of a Habitable Planet   115
JAMES KASTING AND JOSEPH KIRSCHVINK

7 Our Evolving Planet: From Dark Ages to Evolutionary
   Renaissance   132
   ERIC GAIDOS AND ANDREW KNOLL

   Part IV Habitability of the Solar System   155

8 Early Mars – Cradle or Cauldron?   157
   ARMANDO AZUA-BUSTOS, RAFAEL VICUÑA, AND RAYMOND PIERREHUMBERT

9 Large Habitable Moons: Titan and Europa   175
   ATHENA COUSTENIS AND MICHEL BLANC

10 Small Habitable Worlds   201
   JULIE CASTILLO-ROGEZ AND JONATHAN LUNINE

   Part V Exoplanets and Life in the Galaxy   229

11 Searches for Habitable Exoplanets   231
   SARA SEAGER

12 Review of Known Exoplanets   250
   CHRISTOPHE LOVIS AND DANTE MINNITI

13 Characterizing Exoplanet Atmospheres   266
   GIOVANNA TINETTI

14 If You Want to Talk to ET, You Must First Find ET   286
   JILL TARTER AND CHRIS IMPEY

   Index   306

See colour plate section between pp. 148 and 149.

Contributors

Armando Azua-Bustos
Departamento de Genética Molecular y Microbiología
Facultad de Ciencias Biológicas
Pontificia Universidad Católica de Chile
Alameda 340, Santiago
Chile

John Baross
School of Oceanography
University of Washington
Seattle, WA 98195
USA

Steven Benner
Foundation for Applied Molecular Evolution
PO Box 13174
Gainesville, FL 32604
USA

Willy Benz
Physics Institute
University of Bern
Sidlerstrasse 5
CH-3012 Bern
Switzerland

Michel Blanc
École Polytechnique
91128 Palaiseau Cedex
France

Julie Castillo-Rogez

Jet Propulsion Laboratory

4800 Oak Grove Dr.

Pasadena, CA 9110

USA

Shelley Copley

University of Colorado at Boulder

CIRES

216 UCB

Boulder, CO 80309–0216

USA

Athena Coustenis

ESIA

Observatoire de Paris – Section de Meudon

5 Place Jules Janssen

92195 Meudon Cedex

France

Paul Davies

Beyond Institute

Room 2476, ASU Fulton Center

300 E. University Dr.

PO Box 876505

Tempe, AZ 85281

USA

José G. Funes, S.J.

Specola Vaticana

V-00120

Vatican City State

Eric Gaidos

Department of Geology and Geophysics

University of Hawaii at Manoa

1680 East-West Road POST 710

Honolulu, HI 96822

USA

Chris Impey
University of Arizona
Steward Observatory
Tucson, AZ 85721
USA

James Kasting
Department of Geoscience
443 Deike Building
Penn State University Park, PA 16802
USA

Joseph Kirschvink
Division of Geological and Planetary Sciences
California Institute of Technology
Pasadena CA 91125
USA

Andrew Knoll
Department of Earth and Planetary Sciences
20 Oxford Street
Cambridge MA 02138
USA

Cardinal Giovanni Lajolo
V-00120
Vatican City State

Christophe Lovis
Observatoire Astronomique de l'Université de Genève
51 Chemin des Maillettes
CH-1290 Sauverny
Switzerland

Jonathan Lunine
402 Space Sciences Building
Cornell University
Ithaca, NY 14853
USA

Dante Minniti
Departamento de Astronomía y Astrofísica
Pontificia Universidad Católica de Chile
Vicuña Mackenna 4860, 7820436 Macul
Santiago
Chile

Raymond Pierrehumbert
Department of Geophysical Sciences
University of Chicago
5734 S. Ellis Ave.
Chicago, IL 60637
USA

Sean N. Raymond
Laboratoire d'Astrophysique de Bordeaux
2 Rue de l'Observatoire
33270 Floirac
France

Sara Seager
Massachusetts Institute of Technology,
77 Massachusetts Ave.
54–1626,
Cambridge, MA 02139
USA

Roger Summons
Department of Earth, Atmospheric and Planetary Sciences
Massachusetts Institute of Technology
77 Massachusetts Ave.
54–1626
Cambridge, MA 02139–4307
USA

Jill Tarter
SETI Institute
515 N. Whisman Road
Mountain View, CA 94043
USA

Giovanna Tinetti
Department of Physics and Astronomy
University College London
Gower Street
London
WC1E 6BT
United Kingdom

Rafael Vicuña
Departamento de Genética Molecular y Microbiología
Facultad de Ciencias Biológicas
Pontificia Universidad Católica de Chile
Alameda 340, Santiago
Chile

Frances Westall
Centre de biophysique moléculaire
CNRS
Rue Charles-Sadron
F45071 Orleans Cedex 2
France

PART I INTRODUCTION

The volume you hold in your hands is the result not of a workshop or a conference, but of a "Study Week" in astrobiology. The distinction is important, because the term connotes a definite period of time in which participants, literally cloistered in the Pontifical Academy of Sciences within the walled Vatican City, confronted each other's research and together gained new insights by studying the connections. The usual bustle of a scientific conference was absent, as was the distraction of a dramatically seductive natural wonder (beach, mountains, forest) that seems obligatory at scientific workshops. In their place was the classical serenity of the Casina Pio IV, the unobtrusive and continuous maintenance of food and comfort by our gracious Vatican hosts, and a rich interchange of ideas that flowed effortlessly among the participants, making the week seem almost tragically too short. (In the interests of full disclosure, there were memorable visits to the remarkable town of Assisi and to the Vatican Museum, hosted again by the Pontifical Academy). Under such wonderful conditions, we all learned volumes from each other, so that the chapters you are about to read will reflect, we hope, a broader vision than might have been expounded before the Study Week.

As for the subject matter of this book, it speaks for itself. The usual introduction to a book on astrobiology includes a section, sometimes lengthy, defining the field and explaining its scope, and assuring the reader that it is indeed worthy of their attention. We will eschew that approach here, and say simply that the Study Week explored issues related to one of the most profound questions humans can ask in science: Is there life beyond Earth? If that question provokes your interest, then read on. As editors, we have paired authors of differing disciplines or viewpoints and asked them to write their chapters in a seamless, collaborative style. By common assent, the authors have written chapters for a much broader audience than would normally read a volume of conference proceedings. We hope we have been successful, and readers with an active interest but perhaps a modest background in science will find this book an entrée to a vibrant, interdisciplinary field.

It is a pleasure to express our deepest gratitude to the Pontifical Academy of Sciences for organizing and hosting the Study Week on Astrobiology, and especially to His Excellency Msgr. Marcelo Sánchez Sorondo, its Chancellor. We are very grateful as well for the support and presence of His Eminence Giovanni Cardinal Lajolo, emeritus President of the Pontifical Commission for Vatican City State, whose beautiful words of welcome grace the opening pages of this book. Finally, we are grateful to our colleagues who took time away from their research to come together in Vatican City, to engage in the purest form of scientific exchange, and then to write the chapters comprising this book. For all their insights, we thank them.

CI

JF

JL

Inauguration of the Study Week on Astrobiology

CARDINAL GIOVANNI LAJOLO

It is with great pleasure that I welcome all the distinguished scientists convened here, at the invitation of the Pontifical Academy of Sciences, to discuss a theme that is as new as it is difficult and fascinating: Astrobiology. It is a field which requires a range of all but the most profound of scientific knowledge, as well as highly refined research techniques, and it means often proceeding on the basis of scarce evidence and formulating hypotheses requiring strict verification, which in turn can be diversely configured. It means resorting to results of research based on extreme aspects of the possibility of life on Earth, and to study how to verify its presence on other planets or exoplanets. It means – at its limit – studying if and how one could verify the existence of extraterrestrial forms of intelligence and how to enter in contact with them. This is a task that demands scientific integrity, not to be confused with science fiction.

In your study, which represents, I would say, an intense and indispensable case of vast multidisciplinary research, I don't doubt that you will find yourself accompanied and stimulated by that human atmosphere of collegiality and friendship offered by the Pontifical Academy of Sciences.

In research we should not fear the truth. Only the error, which lies in ambush, can cause us fear. But the scientist must also be allowed the possibility to walk paths which do not always lead to positive results, otherwise it would not be research. Nonetheless, even these types of errors are never useless, precisely because, being led by the scientific method, they help us test other paths. And it is thus that the sciences are able to progress, and just as they open humanity to new knowledge, they contribute to the fulfilment of man as man.

Illustrious ladies and gentlemen, at the beginning of your week of study, I am very glad to offer you all a cordial welcome, good wishes for a successful collaboration, and the Blessing of the Holy Father.

Unfortunately, I am not able to participate personally in your lectures and discussions as I would certainly like, not for any competence I have in the field but merely to open myself to new horizons of knowledge. However, to accompany you on behalf of the Vatican, will be our scientists from the Vatican Observatory guided by their Director Jesuit Father José Gabriel Funes.

We will see each other again in Assisi, as scheduled in the programme prepared by His Excellency Monsignor Sanchez Sorondo, Chancellor of the Pontifical Academy of the Sciences, whom I want to thank for his renowned and always cordial hospitality.

I hope you will remember these days as rich in intellectual gratification and benevolent friendship among your scientific colleagues. I wish you very fruitful work.

1

Astrobiology – A New Synthesis

JOHN BAROSS AND CHRIS IMPEY

Introduction

It is easy to envision our earliest ancestors, hands under chins, staring out into starlit space, wondering about their existence and how they came to be. With infinite possibilities it is no wonder that a myriad of gods could show their face in the galaxies and set the stage for the birth of religions, their theology, a moral code of ethics, a special and exclusive position for humans, and eventually a method of inquiry that has led to modern science. Furthermore, the early history of thought about life elsewhere in the universe was linked to more of a theological perception of man's relationship to gods, and Earth as the exclusive domain of life – exemplified in the thirteenth-century Aristotelian–Thomistic synthesis of science and religion.

It is particularly interesting and germane to astrobiology today, that it was astronomers, and particularly those in the fifteenth to seventeenth centuries, that were the first to make quantitative observations that questioned the orthodox view of man's place in the universe. One looked to the heavens for answers to the most profound philosophical questions and in doing so helped establish the scientific method as another method of inquiry. Perhaps a lesson from this early philosophical-based history is that the drive to grow knowledge is a fundamental and perhaps evolutionary characteristic of humans. The implication is that our survival is likely to be dependent on our exploring all possibilities that help us to understand the how, why, and uniqueness of our existence. Perhaps, no topic inspires this need to explore more than the search for life elsewhere with its implications for the origin of life and the possibility of life forms unrelated to Earth life.

Frontiers of Astrobiology, ed. Chris Impey, Jonathan Lunine and José Funes.
Published by Cambridge University Press. © Cambridge University Press 2012.

There are obvious parallels between our explorations for evidence of life in the universe and Darwin's almost five-year voyage on the *HMS Beagle* that spawned his theory of evolution by natural selection. Like Darwin's *Beagle* voyage, the current astrobiology "voyage" in search for life throughout the universe promises to profoundly alter and expand our notions of life, its origin, and a test of Darwin's (1979) notion that "from so simple a beginning endless forms most beautiful and most wonderful have been, and are being evolved." Moreover, as in Darwin's time when there were broad social and philosophical implications to the theory of evolution, there would be even broader implications if extraterrestrial life were found to exist, a second origin of life were discovered, or life very different from Earth life were found that could exist outside the bounds for life on Earth – and, even more profound, extraterrestrial life that we can communicate with.

In *The Origin of Species*, Darwin (1979) opined about natural selection: "I can see no limit to this power, in slowly and beautifully adapting each form to the most complex relations of life." While most of the diversity of life and particularly microscopic life during Darwin's time was mostly unknown, today we are beginning to appreciate the full scope of Darwin's vision of natural selection in the almost limitless diversity of life forms. There are very few environments on Earth that are incapable of supporting life – only extremes in temperature and the availability of liquid water limit Earth life.

In the search for life on other planets and moons both in our Solar System and beyond, we not only look for evidence of "life" as we know it, we also explore the possible evolutionary adaptation of "life" to conditions outside the bounds of Earth environments that are known to exist on other planetary bodies – what has been described as the search for "weird life." Some of these environmental conditions that might support life on other planetary bodies include liquid organic solvents instead of water, temperatures considerably colder than freezing or hotter than boiling where water might remain liquid due to high salt concentrations, high pressure or presence of organic solvents, or the use of energy sources other than light or chemical – the two sources used by Earth life.

Thus, astrobiology questions some of the fundamental tenets of Earth life including their canonical characteristics: Do we fully understand the evolutionary possibilities of the only life we know, and its limits? Is Earth life the best of all possible life forms? Would Darwinian selection always result in life resembling Earth life? Are there multiple ways to make life, even life as we know it? Like the astronomers of the past, we (as astrobiologists) seek the answers to these and other questions through exploration of the universe.

History and goals

The earliest published use of the word astrobiology is ascribed to Lafleur (1941) who defined it as "the consideration of life in the universe elsewhere than

Earth" (reference and quote from Chyba and Hand 2005). In 1957, a symposium described as "the first American symposium in astrobiology" was held in Flagstaff, Arizona, interestingly, just months before the Soviet Union shocked the United States with its launch of Sputnik 1 (see Smith 2004 for more details). By 1958, the National Academy of Science created the Space Studies Board, and shortly afterwards the National Aeronautics and Space Administration (NASA) was also created. While much of the initial US response to the launch of Sputnik 1 was politically motivated, the potential for biological discoveries from space exploration was not lost on the science and lay communities. Smith (2004) quotes from the book, *Science in Space* (Berkner and Odishaw 1961): "Unquestionably, the possibility that some form of living matter exists on other planets is the most exciting prospect: the origin of life under radically different conditions of environment and ecology is a subject of unprecedented significance to fundamental biology."

Before the formation of the NASA Astrobiology Institute (NAI) in 2004, the word exobiology, ascribed to Joshua Lederberg (1960), was widely used by NASA and the science community to describe the scientific study of life beyond Earth's atmosphere. Exobiology is still in use but generally more focused on studies of the origin and evolution of life.

The motivation behind Lederberg's interest in exobiology was his concern about interstellar contamination, both forward contamination by Earth organisms and back contamination that might include a pathogen to which we had no immunity. Lederberg was one of the founding members on the Space Studies Board established in 1958 and was very influential in formulating the Space Studies Board recommendations that all spacecraft be sterilized before launch and that samples returned from other planets be quarantined until determined to be harmless. NASA maintains these recommendations, presently under the control of a Planetary Protection Officer.

The Viking mission to Mars in 1976 was the most ambitious and expensive NASA program to date. At the time, state-of-the-art experiments were designed to detect life in Martian surface samples. Because the mission was unsuccessful in finding evidence of life, interest in astrobiology seemed to wane for almost 20 years. NASA continued to fund laboratory research in exobiology and particularly in origin-of-life related studies. While somewhat dampened by the results of the Viking mission, there remained generally keen interest in the possibility of life on Mars. The issue became more focused on how and where to sample on Mars and what to look for.

Twenty years after the Viking mission, the question of what to look for, or what are definitive signs of life, became the central issue in the claim by McKay *et al.* (1996) of evidence of life including fossil microorganisms in the Martian meteorite ALH84001. The McKay paper stimulated much discussion and debate with the conclusions being scrutinized by the scientific community (see Jakosky *et al.* 2007 for a thorough discussion of ALH84001). However, the seed was sowed that would blossom into then President Clinton's proposed NASA budget that would fund a new initiative in the study of "Origins – of life, planetary systems, stars, galaxies, or the universe" (Smith 2004). Significant funding was available for missions to Mars, Europa, and for the establishment of a program in astrobiology. In 1997, the NASA Astrobiology Institute was established.

It is interesting that the Mars Global Surveyor Orbiter, launched three months after the publication of the McKay *et al.* (1996) paper, captured images that showed the strongest evidence that liquid water did exist on the surface of Mars. The most dramatic example was from the Nanedi Vallis in the Xanthe Terra region. The Mars Odyssey Orbiter, launched in 2001, obtained dramatic evidence from hydrogen spectral measurements that extensive areas of water ice still exist below the surface of Mars. More recently, the Mars Exploration Rovers obtained confirming evidence of extensive quantities of liquid water existing on the surface and in the subsurface of Mars in the past. Moreover, the Mars Rovers identified iron and magnesium silicates and serpentine, and other minerals that also implicate water–rock interactions (Ehlmann *et al.* 2009, Mustard *et al.* 2008). On Earth, these water and mineral interactions, known as serpentinization reactions, produce hydrogen, methane, hydrocarbons, and organic acids, and result in alkaline conditions that support a hydrogen-based microbial community involved in methane production and consumption (Kelley *et al.* 2005). The recent evidence that methane may be escaping into the atmosphere from the subsurface of Mars (Mumma *et al.* 2009), and may be formed by serpentinization reactions, adds to the intrigue that Mars has the potential to support subsurface microbial communities today, or at some time in the past.

Besides the significant results from the Mars Orbiters and Rovers, the time since 1996 has seen many breakthroughs in astrobiology related research. Observations from the Hubble Space Telescope identified protoplanetary disks around young stars that have been hypothesized to lead to planetary systems. Observations from the Galileo spacecraft suggest the presence of a liquid ocean below Europa's icy surface. Cassini/Huygens reveals that Titan has organic solvent lakes suggesting this moon of Saturn may be an exciting laboratory for understanding prebiotic biochemistry. Cassini also made close passes by Enceladus, another moon of Saturn, and found water-rich geysers spewing organic compounds.

The first extrasolar planet was also identified during this period and since then more than 700 extrasolar planets have been reported, plus several thousand candidates awaiting confirmation. NASA's Kepler mission, launched in 2009, has obtained data estimating more than 50 million planets in our Galaxy. Given the rapid advancements in our technology to detect extrasolar planets, and measure and model their atmospheres, it is quite likely that an Earth-like planet in a habitable zone will be discovered during this decade (Impey 2010).

Coupled with these discoveries are advances being made in the biological sciences. For example, the application of new molecular methods has unveiled a huge phylogenetically and physiologically diverse group of previously unknown microorganisms and viruses, referred to as "the unknown biosphere." We now estimate that there may be more than one billion species of microorganisms of which less than 10 000 have been characterized. The physiological diversity of these unknown microbes, and particularly those from extreme environments, may greatly expand the environmental limits of carbon-based life. Another application of molecular tools is to control evolution and construct organisms that have the potential to grow under environmental conditions not found on Earth but on other planetary bodies. This line of research could ascertain the actual environmental limits of carbon-based life; it is also a line of research fraught with ethical issues.

These scientific discoveries are part of the backdrop to the NASA Astrobiology Institute. The guiding principles of this new discipline involve questions and issues that are multidisciplinary and that require an interdisciplinary approach to address. Astrobiology is focused on three basic questions: "How does life begin and evolve? Does life exist elsewhere in the universe? What is the future of life on Earth and beyond?" (Des Marais *et al.* 2008). The astrobiology roadmap identifies seven goals that expand on these three basic questions with the life sciences being the driving theme. Thus, astrobiology will endeavor to make life a central component of future missions to planetary bodies in our Solar System, in our search for Earth-like exoplanets, and in theories and models of the origin of the universe. It will also foster research to better understand life on Earth: its history, evolution, diversity, and limits, with a goal of increasing our chances of detecting life elsewhere. We might discover that life is a natural and inevitable product of cosmic evolution.

Defining life – Is it necessary?

Astrobiology, broadly defined as "understanding the origins, evolution, and distribution of life in the universe," forces us to confront an essential question in science: What is life and how did it originate? The search for life elsewhere begs this question. Even the simple assumption that all life in the universe will resemble Earth life restricts our search for life to Earth-like planets. While we don't yet know how life arose, we assume that it originated some four billion years ago on the Hadean Earth rather than being delivered to Earth from elsewhere. What were the conditions of the Hadean Earth that were essential to the origin of life and how would we be able to determine if other planetary bodies went through Earth-like geophysical and geological transformations that could lead to an independent origin of life or if planets with a different history could spawn a different form of life?

Any definition of life would have to include any living entity even if it were radically different from Earth life. This implies that a definition of life is not a list of canonical characteristics possessed by Earth life, such as being carbon-based, self-replicating, and having the ability to undergo Darwinian evolution. A case can be made that the ability to replicate and undergo Darwinian evolution are essential characteristics of all life even if significantly different from Earth life. However, these requirements would not be part of a definition of life, but instead essential mechanisms to produce progeny and to create diversity and complexity (Baross 2007). What we don't know is how all of the components that make up a living entity become life – the "gestaltian" issue. Cleland and Chyba (2007) approach this dilemma by rightly pointing out that "to answer the question 'What is life?' we require not a definition but a general theory of the nature of living systems." This theory does not yet exist.

In the absence of a theory of the nature of all possible living systems, searching for evidence of past or present life on other planets relies on understanding the physical and chemical characteristics germane to life, and the properties of life itself. The presence of water, essential elements, and sources of carbon and energy define the Earth conditions that allow life to thrive. Living organisms also leave a signature of their existence, either as cellular remnants or as chemical indicators such as the fractionation of carbon, nitrogen, sulfur, and oxygen isotopes, and chemical elements out of normal equilibrium known in lifeless environments. Again, the assumption is that carbon-based life on an Earth-like planetary body will affect chemical transformations in a way analogous to Earth organisms. An example of this line of thinking is the principle of "follow the energy" (Hoehler et al. 2007), that posits that energy sources used by terrestrial life forms would be used by any life forms even if they are radically different biochemically.

While some in the scientific community debate whether or not a definition of life is possible, the astrobiology community has adopted the practical view of looking for life elsewhere that is recognizable as life because of its resemblance to Earth life; that is, it is carbon-based, requires water, and uses either chemical or light energy. Life having these characteristics will also leave some signature from its existence even if it is presently extinct. Astrobiology endeavors to better understand the full range of these biosignatures on Earth and elsewhere, from living organisms or their fossils to remotely measured disequilibrium in atmospheric gas chemistry of extrasolar planets. It is assumed that extraterrestrial life, no matter how different from Earth life, will leave its mark – we just need to be able to read the signs.

Origin and early evolution of life

Many of the questions associated with the search for life in the universe could be answered more readily if it were known how life arose, what environmental conditions are essential for the generation of life, and whether or not there is more than one recipe for making life. Because the answers to these questions aren't known, the search for life elsewhere, even if water and essential chemicals for life are present, could be fruitless if life were not established by having either a *de novo* origin or being contaminated from another planetary body.

Astrobiology takes a broad and practical view of the origin of life to include both a "bottom up" and a "top down" approach. The bottom up approach starts with prebiotic chemistry and the synthesis of organic compounds essential to life. There are many sources of organic building blocks including those delivered from space, such as comets, meteorites, and interplanetary dust particles, or formed on Earth and its atmosphere by volcanoes, ultraviolet light, lightning, and "shock synthesis." Minerals are believed to be critical for the polymerization of these organic compounds into macromolecules such as proteins and nucleic acids. The bottom up research then makes giant leaps of faith to get to the next level: the formation of a "replicator" and its enclosure into a "cell-like" structure. Next, the replicator attains the ability to transfer information, make proteins, and eventually it evolves.

The earliest replicator is widely believed to be ribonucleic acid (RNA), due to the discovery in the 1980s that some RNA molecules have the ability to catalyze chemical transformations including partial self-replication. Eventually, a self-replicating RNA attained the ability to transfer information and make proteins. This period is referred to as "RNA world." Next, there was a takeover of the RNA world by deoxyribonucleic acid (DNA) and this led to greater diversity and complexity of life. There is considerable data that supports this model that RNA preceded DNA and that the genetic code and protein synthesis evolved in the RNA world (Goldman *et al.* 2010). The problem is in the details: How was RNA first synthesized, and an even more difficult challenge, how did a self-replicating RNA get synthesized? The steps leading to RNA that can "translate" a genetic code and "transcribe" proteins are also not understood.

The top down approach asks the question: What can be inferred about the origin of life and the most ancient microbial communities from the biochemistry and physiology of extant organisms? This approach is possible because all life as we know it possesses the same underlying biochemistry and molecular biology, even though there is great variation in the physiology of organisms. This concept, referred to as "the unity of biochemistry," dates back almost a century (Kluyver and Donker 1926), before the discovery that genetic information resides in DNA, the elucidation of the genetic code, and translation of the code to synthesize proteins. The unity of biochemistry is one of the most compelling arguments for the importance of Darwinian evolution even at the very onset of life's emergence. Thus, extant life is the result of both the selection of the most "fit" genetic lineages and the homogenization of these fit genetic lineages into a common ancestral community referred to as the Last Universal Common Ancestor (LUCA). All terrestrial forms of life, represented by three separate domains – Bacteria, Archaea, and Eukarya – share the same molecular information strategies and energy-yielding pathways among what would have been a wide range of possibilities before the emergence of LUCA.

A list of examples of how the top down approach has helped shape the origin of life research includes: insight from molecular phylogeny studies into the most ancient of extant organisms, the discovery of catalytic RNA, autotrophic metabolic pathways (carbon dioxide as the primary carbon source) inferred to be ancient from biochemical first principles and from chemical and isotopic evidence in the Archaean rock record, the importance of metal–sulfur minerals in catalyzing chemical reactions before the emergence of proteins, and a better understanding of the environmental conditions of the early Earth that likely contributed to the origin of life.

The "bottom up" and "top down" approaches have yet to merge into a single approach. The bottom up approach remains strongly rooted in what has been referred to as the primordial soup model stemming from Darwin's "warm pond" setting for the origin of life. The soup model implies a setting that contains perhaps millions of different organic compounds, not just those found in life. The demonstration by Stanley Miller in 1953 that amino acids could be synthesized by applying a spark discharge (mimicking lightning) to water and gases thought to be dominant on the early Earth supported the soup setting for the origin of life. Dry and wet cycles could allow organic compounds to condense into macromolecules that could lead to a replicating RNA. The criticism of the soup model is that the replicator would have to form by chance, a formation that would have a very low probability of occurrence.

An alternative model that could possibly increase the probability of abiotic synthesis of a replicating RNA is to establish proto-metabolic networks that would result in production of high concentrations of key organic compounds including those with stored energy, rather than a soup with key organic compounds in great dilution (Cody and Scott 2007, Copley et al. 2007, Shapiro 2007). De Duve (1991) makes the argument that high-energy organic compounds, such as thioesters (organic sulfur compounds that are a key intermediate in the metabolic pathway used by methane-producing microorganisms), are necessary to establish an RNA world. Many protein enzymes involved in metabolism and biosynthesis (anabolism) of organic compounds have metal−sulfur cores that in some instances are the sites of catalysis. Minerals such as pyrite (metal sulfides) and olivine (iron, magnesium silicates) do catalyze the synthesis of organic compounds, while others, such as clays, can catalyze the polymerization of organic compounds into macromolecules. These and other minerals preceded protein enzymes as the early catalysts involved in the origin of life and perhaps early life. Metal−sulfur proteins are relics of this early catalysis.

It is very likely that future studies will result in the merging of the replicator-first and metabolism-first hypotheses that could result in a model that integrates catalytic minerals, peptides, and RNA into a structure that might resemble a proto-ribosome rather than a replicating RNA. The proto-ribosome could be self-replicating, be the site of development of the genetic code, and it would have a higher probability of Darwinian selection than a replicating RNA.

Possible settings for the origin of life

Many of the origin-of-life models include a favorable setting. The various settings hypothesized for an origin of life all have the common characteristic of fostering one or more reaction steps believed to be essential for life, such as the abiotic synthesis of organic compounds, or the formation of proto-metabolic pathways (see Deamer 2007 for a discussion of the different settings). Is it unrealistic to think that all of the steps that could lead to life could occur in one setting, even a setting that is dynamic and energetic? The answer is probably yes, since it seems more reasonable to assume that there were numerous steps from organic precursor compounds to a living entity and that these steps would require a wide range of environmental conditions. The result is that the conditions that may have led to the origin of life were likely uncoupled from life itself − life may initially have had a very narrow range of environmental conditions for growth and survival.

Why would multiple settings be necessary for the origins of life as we know it? A close examination of the requirements for life provides the clues. The essential elements for life include carbon, hydrogen, oxygen, nitrogen, phosphorus, and sulfur. In addition, life uses a wide range of transition metals, such as iron, manganese, copper, cobalt, nickel, molybdenum, tungsten, and zinc, for protein structure and function, as energy sources, and as electron acceptors and donors. The dynamic nature of Earth provides environmental settings that provide these elements including those that form catalytic minerals essential for the early stages in the origin of life. The reaction of water with hot basaltic rock, the mark of submarine hydrothermal systems, provides all of the essential transition elements in addition to high concentrations of carbon dioxide, hydrogen sulfide, hydrogen, and methane. Catalytically active metal−sulfur minerals are also produced.

Hydrothermal vents are cracks along ridges on the ocean floor and are usually located in areas of seafloor spreading, the result of plate tectonics. At sites of seafloor spreading, new crust is formed and components of this young crust react with seawater to produce hydrogen, organic compounds, and other minerals that may have been important in the origin of life.

What is clear is that plate tectonics is the key driver in creating a dynamic Earth and the array of environmental settings that could contribute chemical reactions and products essential for an origin of life to occur. It is interesting that there is evidence for life in hydrothermally altered rocks that date back to 3.8 billion years ago. The isotopic and mineralogical evidence points to hydrogen- and sulfur-based microbial ecosystems including anoxygenic photosynthesis (Sleep 2010).

How can we use this information in our search for life elsewhere? It may be more important to understand the geophysical properties of a planetary body than to understand how life originated. Life as we know it could probably not have originated without tectonics and the environmental conditions that can only result from tectonics. Any rocky planetary body with water and active tectonics would have the potential for a *de novo* origin of life and for creating the highly varied chemical and physical environments that could lead to the great diversity of life analogous to Earth life.

Sites for life – near and far

The search for life in the universe embodies a tension between expectations sculpted by the nature of terrestrial biology and the sense that there may be some major surprises awaiting us in a universe of roughly 10^{23} stars. In astrobiology, we have to draw as many inferences as possible from a single example of life – terrestrial life branched off from a Last Universal Common Ancestor, and we still have no general theory of living systems. Yet life on Earth has been shaped by the geological and chemical context of this planet, and life elsewhere would naturally be subject to a different set of constraints. A microcosm of this tension is seen in the search for life on Mars. Astrobiologists set the framework for life detection on or near the surface of Mars using the closest analogous environments on the Earth (Doran *et al.* 2010), but from the Viking landers through to the upcoming Mars Science Laboratory, it is very difficult to gather evidence for, or decisive evidence against, Mars life based on terrestrial biology (Klein and Levin 1976). The obvious concern is that these experiments are only designed to detect organisms with nucleic acid information storage and familiar metabolic pathways. If life does or ever has existed on Mars it may be at or beyond the envelope of conditions tolerated by terrestrial extremophiles (DasSarma 2006).

If we cannot answer the question "Is there life elsewhere?" it seems premature to ask the question "How strange can life be?" However, the question of alternative biochemistry arises within the Solar System, on Saturn's large and enigmatic moon Titan. In our Solar System and others, the real estate in the "cryogenic" biosphere is much larger than the real estate in a conventional habitable zone – defined as the range of distances where water can be liquid on the surface of a planet or moon – and the outer Solar System is potentially home to as many as a dozen worlds with liquid water (Hussmann *et al.* 2006). Titan is exceptional because it is large enough to retain a nitrogen atmosphere thicker than Earth's, it has weather and erosion, it has shallow seas of ethane and methane, and it has an interior magma layer that drives episodic cryovolcanism involving water and ammonia. This is reminiscent (at lower temperature and with a very different chemical basis) of the primordial Earth. Depletion of hydrogen and acetylene at the surface has been taken as evidence for methanogenic life forms, but the evidence so far is not compelling (McKay and Smith 2005).

If the necessary and sufficient requirements for biology are water, organic material, and free energy, then those minimum requirements are likely to be met in diverse planet and moon settings in our Solar System and beyond. Extremophiles tell us that biology is possible with only episodic access to water, or with water that veers above the boiling point and below the freezing point. There are many sources of energy in astrophysics: chemical, geological, gravitational, and magnetic, to give an incomplete list. Tectonic activity may be needed to guarantee biological viability but we should not gainsay the possibility of biology with less dynamic sources of energy. These more subtle environments will be extremely difficult to identify as we turn our attention to the universe beyond our backyard.

The "explosion" in the number of extrasolar planets is perhaps the most exciting phenomenon in all of science. Two decades ago, no planets were known beyond the Solar System, and more than a few researchers had been burned by claims of detections that did not hold up, and many others had given up on the chase. When a planet with half the mass of Jupiter was found whipping around the star 51 Peg every four days, it was a stunning surprise (Mayor and Queloz 1995). We should, however, spare some surprise for the earlier discovery of planets around a pulsar, demonstrating that expectations are meant to be defied in astrobiology (Wolszczan and Frail 1992). Since 1995, the number of extrasolar planets has had a doubling time of 30 months, with a total of about 780 as of early July 2012. For an updated list see http://exoplanet.eu. Alongside the growing numbers is the steady downward march of the detection limit from Jupiter mass to Neptune mass to near-Earth mass. Currently there are roughly 30 extrasolar planets in the range 2–10 Earth masses, colloquially referred to as "super-Earths."

The bulk of the heavy lifting in extrasolar planet research has used, and continues to use, the indirect Doppler method. In the past decades, eclipses (or transits) have given the extra information on size, and so a constraint on mean density, while direct imaging has become effective with space-based observations and nulling interferometry on the ground. About 10% of Sun-like stars have planets, with indications that the true fraction may be much higher and that rocky terrestrial planets may outnumber gas giants (Marcy *et al.* 2005).

NASA's Kepler mission has blown the lid off the search for low-mass planets. This modest telescope in a stable space environment can readily detect the 0.01% depth of eclipse cause by an Earth-like planet transiting a Sun-like star. The team announced over 1200 candidates in early 2011, over 50 of which were in their habitable zones, among which five are probably less than twice the Earth's size (Borucki *et al.* 2011). Simple arguments suggest that the habitable "real estate" around dwarf stars exceeds that around Sun-like stars, motivating new wide-field surveys for transits associated with stars much nearer and brighter than Kepler's targets. Extrasolar planet research is a burgeoning but young field, with many theoretical puzzles to solve before we can confidently project the number of habitable worlds (Baraffe *et al.* 2010).

Signs of life – simple and advanced

The Copernican Principle has been robust enough to bear our weight at every turn in the long history of astronomy. Our situation on a rocky planet that orbits a middle-weight star on the outskirts of an unexceptional spiral galaxy appears not be unusual or unique. A conservative estimate might be a billion habitable "spots" – terrestrial planets in conventionally defined habitable zones, plus moons of giant planets harboring liquid water – in the Milky Way alone. That number must be multiplied by 10^{11} for the number of "Petri dishes" in the observable cosmos. Do we imagine that they are all stillborn and inert? Or do we imagine that a significant fraction of them host biological experiments, either like or unlike the experiment that took place on Earth? That is the central question of astrobiology.

Biomarkers are required to take the huge step forward from habitability to the first detection of life beyond Earth. That detection – keenly anticipated by all astrobiologists and by members of the general public with an interest in science – might come in the form of a shadow biosphere on our planet, or from trace fossils in a Martian rock, or from future diagnostics of targets in the outer Solar System, or from a spectral signature in the atmosphere of an extrasolar planet, or from success in the campaign to detect signals from remote civilizations. Each of these possibilities implies a different type of evidence, which must be matched against uncertain criteria for the definition of success.

Mars gives us indications of the challenges in life detection. Geochemical traces in ALH84001 and the more recent remote sensing of methane seemed to implicate biological activity, but in both cases we're left with the Scottish verdict "not proven" (McKay *et al.* 1996, Mumma *et al.* 2009). Titan presents a different issue. We simply do not have a basis in lab experiments of a general theory of biochemistry to predict what we should look for. Extrasolar planets simplify the problem because the bar is set at the global alteration of atmospheric composition by metabolic processes. Oxygen, and its photolytic product, ozone, are the "gold standards" of biomarkers because their reactivity means they are rapidly depleted on any Earth-like planet without continual replenishment by photosynthesis. Methane and nitrous oxide are also good biomarkers. In practice, a suite of biomarkers will be needed to confidently assert microbial life on another planet, bolstered by simulations and lab experiments (Kaltenegger *et al.* 2010).

Terrestrial extrasolar planets reflect just a billionth of the host star's light, a contrast that improves by a factor of 10 by moving into the thermal infrared, and they are projected so close to the star that large telescopes and adaptive optics are needed to pry them apart. Even with NASA's future flagship James Webb Space Telescope, searching for biomarkers on nearby super-Earths will be extremely challenging.

At the beginning of this chapter, we speculated that the drive to understand our place in the universe is a fundamental, evolutionary characteristic of humans. On this planet, natural selection did not stand still after microbes had radiated into every conceivable niche. The restless nature of evolution led to multicellular life, central nervous systems, brains, and eventually, one species with the capacity to alter its global environment and escape the grip of its planet. If microbial life is not vanishingly rare in the Milky Way, and if Darwin's insight applies beyond the Earth, then microbes may represent a way-station rather than an endpoint of evolution. It happened here; it could have happened on some fraction of the habitable worlds we are discovering. The search for advanced organisms is definitely well-motivated.

We have no version of Darwin's *Beagle* with which to voyage on interstellar seas. Our technology is too immature and we are too finite and frail to make the trip. So the Search for Extraterrestrial Intelligence (SETI) uses remote sensing to vault over the uncertainties in the existence and evolution of life elsewhere, and look directly for analogs of us (Tarter 2004). As its practitioners readily acknowledge, the activity is really a search for remote technology, since intelligent life forms elsewhere might not have technology, and even if they did, might have many reasons for not making their presence known or for not actively exploring their cosmic environment.

SETI has been met with 50 years of the "great silence" as researchers detect only radio static from stars that may host habitable planets. The practitioners are not discouraged, however, since advances in computation and the detection of narrow-bandwidth radio signals mean that the exploration of the domain where signals might be detected is in its infancy. The Allen Array promises to place SETI in a regime where the sensitivity to equivalent versions of our technology will span millions of Sun-like stars and any continued non-detection of artificial signals will mean that technological civilizations like ours are rare (Welch *et al.* 2009). To many astronomers, SETI is an essential "side bet" in astrobiology, worth the effort and the null results because we know so little about the outcomes of biological experiments and because the implications of making contact would be so huge.

Astrobiology and human culture

Astrobiology is tackling one of the most profound questions we can ask about our place in the universe: Is the 4-billion-year biological experiment that has taken place on the Earth unique? After centuries of speculation, our technical tools are finally up to the challenge of detecting habitable planets and inspecting them for alteration by biology. As Arthur C. Clarke said: "Two possibilities exist: either we are alone in the universe or we are not. Both are equally terrifying" (as quoted in Kaku 1999). The search for life in the universe has profound philosophical implications; knowledge that life on Earth was not unique would also have impacts on human culture and religion. We are careless stewards of our planet. We may find Earth "clones" in due course, but they will be unreachable by our technology for some time to come, and it would cost trillions of dollars and take centuries to terraform Mars as a safe haven (Lovelock and Allaby 1985). The prospect and nature of life beyond Earth is one of the few topics in science that can grab the attention of any public audience, from pre-school to retirement age. Those who work in astrobiology are perfectly placed to improve public science literacy by conveying the research that addresses this fundamental question.

Astrobiology and humility

One of the privileges of being a scientist is that we have the opportunity to make new discoveries through exploration of what is unknown in our scientific discipline. Astrobiology, referred to earlier as the new Darwinian voyage, is very much a discovery-based science, but with one overriding unique goal – to find a second example of life on a living planet. Perhaps we search for life elsewhere because we are really exploring the nature of our own existence. Why us and not a simple microbial community free of thought and wonderment, or why life at all? What are the philosophical implications if there are multiple ways to make life, or, that life elsewhere is different or exactly the same as Earth life? The origin of life, the result of both contingent and deterministic emerging properties, is likely to have occurred, or is still to occur on many other planetary bodies. Yet we can't define life or know how to devise a theory about the nature of a living organism. A vast and expanding universe, with stars and galaxies beyond the grasp of our telescopes, adds more questions and issues that are unexplained. We can only express great humility in our limitations as we try to understand the mysteries associated with the origins and evolution of life and the universe. Humility dictates that we should remain open to explanations that might be outside the bounds of current scientific scrutiny.

References

Baraffe, I., Chabrier, G., and Barman, T. (2010). "The physical properties of extra-solar planets," *Reports on Progress in Physics*, Vol. 73, 016901.

Baross, J. A. (2007). "Evolution: a defining feature of life," in *Planets and Life: The Emerging Science of Astrobiology*, W. T. Sullivan, III and J. A. Baross (eds.). Cambridge: Cambridge University Press, pp. 213–221.

Berkner, L. V. and Odishaw, H. (1961). "Dimensions and problems," in *Science in Space*, L. V. Berkner and H. Odishaw (eds.). New York: McGraw-Hill Book Company, pp. 3–47.

Borucki, W. J. *et al.* (2011). "Characteristics of planetary candidates observed by Kepler, II: analysis of the first four months of data," *Astrophysical Journal*, Vol. 736, pp. 19–40.

Cleland, C. E. and Chyba, C. F. (2007). "Does 'Life' have a definition?," in *Planets and Life: The Emerging Science of Astrobiology*, W. T. Sullivan, III and J. A. Baross (eds.). Cambridge: Cambridge University Press, pp. 119–131.

Chyba, C. F. and Hand, K. P. (2005). "Astrobiology: the study of the living universe," *Annual Reviews of Astronomy and Astrophysics*, Vol. 43, pp. 31–74.

Cody, G. D. and Scott, J. H. (2007). "The roots of metabolism," in *Planets and Life: The Emerging Science of Astrobiology*, W. T. Sullivan, III and J. A. Baross (eds.). Cambridge: Cambridge University Press, pp. 174–186.

Copley, S. D., Smith, E., and Morowitz, H. J. (2007). "The origin of the RNA world: co-evolution of genes and metabolism," *Bioorganic Chemistry*, Vol. 35, pp. 430–443.

Darwin, C. (1979). *The Origin of Species*. New York: Random House (original edition 1859).

DasSarma, S. (2006). "Extreme halophiles are models for astrobiology," *Microbe*, Vol. 1, pp. 120–126.

De Duve, C. (1991). *Blueprint for a Cell: The Nature and Origin of Life*. Burlington, VT: Portland Press.

Des Marais, D. J. *et al.* (2008). "The NASA astrobiology roadmap," *Astrobiology*, Vol. 8, pp. 715–730.

Deamer, D. W. (2007). "The origin of cellular life," in *Planets and Life: The Emerging Science of Astrobiology*, W. T. Sullivan, III and J. A. Baross (eds.). Cambridge: Cambridge University Press, pp. 187–209.

Dick, S. J. (2007). "From exobiology to astrobiology," in *Planets and Life: The Emerging Science of Astrobiology*, W. T. Sullivan, III and J. A. Baross (eds.). Cambridge: Cambridge University Press, pp. 46–65.

Doran, P. T., Lyons, W. B., and McKnight, D. M. (eds.) (2010). *Life in Antarctic Deserts and Other Cold Dry Environments: Astrobiological Analogs*. Cambridge: Cambridge University Press.

Ehlmann, B. L. *et al.* (2009). "Identification of hydrated silicate minerals on Mars using MRO-CRISM: geologic context near Nili Fossae and implications for aqueous alteration," *Journal of Geophysical Research*, Vol. 114, E00D08, doi:10.1029/2009JE003339.

Goldman, A. D., Samudraia, R., and Baross, J. A. (2010). "The evolution and functional repertoire of translation proteins during the origin of life," *Biology Direct*, Vol. 5, pp. 1–15.

Hoehler, T. M., Amend, J. P., and Shock, E. L. (2007). *Astrobiology*, Vol. 7, pp. 819–823.

Hussman, H., Sohl, F., and Spohn, T. (2006). "Subsurface oceans and deep interiors of medium-sized outer planets satellites and large trans-Neptunian objects," *Icarus*, Vol. 185, pp. 258–273.

Impey, C. D. (ed.) (2010). *Talking About Life: Astrobiology Conversations*. Cambridge: Cambridge University Press.

Jakosky, B. M., Westall, F., and Brack, A. (2007). "Mars," in *Planets and Life: The Emerging Science of Astrobiology*, W. T. Sullivan, III and J. A. Baross (eds.). Cambridge: Cambridge University Press, pp. 357–387.

Kaku, M. (1999). *Visions: How Science Will Revolutionize the Twenty-First Century*. New York: Oxford University Press.

Kaltenegger, L. *et al.* (2010). "Deciphering spectral fingerprints of habitable planets," *Astrobiology*, Vol. 10, pp. 89–102.

Kelley, D. S. *et al.* (2005). "A serpentinite-hosted ecosystem: the Lost City hydrothermal field," *Science*, Vol. 307, pp. 1428–1434.

Klein, H. P. and Levin, G. V. (1976). "The Viking biological investigation: preliminary results," *Science*, Vol. 194, pp. 99–105.

Kluyver, A. J. and Donker, H. J. (1926). "Die Einheit in der Biochemie," *Chemie dev Zelle und Gewebe*, Vol. 13, pp. 134–199. (Reprinted in A. F. Kamp, J. W. M. La Rivière and W. Verhoeven (eds.). *Albert Jan Kluyver, His Life and Work*, Amsterdam: North-Holland Publishing Co., 1959, pp. 211–267).

Lafleur, L. J. (1941). "Astrobiology," *Astronomical Society of the Pacific Leaflets*, Vol. 3, pp. 333–340.

Lederberg, J. (1960). "Exobiology: approach to life beyond the Earth," *Science*, Vol. 132, pp. 393–400.

Lovelock, J. and Allaby, M. (1985). *The Greening of Mars*. Clayton, Victoria: Warner Books.

Marcy, G., Butler, R. P., Fischer, D. *et al.* (2005). "Observed properties of exoplanets: masses, orbits, and metallicities," *Progress in Theoretical Physics Supplement*, Vol. 158, pp. 24–42.

Mayor, M. and Queloz, D. (1995). "A Jupiter-mass companion to a solar-type star," *Nature*, Vol. 378, pp. 355–359.

McKay, C. P. and Smith, H. D. (2005). "Possibilities for methanogenic life in liquid methane on the surface of Titan," *Icarus*, Vol. 178, pp. 274–276.

McKay, D. S., Gibson, E. K., Jr., Thomas-Keprt, K. L. *et al.* (1996). "Search for past life on Mars: possible relic biogenic activity in Martian meteorite ALH 84001," *Science*, Vol. 273, pp. 924–930.

Mumma, M. J. *et al.* (2009). "Strong release of methane on Mars," *Science*, Vol. 323, pp. 1041–1045.

Mustard, J. F. *et al.* (2008). "Hydrated silicate minerals on Mars observed by the Mars reconnaissance orbiter CRISM instrument," *Nature*, Vol. 454, pp. 305–309.

Shapiro, R. (2007). "Origin of life: the crucial issues," in *Planets and Life: The Emerging Science of Astrobiology*, W. T. Sullivan, III and J. A. Baross (eds.). Cambridge: Cambridge University Press, pp. 132–153.

Sleep, N. H. (2010). "The Hadean-Archaean environment," in *The Origins of Life*, D. Deamer and J. W. Szostak (eds.). New York: Cold Spring Harbor Laboratory Press, pp. 35–48.

Smith, D. H. (2004). "Astrobiology in the United States: a policy perspective," in *Astrobiology: Future Perspectives*, P. Ehrenfreund *et al.* (eds.). Dordrecht: Kluwer, pp. 445–466.

Sullivan, W. T. III and Carney, D. (2007). "History of astrobiological ideas," in *Planets and Life: The Emerging Science of Astrobiology*, W. T. Sullivan, III and J. A. Baross (eds.). Cambridge: Cambridge University Press, pp. 9–45.

Tarter, J. C. (2004). "Astrobiology and SETI," *New Astronomy Reviews*, Vol. 48, pp. 1543–1549.

Welch, J. *et al.* (2009). "The Allen Telescope Array: the first wide-field, panchromatic, snapshot radio camera for radio astronomy and SETI," Special Issue of *Proceedings of the IEEE*, "Advances in Radio Telescopes," J. Baars, R. Thompson, and L. D'Addario (eds.). Vol. 97, pp. 1438–1447.

Wolszczan, A. and Frail, D. (1992). "A planetary system around the millisecond pulsar PSR 1257+12," *Nature*, Vol. 355, pp. 145–147.

PART II ORIGIN OF PLANETS AND LIFE

2

Towards a Theory of Life

STEVEN BENNER AND PAUL DAVIES

Introduction

What is this thing called "life" that seems to interest everyone? And why are we looking for it so hard? Why, for example, are we seeking "Earth-like planets" in "habitable zones," or the spectroscopic signatures of amino acids and carbohydrates in the interstellar dust? Why not just accept the perspective of Dr. Manhattan from *The Watchmen*: most of the universe appears uninhabitable, yet seems to get along quite well.

Certainly, it is easy to understand the motivation of SETI astronomers who seek *intelligent* life. They seek someone to talk to, someone who can teach us something, if it does not consume us, or destabilize our society with technological wizardry. If we *physically* encounter life of the type that SETI seeks, we will almost certainly recognize it as life. SETI-style life will have the attributes that we most value in life, whether it explains the secrets of dark matter or simply applies 40 of us to a recipe for Rigelian stew. But the higher probability seems now that we will not find anything so complex. The life that we seem most likely to encounter will be "primitive," single-celled organisms visible only under the microscope. And the life that we now seem most likely to encounter will be detectable only by its chemical signatures, no longer able to grow after its first encounter with us.

Given these likelihoods, it is less clear what we are seeking. Witness the difficulty that people have had crafting definitions of "life" as a universal concept as it would apply to simpler, unintelligent organisms. For example, Daniel Koshland, Distinguished Professor of Biochemistry at the University of California in Berkeley, and then President of the American Association for the Advancement of Science,

Frontiers of Astrobiology, ed. Chris Impey, Jonathan Lunine and José Funes.
Published by Cambridge University Press. © Cambridge University Press 2012.

recounted his own experience attempting to define life as universal (Koshland 2002):

> What is the definition of life? I remember a conference of the scientific elite that sought to answer that question. Is an enzyme alive? Is a virus alive? Is a cell alive? After many hours of launching promising balloons that defined life in a sentence, followed by equally conclusive punctures of these balloons, a solution seemed at hand: "The ability to reproduce – that is the essential characteristic of life," said one statesman of science. Everyone nodded in agreement that the essential of life was the ability to reproduce, until one small voice was heard. "Then one rabbit is dead. Two rabbits – a male and female – are alive but either one alone is dead." At that point, we all became convinced that although everyone knows what life is, there is no simple definition of life.

And yet we are looking for it.

However, Koshland and his "elite" did not make a good run at the problem. They confused up front the concept of "being alive" with the concept of "life." This is not just a confusion of an adjective with a noun. Rather, the attempt confuses a *part* of a system with the *whole* of a system. Parts of a living system might themselves be alive (a cell in our finger may be alive), but those parts need not be coextensive with a living system. One rabbit may be alive, but he (or she) need not be a living system, and need not be "life." In another effort, Koshland himself proposed a "list" definition of life. His definition combined thermodynamic, genetic, physiological, metabolic, and cellular features of terran life that he thought to be important. Koshland even created an ungainly acronym for his definition, PICERAS, which combined the first letters of the words: program, improvization, compartmentalization, energy, regeneration, adaptability, and seclusion.

Other committees of scientific statesmen have defined life with better clarity without relying explicitly on lists. For example, a committee was assembled in 1994 by NASA to define what it should be seeking in its life-detection missions. The committee eventually settled on a definition of life as a "self-sustaining chemical system capable of Darwinian evolution" (Joyce 1994). By using the word "system," the committee recognized that entities can be alive (including a cell, virus, or male rabbit) without themselves being "life." The phrase "self-sustaining" requires that a living system should not need continuous intervention by a higher entity (a graduate student or a god, for example) to continue as "life."

The phrase "Darwinian evolution" made specific reference to a process that involves a molecular genetic system (DNA in terran life) that can be replicated imperfectly, and where mistakes arising from imperfect replication can themselves be replicated. Thus, "Darwinian evolution" implies more than "reproduction," a trait that ranks high in many list definitions of life. The requirement for reproduction with errors, where the errors are themselves reproducible, excludes a variety of non-living chemical systems that can reproduce but not evolve. For example, a crystal of sodium chlorate can be powdered and used to seed the growth of other sodium chlorate crystals, $NaClO_3$ (Kondepudi *et al*. 1990). In this way, these crystals can be said to reproduce. Further, features of the crystal, such as whether its atoms are arranged in left-handed spirals or right-handed spirals (the crystals are "chiral") can be passed to its descendants.

However, the replication is imperfect; a real crystal contains many defects. Indeed, to specify all of the defects in any decent-sized crystal would require an enormous amount of information, easily exceeding the 10 billion bits of information contained in a human genome. But the information in these defects is not itself heritable: the defects in the parent crystal are not reproduced in the descendent crystals. Thus, the information held in the defects in the descendent crystal is entirely independent of the information held in the defects in the parent. Therefore, the crystal of sodium chlorate cannot support Darwinian evolution, meaning that a system of sodium chlorate crystals does not qualify as life.

Insights from science fiction

Perhaps the most important feature of the NASA definition was its statement about what kinds of life its framers believed are possible. Thus, the NASA definition embodied a *theory* of life, something necessary for any good definition (Cleland and Chyba 2002). And, like any good theory, it *excluded* certain systems that are *conceivable* as forms of life simply because the theory holds them to be impossible in the real world.

We can turn productively to science fiction to illustrate how easily life forms may be conceived that are not chemical systems capable of Darwinian evolution. For example, the crew of *Star Trek: The Next Generation* encountered nanites that infected the computer of the next-generation starship, the Enterprise, in Episode 50 ("Evolution"). These were informational or, perhaps, electromechanical, but in any case not chemical; their evolution was not tied to an informational molecule like DNA (although they did require a chemical matrix to survive). Likewise, the Crystalline Entity of Episode 18 ("Home Soil") appeared to be chemical, but not obviously Darwinian. The Calamarians from Episode 51 ("Déjà Q") were made of "pure energy" (whatever that means), not chemicals. And in Episode 1, Q himself ("Encounter at Farpoint") appeared to be neither matter nor energy, moving instead in and out of the so-called Continuum without the apparent need of either.

Off screen, Fred Hoyle developed a story about a black cloud, a fictional entity that floated into our Solar System and blocked our sunlight, placing Earth in distress (Hoyle 1957). After the black cloud realized that the Earth held self-aware forms of life, it politely moved out of the way and apologized. The cloud seemed to have no children, however, hence incapable of Darwinian evolution, and therefore to not be "life" under the NASA definition.

If we were to encounter Q, the Calamarians, or any of these other conjectural entities during a real, not conceptual, trek through the stars, we would be forced to concede that they *do* represent living systems. We would also agree that they do not fall within the NASA definition of life. We would therefore be forced to agree that we need a new definition for life. Similarly, if a black cloud floats into our Solar System and apologizes to us, we will certainly reject any definition that does not include it as life, especially if we want to encourage particularly polite life forms (Benner 2009). We submit that we do not change our definition of life now so as to accommodate Q, the Black Cloud, or other examples of "weird" life from Hollywood because *most of us do not constructively believe that this kind of life is possible.*

Approaches to a theory of life

Science is an intellectual activity that embodies a mechanism to prevent scientists from necessarily reaching the conclusions they set out to achieve (Benner 2009). As one such mechanism, scientists are trained to "flag" situations where they find themselves arriving at results that they want and expect. One such flag is raised when we note that all of the members of the NASA panel that defined life as "a self-sustaining chemical system capable of Darwinian evolution" were, themselves, self-sustaining chemical systems capable of Darwinian evolution. Likewise, before we accept Koshland's PICERAS definition of life, we might worry that Professor Koshland himself is characterized by a program, improvization, compartmentalization, energy, regeneration, adaptability, and seclusion. It may be a little too convenient that every definition of life appears to correspond exactly to our form of life.

All life may indeed be a self-sustaining chemical system capable of Darwinian evolution, just like us. However, before settling on this definition, we should at least make an effort to develop a theory of life (and its corresponding definition) that does not just recite features that we find in ourselves. Science is experienced to consider the "universal" from specifics. How might this experience be used to explore more broadly theory-based definitions of life?

Observation, analysis, exploration, and synthesis

Observation, analysis, exploration, and synthesis have been the mainstays of scientific activity for centuries. How might we apply these to develop a theory-based definition of life? Obviously, observation will be limited by available technology and circumstance. We are not yet able to observe remotely Klingons. Their planet, Qo'noS, would be too distant, even if we knew which direction to look. Nor can we necessarily observe everything within even the terran life that we

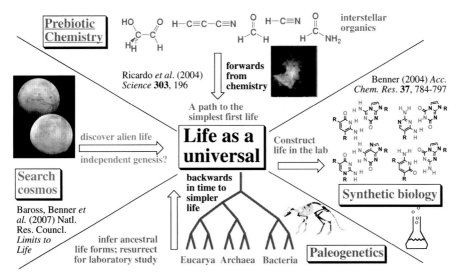

Figure 2.1 Four approaches indirectly adumbrate the character of "life," represented by the "black box" at the center of this figure. Natural history (the bottom wedge) exploits observation of the one exemplar of life that we surely have, using geology, paleontology, and molecular analysis to infer ancient states of life, and paleogenetics to resurrect ancient bits of extinct life for study in the lab. Exploration (the left wedge) hopes to expand on our understanding of life by finding a second exemplar. Prebiotic chemistry (the top wedge) attempts to understand the intrinsic nature of life by understanding chemical constraints for its formation. Synthesis (the right wedge) attempts to understand life as a universal by deliberately creating, in the lab, a new form of life (courtesy Steven Benner).

have in hand. For example, the complex and heterogeneous molecular aggregates that form cellular membranes, crucial to the seclusion and compartmentalization of PICERAS life, remain beyond observation with today's technology. With current technology, we can observe strands of DNA and we can observe chromosomes, but the bit in between – the mesoscopic, complex, amorphous, dynamic, multi-folded, partially ordered, molecular tangle – is largely uncharted, even though the living systems that use these bits are in hand.

Synthesis, the creation of new forms of the matter that we think we understand, is also limited by technology. While we can design and fabricate many individual biomolecules from scratch, we do not have full control over the synthesis of their aggregates, including higher-order structures such as membranes, let alone molecular motors. Many of the distinctive features of life which we label as "emergent" derive their remarkable properties precisely from the organized assembly of those aggregates. In spite of these shortcomings, by the mid-1980s technological advances had enabled us to identify ways to use observation and analysis to constrain a theory-based definition of life. These are summarized in Figure 2.1.

The first approach (represented by the bottom wedge) exploits observations and analyses based on evolutionary theory as a scientific method central to natural history. By 1985, fossil and geological records, richly elaborated by nearly two centuries of observation, began to be augmented by molecular records that included gene and protein sequences. While DNA sequencing was in its infancy, it was clear that this and other analytical techniques would soon yield an avalanche of molecular data relevant to natural history, including crystal structures of many gene products and an advanced understanding of the "parts list" of living systems.

About the same time, it became clear that planetary exploration, which had slowed following the 1976 Viking mission to Mars, might again accelerate, opening opportunities for discovery that might refine our understanding of life. Even if alien life itself were not encountered, we could reliably expect that exploration would place constraints on possible environments for life in the Solar System, including "weird" environments (Baross *et al.* 2007). Planetary exploration is represented by the left-hand wedge in Figure 2.1. Similarly, researchers in prebiotic chemistry (the top wedge in Figure 2.1) had, by the mid-1980s, apparently run out of ideas. However, technology in astrochemistry and planetary science began to revitalize the field, promising additional constraints on possible environments where life might have emerged. The detection of ever more complex organic molecules in interstellar clouds and even comets began to expand rapidly. Reliable organic inventories of meteorites became available (Pizzarello 2004), providing insights into organic chemistry in planetesimals. And planned missions to Titan and other outer Solar System objects offered the prospect of studying a natural prebiotic soup in all its multi-billion-year richness.

The final, right-hand, wedge in Figure 2.1 represents the deliberate synthesis of new forms of organic matter. In the mid-1980s, synthetic organic chemistry was mainly focused on streamlining the complete synthesis of certain commercially valuable natural products. However, a spin-off from this technology was the possibility of creating entirely new life forms from scratch, including artificial genetic and catalytic systems. Such studies serve to identify the range of potential chemical pathways to life. Although "life" is still a largely mysterious "black box" at the center of Figure 2.1, the foregoing four approaches have at least cast some light on how to define it. In the following sections we review these insights separately.

Life in the context of its history

As long as we agree that we ourselves are alive, it makes sense to appeal to life on Earth as an exemplar of a universal. This is, of course, exactly what biologists have been doing even when they called themselves "natural historians." This activity encompasses observation (of systems that are alive) and analysis (dissection, fractionation, biomolecule isolation).

More recent has been the dissection of terran biology at the molecular level, including sequences of genes and their encoded proteins. Databases of these sequences have allowed natural history to be applied to biomolecular structures. DNA and protein sequences can be aligned to make their similarities most perspicuous. In many cases, the sequences of genes and proteins from different organisms are too similar to have arisen by random chance. This implies that biomolecules from different organisms share common ancestry, just as the organisms themselves. For each alignment, a tree is constructed showing the familial relationships of the genes. By the turn of the twenty-first century, catalogs of trees and alignments had been compiled for hundreds of gene families (Benner 2009). This led to the creation of a new field known as "paleogenomics," in which entire genomes are compared to elucidate the evolution of organism-wide behavior over the history of the Earth. Paleogenomics is now casting light on the origin and evolution of human diseases, some of which, such as alcoholism, hypertension, and diabetes, turn out to have quite complicated histories (Johnson *et al.* 2008).

Further, in the process of constructing trees from alignments to interrelate members of gene families, models for the sequences of *ancestral* genes and their encoded proteins can be constructed. These create the possibility of doing paleogenetic experiments, in which ancient genes and proteins are resurrected using biotechnology for study in the laboratory. Before paleogenetics, it was widely supposed that historical hypotheses are intrinsically untestable. However, by studying the behavior of ancestral genes and proteins from species that have long gone extinct, competing historical narratives could be tested in the laboratory.

Furthermore, experiments with ancestral biomolecules sometimes revealed unexpected ancient biological functions. A good example concerns so-called elongation factors (EFs), one of many proteins that are examined using paleogenetic techniques. EFs participate in the biosynthesis of proteins, bringing charged transfer RNA carrying individual amino acids to the ribosome, which adds those amino acids to a growing protein chain. Genomic sequencing allowed Gaucher *et al.* (2003) to infer the sequences of ancestral EFs that lived in very ancient eubacteria, perhaps as long as two billion years ago. Biotechnology was then used to resurrect those ancestral EFs. The most ancient of these were found to work best at a temperature of $65 \pm 5\,°C$. This implied that the most ancient eubacteria lived at $65\,°C$.

Many lines of evidence, combined with the analysis of ribosomal RNA sequences, have helped reconstruct a picture of primitive life and its environment. In these and many other ways paleogenetics has steadily transformed natural history into an experimental science. By 2010, paleogenomic analyses had joined with other disciplines to make three conclusions inescapable. The first is historical: *All life that we have so far encountered on Earth appears to be related by common ancestry*. This conclusion forced a second: The terran biosphere appears to hold only one "example" of life. That example, of course, has been edited and revised by evolution as a consequence of historical accident, adaptation to different environments, and structural drift. But recounting the long history of adaptive changes that permitted a rich diversity of terran life forms is unlikely to cast much light on the question of life universally. The range of environments on Earth is quite narrow. Even environments that are considered "extreme" (as in Antarctica, or in hot springs, or metal-contaminated rivers such as the Rio Tinto) are only moderately different from environments that humans find comfortable, especially when compared to Titan, with its hydrocarbon oceans at 91 K, or the hot, acidic clouds above Venus, or the cold near-vacuum of the Martian surface.

However, a third conclusion offers some hope. In reconstructing the history of terran life, we learned something about how "primitive" life was structured (Benner *et al.* 1989). Viewing the evolutionary history of life backwards in time suggested that very ancient terran life was far more dependent on RNA, and far less dependent on proteins, than contemporary terran life. Extrapolating this finding suggests that a very early form of life may have used no encoded proteins at all, but rather employed ribonucleic acid (RNA) to fulfill both genetic and catalytic roles (Gilbert 1986, Benner *et al.* 1989). That is, contemporary "DNA-RNA-protein"-based life descended from an ancient form that inhabited a so-called "RNA world" (Benner *et al.* 1989). RNA life is not really a second sample of life, but it is perhaps sufficiently different from life as we know it to qualify as a "half example."

So far, the exploration of Earth has not uncovered any surviving "RNA world" organisms, though they may yet be found in some sort of "shadow biosphere" (Cleland and Copley 2005, Davies and Lineweaver 2005). Therefore, much of our model for this "weird" RNA-world life form is based on inference. Nevertheless, RNA may turn out to be a unique component in all chemical systems capable of undergoing Darwinian evolution, in which case it will be universal. If a primitive form of life exists elsewhere in the Solar System, then it might be RNA-based life, and recognizable as such.

Exploration and the example of Mars

Our awareness of only one (perhaps one and a half, if we count a predecessor RNA-world life) form of life on Earth creates a need for another staple of science, exploration, represented by the left wedge of Figure 2.1. Here, we seek a second exemplar of biology (and further). There is a certain cyclicity involved here: exploration is used to develop a definition-based theory of life, but a definition-based theory of life is necessary for exploration to be meaningful in the first place. The cycle defines, then explores, then revises the definition, and then repeats.

The Viking mission

This strategy is illustrated by efforts to find life on Mars, efforts that began (in a real sense) with the 1976 Viking mission. The Viking spacecraft carried three explicit life-detection experiments duplicated on two separate landers (Levin and Straat 1977, 1981). One experiment added Martian soil to an aqueous solution of seven organic compounds labeled with carbon-14. Any life in the Martian soil was expected to oxidize these compounds and release radioactive carbon-14 dioxide. A second experiment also placed a sample of Martian soil into a nutrient broth, and exposed it to sunlight. If life were present and capable of photosynthesis, molecular oxygen should be given off. The third experiment exposed the Martian surface to radioactive carbon dioxide and carbon monoxide in the presence of sunlight. If life were present on the surface with the ability to "fix" carbon from the atmosphere (like terran plants), radiolabels from these gases should be fixed into organic compounds in the soil. Observation of carbon fixation was to be interpreted as a positive sign of Martian life.

Although the Viking scientists did not clearly articulate their theory of life, the design of their tests reveals what definition-based theory they constructively held. Because all of their life-detection tests looked for products of metabolism, mission designers tacitly placed "metabolism" high on their list of criteria for life. More than that, the tests constructively assumed that Martian life had terran-like metabolism. Had the NASA panelists who defined life in 1994 themselves consumed the seven radioactive organic molecules in the Viking life-detection experiment and then exhaled into the Viking detector, that detector would have concluded that the panelists were alive because it would have measured radioactive CO_2 in their breaths. In similar vein, the other Viking tests for life were based on terran photosynthesis. On Earth, if water and nutrients are added to soil exposed to sunlight, any resident photosynthetic organisms will emit oxygen. If you expose a terran plant to radioactive carbon dioxide and sunlight, the plant will fix the CO_2 into organic compounds in the plant, which will then become radioactive.

So what happened on the surface of Mars when the life-detection experiments were run? Without dwelling on details, all three experiments gave positive results. Radiolabeled CO_2 was released when the radiolabeled organic compounds were added with water to the Martian soil. Release was not observed if the Martian soil was heated, as expected if the heat had killed something living in the soil. Molecular oxygen was released when water was added to Martian soil, just as if the soil contained photosynthetic organisms. Radioactive carbon was fixed on the Martian soil when exposed to radiolabeled carbon monoxide and carbon dioxide, consistent with the presence of photosynthetic organisms.

On the face of it, the Viking experiments detected life on Mars – indeed, life of a similar nature to terran life, at least as far as its metabolism and upper temperature limit are concerned. To be sure, the presence of life would have been more clearly indicated if *more* radioactive carbon had been fixed from the atmosphere, rather than the small amounts that were observed. The pattern of oxygen release was also perplexing upon heating and cooling. Nevertheless, Gilbert Levin, who designed the labeled release experiment, still argues that his test found life. And why not? The results fulfilled the criteria adopted by the mission as a signature of life.

These results notwithstanding, almost all scientists concluded that the Martian surface held no life. Why? Briefly, the positive indicators of life were dismissed following results from yet another instrument delivered to Mars, a gas chromatograph-mass spectrometer (GC-MS). Here, a sample of Martian soil was first delivered to a cup, which was then sealed and heated. Vapors emerging were blown by a stream of dihydrogen gas into the gas chromatograph column, where they were separated. From there, the separated compounds were injected into the mass spectrometer to determine their abundances and masses (Biemann *et al.* 1977).

Surprisingly, given the results of the life-detection test, the GC-MS instrument did not detect any organic molecules, to the limit of its sensitivity, other than a trace of chlorinated species and some benzene, which were interpreted as contaminants in the equipment brought from Earth. The null GC-MS result drove the community to conclude that the Martian surface contained no life at all, even though all three experiments designed under the metabolism-based theory of life were positive at some level.

The response to the Viking mission thus tells us something interesting about definition-based theories of life. In the planning stage, life was defined by metabolism. After the positive results were obtained, however, the scientific community shifted to a *reduced-carbon* definition of life, which places a higher value on organic composition than on metabolism. Putting on our anthropologist hats, we can say that the scientists interpreting the Viking results constructively felt that the reduced-carbon theory of life trumped the metabolism theory of life.

Indeed, the Viking experience is remarkable in the speed with which people invented non-life explanations for the metabolism-targeted life-detection tests, as soon as the failure to find reduced carbon trumped the positive metabolism results. Literally within hours, it was noted that the surface of Mars was exposed to solar ultraviolet rays of sufficient energy to split water and produce hydroxyl radicals. These would, over time, render the surface of Mars oxidizing. The oxidant in the soil would *also* serve to release radioactive carbon from the seven

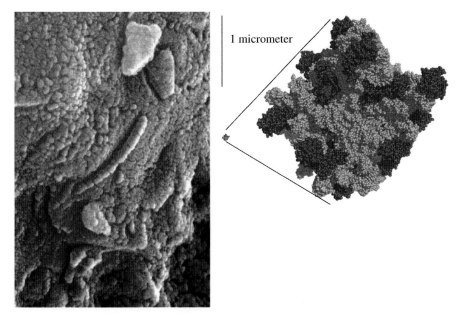

1 micrometer

Figure 2.2 The small, cell-like structures in the Allan Hill meteorite, with a ribosome drawn to scale (courtesy Steven Benner).

labeled substances *non*-biologically. It might also release dioxygen upon wetting. Further, the ultraviolet light might cause the radiolabeled carbon monoxide and dioxide to form organic materials. In other words, the intense ultraviolet radiation on Mars effectively mimics the results of organic metabolism. You would be forgiven for supposing this was a surprise, but in fact it was known from laboratory experiments conducted before Viking landed. They were not considered an issue, however, until the GC-MS results compelled the community to adopt a definition-based theory of life with a specified hierarchical order.

As a consequence of this collection of constructive views, "non-life" explanations for the Viking results soon came to dominate within the community. For the following 20 years, most scientists thought the Martian surface to be highly oxidizing and devoid of life. This opinion undoubtedly colored further exploration plans; in fact biologically directed experiments all but disappeared from Mars missions for many years.

The Allan Hills meteorite

The situation changed yet again when David McKay and his colleagues at NASA's Johnson Space Center drew attention to a meteorite, dubbed ALH84001, found on an ice field in Antarctica near the Allan Hills. The meteorite was determined without doubt to have originated on Mars. McKay and his team published images of small, cell-like structures in the rock (Figure 2.2), accompanied by an analysis suggesting, but by no means proving, that these features could be the fossilized remnants of microbial life (McKay *et al.* 1996). Fuelled by a press conference hosted by NASA's Administrator and accompanied by a statement from the US President, the announcement created an international sensation.

The claim of McKay and his colleagues is of interest to our discussion here because it draws on yet *another* theory-based definition of life, known as the "Cell Theory of Life." Cell Theory can be traced back to observations by Robert Hooke (1635–1703), who observed cells in slices of cork using a microscope. Later studies showed that animal tissues were also made from cells. In 1847, Theodor Schwann and M. J. Schleyden suggested that animal and vegetable biology could be unified under the simple theory that *all* living systems are built from cells. Because of McKay's observations, the community constructively acquired once again an appreciation of the potential for life on Mars. Anthropologists of science might therefore conclude that the cell definition-theory of life must trump the reduced-carbon definition-theory, which in turn trumps the metabolism definition-theory. In fact, many definitions of life conform to this hierarchy. For example, in Koshland's PICERAS definition, the C (compartmentalization) and the S (seclusion) components of the acronym come from cell theory.

The reader can decide the extent to which the structures discovered in the Allan Hills meteorite resemble Earthly microbes (Figure 2.2). But in the decade following the initial publication of the meteorite's discovery, some in the biological community began to argue that the intriguing microbe-like structures could not be remnants of cells because of their small size: only 100 nanometers across. This was "too small" to be life. But too small for *what* form of life? The most frequently cited arguments against McKay's cell-like structures as the remnants of life compared their size to the size of the ribosome, the molecular machine used by terran life to make proteins (Figure 2.2). The ribosome is approximately 25 nanometers across. This means that the "cells" in Allan Hills 84001 can hold only about four ribosomes – too few, supposed many biologists, for a viable organism. The syllogism is clear. Proteins are necessary for life. Ribosomes are necessary to make proteins. Martian cells are too small to hold ribosomes. Hence, Martian cells are too small to make proteins. Therefore, Martian cells are too small to be life.

This criticism shows that the critics of the claims of McKay *et al.* (1996) constructively held yet a *different* definition-theory of life, a *protein* theory of life. Furthermore, within the hierarchy of theory-based definitions, the "protein" theory of life evidently trumps the "cell" theory, which in turn trumps the "carbon" theory, which trumps the "metabolism" theory of life, at least to the extent to which the Viking and Allan Hills observations were connected.

Why should proteins be universally necessary components of life? Could it be that Martian life has no proteins? Here, our earlier observations about the possibility of ancestral RNA-based life *on Earth* are relevant. Life forms in the putative RNA world (by definition) survived without encoded proteins and the ribosomes needed to assemble them. Thus, our additional "half example" of life on Earth refutes the "protein theory of life," and undermines the argument that the structures observed in the Allan Hills meteorite are too small to be remnants of living cells. If those structures represent a trace of an ancient RNA world on Mars, they would not need to be large enough to accommodate ribosomes (Benner 1999). The shapes in meteorite ALH84001 just *might* be fossil organisms from a Martian "RNA world."

Earth and Mars as a system

Unfortunately, we cannot stop here, as any life found on Mars need not represent a "second" form of life, simply because Earth and Mars are not biologically quarantined. From time to time, both planets are struck by asteroids and comets with enough force to splash rocks into solar orbit, from where they can migrate to other planets. Thus, dozens of rocks from Mars can be found among meteorites collected on Earth, and traffic of material has undoubtedly occurred (with lower efficiency due to the influence of the Sun and the higher gravity and thicker atmosphere of the Earth) in the other direction.

Thus, through the multi-billion-year history of the Solar System, trillions of tons of rocks have been traded between these two planets. Microorganisms could hitch a ride on this material and be safely transported from Mars to Earth (or, less commonly, vice versa), a phenomenon known as lithopanspermia or transpermia. Cocooned inside a large rock, a bacterium could easily withstand space conditions for sufficient time in a dormant but still viable form, shielded by the rock from ultraviolet radiation, solar flares, and all but the highest-energy cosmic rays. This scenario, greeted with skepticism when originally presented (Davies 1996), is now widely accepted (Melosh 2003). If correct, then Mars and Earth form a weakly coupled combined ecosystem. This makes it possible that life originating on Earth spread to Mars, or that there is a Martian site for the origin of terran life.

Exploration and the example of Earth

It would be disappointing to go all of the way to Mars only to discover its life is merely an unexceptional branch on the terran tree of life, and not a second independent genesis. There is, however, another possibility for finding life derived from a second genesis that does not require expensive space exploration. If life really *does* emerge readily in Earth-like conditions, then perhaps it emerged more than once on Earth. After all, no planet is more "Earth-like" than Earth itself.

Perhaps our exploration should be directed more intensely to find non-standard life on Earth.

What are the odds that deeper exploration on Earth might turn up life that does *not* belong to the same tree as the terran life already known? Biologists have only scratched the surface of the microbial realm. As the vast majority of terrestrial species are microbes, we cannot rule out that intermingled with standard terran life that we know are denizens of a completely different tree. One or more radically different form of life might lurk all around us (Davies and Lineweaver 2005), forming a sort of "shadow biosphere" (Cleland and Copley 2005). The discovery of a microbe with an alternative biochemical and genetic scheme – an alien organism – living right under our noses would be one of the greatest advances in biology since the theory of evolution.

How could we identify microbial aliens on Earth? Two possibilities suggest themselves. First, non-standard life might be adapted to somewhat different habitats than standard terran life. It is of course well known that standard life inhabits a remarkably wide range of habitats. Some "extremophiles" thrive in temperatures well above the normal boiling point of water; others live in dry and cold conditions far below freezing. Still others withstand high concentrations of salt or toxic metals, or extremely high or low pH, or even intense radiation. Nevertheless, limits exist. In the deep ocean volcanic vent ecosystems, organisms have been found alive at temperatures above 120 °C (Baross *et al.* 2007). But 130 °C seems to present an upper limit, and there are good chemical reasons for this relating, in particular, to the stability of DNA. If microbes were found living in a temperature range of, say, 170 to 190 °C, the discontinuity in the temperature would make these organisms stand out as candidates for non-standard life, perhaps ones utilizing an informational molecule more stable than DNA.

Alternatively, we might explore for non-standard life at depth. It is well established that microorganisms inhabit the continental and ocean bed subsurface, to a depth of one kilometer or even more, a realm sometimes dubbed "the deep hot biosphere." A radically alternative form of life might lurk at even greater depths, awaiting the prospector's drill.

If standard and putative non-standard life occupied broadly the same habitats, non-standard life would be much harder to identify. Since one cannot generally tell by observation what biochemistry supports a microbe, we are presented with the problem of how to "separate the sheep from the goats" in any exploration of standard habitats. Two approaches suggest themselves.

The first would exploit an educated guess as to what alternative biochemistry might be utilized, and then target a feature of it. Returning to our experience with Mars, the fact that standard life forms all use ribosomes to make proteins, standard living cells cannot be smaller than a certain size. It is relatively easy to explore for cells below this size limit, even if they are intermingled with larger microbes. In fact, there are persistent claims that such "nanobacteria" or "nanobes" exist in the environment (Benner 1999).

Another method would devise some sort of filter that would disable the metabolism of standard life, and then attempt to culture organisms from a wide range of environments to see whether anything stands out as thriving in spite of the filter. One experiment along these lines sought to explore for bacteria able to grow in "mirror soup" – a nutrient medium having a chirality opposite to that used by standard life, and therefore unpalatable to standard life, but nutritious to hypothetical "mirror life" that employs sugars and amino acids of the opposite chirality. However, this approach is complicated by many factors, and is not a decisive discriminator when used in the simple experiment just described. A more sophisticated version of the experiment might however yield interesting results.

Yet another possibility might use antibiotics to disable any organisms using the standard ribosome. Anything that thrives in a culture containing these antibiotics would be a candidate for a life form having a different mechanism for making biopolymers, perhaps even using different biopolymers. A final possibility is that all life on Earth had a common origin, but split early into branches that diverged very far biochemically. This creates the possibility that exploration will find "living fossils" – primitive precursors of standard terran life that cling on in certain special niches. For example, the familiar triplet code might have been preceded by a doublet or even a quadruplet code (Neumann *et al.* 2010). Could microbes still exist using this ancient code? Another intriguing idea is that RNA-world life is still out there somewhere (Benner 1999). All these possibilities offer well-defined targets for scientists interested in trawling the environment for "forms most wonderful" even by Darwin's standards.

Working forwards in time from chemistry

The back-and-forth between observation and definition-based theories of life as we explore for life illustrates the limitations of exploration as a strategy to define life as a universal. What is interpreted as a biosignature is conditioned on the community's theory of life. As a practical matter, this makes it difficult to decide in advance precisely what to observe, which makes it challenging to allocate scarce payload to life-detection experiments. We can, of course, constrain our view of the kinds of life to seek by recognizing that a form of life is possible only if it can originate. If the universe had a beginning, then so must life. Excluding supernatural causes, we might ask: What kinds of chemical systems capable of self-sustaining support of Darwinian evolution could also emerge from non-Darwinian chemistry? This would narrow the scope of our search by excluding self-sustaining chemical systems capable of Darwinian evolution that could not have emerged without an intelligent designer, i.e. divine, human, or alien intervention. This approach is illustrated by the top wedge in Figure 2.1.

What Darwinian chemical system might emerge from non-Darwinian chemistry? One hypothesis for the origin of life holds that life on Earth began with RNA that played both genetic and catalytic roles. This "RNA first" hypothesis was advanced by Alex Rich a half century ago (Rich 1962). It solves, at least to some extent, the so-called "chicken-or-egg" problem. In modern terran life, proteins are needed to make DNA, and DNA is needed to make proteins. Both DNA and proteins almost certainly could not have emerged spontaneously from inanimate matter at the same time. And certainly not in the form where the DNA that spontaneously emerged just happened to encode proteins, while the proteins that spontaneously emerged just happened to catalyze the synthesis of DNA.

But if RNA originally played both the genetic role now done by DNA and the catalytic role now done by proteins, and if RNA could have supported Darwinian evolution via mutations and selection to give better catalysts (where those mutations were heritable by copying RNA as a genetic molecule), then life could have perhaps originated somehow by the abiotic emergence of RNA. This is the familiar "RNA-first hypothesis" for the origin of life. The vestige of this origin hypothesis is the role of messenger RNA as the intermediate in modern terran life between DNA and proteins, and the catalytic role of RNA in the biosynthesis of proteins.

Given this hypothesis, prebiotic chemists start with a list of organic compounds that were plausibly present on early Earth. Then, experiments are done in the laboratory to understand how rules of chemical reactivity, presumed invariant over time and space, might have allowed RNA to emerge spontaneously from these compounds. The literature attempting to generate oligomeric RNA from plausible prebiotic organic compounds is large (e.g. Powner *et al.* 2009), involving minerals (Ricardo *et al.* 2004), evaporating lipids (Rajamani *et al.* 2008), and exotic solvents such as formamide (Costanzo *et al.* 2009). Many experiments have been controversial; some have been difficult to reproduce. Experts in the community have doubted that such efforts will ever produce a compelling set of transformations that proceed from prebiotic compounds to oligomeric RNA without excessive numbers of steps requiring inordinate human supervision (Shapiro 2007).

Furthermore, prebiotic chemistry occurring billions of years ago in the history of Earth is unlikely to have left any discernible traces; the specific events leading to the origin of life on Earth may be inherently unknowable. More seriously, no research in prebiotic chemistry has yet managed to resolve a paradox that is well known to all chemists. When energy is applied to large collections of organic molecules having access to Darwinian evolution, the molecules reproduce themselves, grow, and adapt. However, when energy is applied to large collections of organic molecules not having access to Darwinian evolution, they are transformed into complex mixtures that bear an ever-increasing resemblance to tar and an ever-decreasing resemblance to life. Further application of energy leads to tar that resembles asphalt, and appears to be even less able to sustain Darwinian evolution. Until this paradox is resolved by some appropriate combination of chemistry and environment, the "origins" problem will not be solved.

Synthesizing life from scratch

Of course, even if a laboratory experiment does generate RNA from plausible prebiotic precursors, the result need not offer us a constraint-by-origin view of what forms life might take universally. After all, molecules other than RNA might be able to simultaneously support genetics and catalysis. These too might be able to straddle non-Darwinian and Darwinian chemistry in a way that leads to their own form of life, possibly quite different from the life we know on Earth, and possibly in environments quite different from those found on early Earth.

This observation suggests a fourth approach to understanding life as a universal. This approach, represented by the right wedge in Figure 2.1, comes under the title "synthetic biology." Synthesis, especially in chemistry, reflects first a definition of understanding as an ability to create. If we truly understand chemical systems that support Darwinian evolution, we should be able to construct one of our own in the laboratory. If the NASA definition-theory of life is on point, this artificial system should be able to re-create all of the properties that we value in life. Conversely, if we cannot build life from scratch, then we must not understand completely what life is. Further, *de novo* synthesis provides us with candidate Darwinian molecules that are different in structure from those found in terran biology. Through synthesis, we can ask: What kinds of molecules other than standard DNA and RNA might support Darwinian evolution? This takes us in the direction of universality, at least as far as our imagination and synthetic technology allow.

Underlying the famous double helix of Watson and Crick (Watson and Crick 1953) are two simple rules for genetics, rules that describe complementarity between two DNA strands. The first rule, size complementarity, pairs large purines in one strand with small pyrimidines in the other. The second, hydrogen bonding complementarity, pairs hydrogen-bond donors from one nucleobase with hydrogen-bond acceptors from the other (Figure 2.3). In the first-generation model for the double helix, the nucleobase pairs were central. In contrast, the backbone, made of alternating sugar and negatively charged phosphate groups, was viewed as being largely incidental to the molecular recognition event at the center of natural genetics and Darwinian evolution. It provided a scaffold, and little more. If this simple "first-generation" model for the double helix were correct and complete,

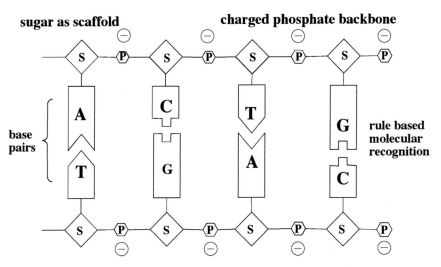

Figure 2.3 The two rules of Watson–Crick complementarity, illustrated in cartoon form (below) and molecular form (above).

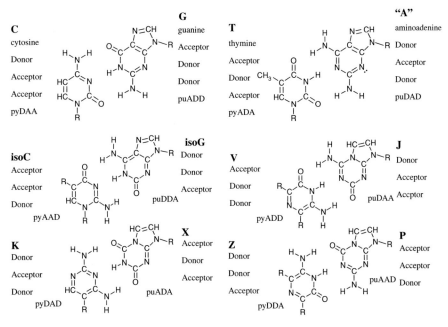

Figure 2.4 Rearranging the hydrogen-bond donor and acceptor groups in the laboratory allowed the preparation of eight new nucleotides that form four, non-standard, base pairs, in addition to the standard four nucleotides that form the two standard base pairs.

then we should be able to synthesize a different molecular system with different nucleobases to get an unnatural synthetic system that could mimic the molecular recognition displayed by natural DNA and RNA. We might even be able to get this artificial genetic system to have children and, possibly, evolve.

How might this be done? On paper, if we shuffled the hydrogen-bond donor and acceptor groups in the A:T and G:C pairs (Figure 2.4), we can write down eight new nucleobases that fit together to give four new base pairs having the same geometry as the A:T and G:C pairs. Photocopy the page from this book, cut out the non-standard base pairs, and fit them together yourself as a modern James Watson. As with the four standard nucleobases examined by Watson and Crick, the new nucleobases were predicted to pair with size complementarity (large with small) and hydrogen-bond complementarity (hydrogen-bond donors with acceptors), *if the theory behind the pairing was so simple.*

Using the synthetic technology developed and enjoyed by chemists over the previous century, the Benner laboratory was able to synthesize all of the synthetic components of our new artificial genetic alphabet (Benner 2004). Further, these non-standard synthetic nucleotides actually worked in synthetic DNA and RNA strands. Artificial synthetic DNA sequences containing the eight new synthetic nucleotides formed double helices with their complementary synthetic sequences. Complementation followed simple rules; just as A pairs with T and G pairs with C, P pairs with Z, V pairs with J, X pairs with K, and isoG pairs with isoC. The non-standard large nucleotides paired only with the correct synthetic small nucleotide. The artificial non-standard DNA worked as well as natural DNA, at least in its ability to pair following simple rules (Benner 2004). But why stop here? The next challenge in assembling a synthetic biology requires us to have our synthetic genetic system support Darwinian evolution. For this, we need technology to replicate synthetic DNA. Of course, replication alone would not be sufficient. The copies must occasionally be imperfect and the imperfections must themselves be replicable.

DNA polymerases, enzymes that copy standard DNA strands by synthesizing new strands that pair A with T, T with A, G with C, and C with G, might be called upon to replicate non-standard DNA containing extra nucleotide "letters" (Yang *et al.* 2007). Unfortunately, natural polymerases have evolved for billions of years to accept the G, A, T, and C of natural genetic systems, not non-standard genetic systems. In most cases, non-standard DNA differs too much from natural DNA to be accepted by most natural polymerases. Fortunately, the polymerases themselves can be changed. Their amino acids can be replaced, leading to non-standard polymerases that can accept non-standard DNA. Several examples of successful replication of non-standard DNA are now known with non-standard polymerases (Sismour *et al.* 2004, Yang *et al.* 2010). The non-standard polymerases copy the non-standard DNA, and then copy the copies, and then copy the copies of the copies. But do these polymerases also occasionally make mistakes (mutations)? And can these mistakes themselves be copied? Could non-standard DNA support Darwinian evolution?

Here, careful experiments were done to determine whether non-standard nucleotides could participate in mutation processes (Yang *et al.* 2010). This work showed that Z and P could indeed mutate to C and G and, more surprisingly, that C and G could mutate to give Z and P. The details of the mutation process were studied. Sometimes Z is incorporated opposite G instead of C. Sometimes C is incorporated opposite P instead of Z. Sometimes P is incorporated opposite C instead of G. Sometimes G is incorporated opposite Z instead of P. This low level of mutation is just a few percent per copy. But once mutations are introduced into the children DNA, they themselves can be copied and therefore propagated to the next generation. Thus, the synthetic genetic system built from G, A, C, T, Z, and P (GACTZP) is indeed capable of supporting Darwinian evolution. A GACTZP synthetic six-letter genetic system is clearly not homologous to the genetic system that we find naturally on Earth. It does, of course, share many structural features with natural genetic systems. Some of these we believe to be universal based on theories like the polyelectrolyte theory of the gene (Benner and Hutter 2002). The repeating backbone phosphates are not, according to that theory, dispensable.

However, GACTZP DNA can support Darwinian evolution. Is it therefore artificial synthetic life? Our theory-based definition holds that life is a *self-sustaining* chemical system capable of Darwinian evolution, and GACTZP DNA is not self-sustaining. For each round of evolution, a graduate student must add something by way of food; the system cannot go out to lunch on its own. Thus, GACTZP DNA is not yet a "second example" of life. We are not ready to use our system to see whether it can spontaneously generate traits that we recognize from natural biology. Nevertheless it is a start.

References

Baross, J., Benner, S. A., Cody, G. D. *et al.* (2007). *The Limits of Organic Life in Planetary Systems.* Washington, DC: National Academies Press.

Benner, S. A. (1999). "How small can a microorganism be? Size limits of very small microorganisms," Proceedings of a Workshop, Steering Group on Astrobiology of the Space Studies Board. National Research Council, pp. 126–135.

Benner, S. A. (2004). "Understanding nucleic acids using synthetic chemistry," *Accounts of Chemical Research*, Vol. 37, pp. 784–797.

Benner, S. A. (2009). *Life, the Universe and the Scientific Method.* Gainesville, FL: FAME Press.

Benner, S. A. and Hutter, D. (2002). "Phosphates, DNA, and the search for nonterran life: a second generation model for genetic molecules," *Bioorganic Chemistry*, Vol. 30, pp. 62–80.

Benner, S. A., Ellington, A. D., and Tauer, A. (1989). "Modern metabolism as a palimpsest of the RNA world," *Proceedings of the National Academy of Science*, Vol. 86, pp. 7054–7058.

Benner, S. A., Devine, K. G., Matveeva, L. N., and Powell, D. H. (2000). "The missing organic molecules on Mars." *Proceedings of the National Academy of Science*, Vol. 97, pp. 2425–2430.

Biemann, K., Oro, J., Toulmin, III, P. *et al.* (1977). "The search for organic substances and inorganic volatile compounds in the surface of Mars," *Journal of Geophysical Research*, Vol. 82, pp. 4641–4658.

Cleland, C. E. and Chyba, C. F. (2002). "Defining 'life,'" *Origins of Life and Evolution of the Biosphere*, Vol. 32, pp. 387–393.

Cleland, C. E. and Copley, S. D. (2005). "The possibility of alternative microbial life on Earth," *International Journal of Astrobiology*, Vol. 4, pp. 165–173.

Cooper, G., Kimmich, N., Belisle, W. *et al.* (2001). "Carbonaceous meteorites as a source of sugar-related organic compounds for the early Earth," *Nature*, Vol. 414, pp. 879–884.

Costanzo, G., Pino, S., Ciciriello, F., and Di Mauro, E. (2009). "Generation of long RNA chains in water," *Journal of Biological Chemistry*, Vol. 284, pp. 33206–33216.

Davies, P. C. W. 1996. "The transfer of viable micro-organisms between planets," in *Evolution of Hydrothermal Ecosystems on Earth (and Mars?)*: Proceedings of the CIBA Foundation Symposium No. 20. G. Brock and J. Goode (eds). New York: Wiley.

Davies, P. C. W. and Lineweaver, C. H. (2005). "Hypothesis paper – finding a second sample of life on Earth," *Astrobiology*, Vol. 5, pp. 154–163.

Gaucher, E. A., Thomson, J. M., Burgan, M. F., and Benner, S. A. (2003). "Inferring the paleoenvironment of ancient bacteria on the basis of resurrected proteins," *Nature*, Vol. 425, pp. 285–288.

Gilbert, W. (1986). "Origin of life: the RNA world," *Nature*, Vol. 319, p. 618.

Hoyle, F. 1957. *The Black Cloud*. London: Heinemann.

Huang, Z., Schneider, K. C., and Benner, S. A. (1991). "Building blocks for analogs of ribo- and deoxyribonucleotides with dimethylene-sulfide, -sulfoxide and -sulfone groups replacing phosphodiester linkages," *Journal of Organic Chemistry*, Vol. 56, pp. 3869–3882.

Johnson, R. J., Gaucher, E. A., Sautin, Y. Y. *et al.* (2008). "The planetary biology of ascorbate and uric acid and their relationship with the epidemic of obesity and cardiovascular disease," *Medical Hypotheses*, Vol. 71, pp. 22–31.

Joyce, G. F. (1994). From foreword of *Origins of Life: The Central Concepts*, by D. W. Deamer and G. R. Fleischaker. Boston, MA: Jones and Bartlett.

Kondepudi, D. K., Kaufman, R. J., and Singh, N. (1990). "Chiral symmetry breaking in sodium chlorate crystallization," *Science*, Vol. 250, pp. 975–976.

Koshland, D. E. (2002). "The seven pillars of life," *Science*, Vol. 295, pp. 2215–2216.

Levin, G. V. and Straat, P. A. (1977). "Recent results from the Viking labeled release experiment on Mars," *Journal of Geophysical Research*, Vol. 82, pp. 4663–4667.

Levin, G. V. and Straat, P. A. (1981). "A search for a nonbiological explanation of the Viking labeled release life detection experiment," *Icarus*, Vol. 45, pp. 494–516.

McKay, D. S., Gibson, E. K., Thomas-Keprta, K. L. *et al.* (1996). "Search for past life on Mars: possible relic biogenic activity in Martian meteorite ALH84001," *Science*, Vol. 273, pp. 924–930.

Melosh, H. J. (2003). "Exchange of meteorites (and life) between stellar systems," *Astrobiology*, Vol. 3, pp. 207–215.

Neumann, H. *et al.* (2010). "Encoding multiple unnatural amino acids via evolution of a quadruplet-decoding ribosome," *Nature*, Vol. 464, pp. 441–444.

Pizzarello, S. (2004). "Chemical evolution and meteorites: an update," *Origins of Life and Evolution of Biospheres*, Vol. 34, pp. 25–34.

Powner, M. W., Gerland, B., and Sutherland, J. D. (2009). "Synthesis of activated pyrimidine ribonucleotides in prebiotically plausible conditions," *Nature*, Vol. 459, pp. 239–242.

Rajamani, S., Vlassov, A., Benner, S. *et al.* (2008). "Lipid-assisted synthesis of RNA-like polymers from mononucleotides," *Origins of Life and Evolution of Biospheres*, Vol. 38, pp. 57–74.

Ricardo, A., Carrigan, M. A., Olcott, A. N., and Benner, S. A. (2004). "Borate minerals stabilize ribose," *Science*, Vol. 303, p. 196.

Rich, A. (1962). "On the problems of evolution and biochemical information transfer," in *Horizons in Biochemistry*, M. Kasha and B. Pullmann (eds.). New York: Academic Press.

Shapiro, R. (2007). "A simpler origin for life," *Scientific American*, Vol. 296, pp. 46–53.

Sismour, A. M., Lutz, S., Park, J.-H. *et al.* (2004). "PCR amplification of DNA containing non-standard base pairs by variants of reverse transcriptase from human immunodeficiency virus-1," *Nucleic Acids Research*, Vol. 32, pp. 728–735.

Watson, J. D. and Crick, F. H. C. (1953). "Molecular structure of nucleic acids: a structure for deoxyribose nucleic acid," *Nature*, Vol. 171, pp. 737–738.

Yang, Z., Sismour, A. M., Sheng, P. *et al.* (2007). "Enzymatic incorporation of a third nucleobase pair," *Nucleic Acids Research*, Vol. 35, pp. 4238–4249.

Yang, Z., Hutter, D., Sheng, P. *et al.* (2006). "Artificially expanded genetic information system: a new base pair with an alternative hydrogen bonding pattern," *Nucleic Acids Research*, Vol. 34, pp. 6095–6101.

Yang, Z., Chen, F., Chamberlin, S. G., and Benner, S. A. (2010). "Expanded genetic alphabets in the polymerase chain reaction," *Angewandte Chemie*, Vol. 49, pp. 177–180.

3

Terran Metabolism

The First Billion Years

SHELLEY COPLEY AND ROGER SUMMONS

Introduction

Microbial life on Earth occupies an astonishingly wide range of habitats. Microbes proliferate in hydrothermal vents at the bottom of the ocean, in the intensely cold and dry valleys of Antarctica, in acid mine drainage streams laden with toxic metals, and in anoxic sediments in lakes, rivers, and the bottom of the sea. They occur in vast numbers in environments as diverse as the nutrient-poor waters of the open ocean and the guts of animals. The sources of carbon, nitrogen, phosphorous, sulfur, and energy for building biomass vary widely in these environments. Microbes have evolved an incredibly diverse range of strategies that take advantage of the resources available in particular environmental niches. Some, termed autotrophs ("self-feeding"), can synthesize all organic compounds needed for cell growth from CO_2, H_2, and inorganic sources of the phosphorous, sulfur, and trace metals they need. Others, termed heterotrophs ("other"-feeding), derive carbon and energy from organic molecules synthesized by other life forms. Figure 3.1 summarizes these two approaches to metabolism.

Building biomass requires energy. Consequently, living organisms must harness energy from the environment and store it in a chemical or physical form. Energy can be harnessed either from light or from oxidation of electron-rich compounds. Some microbes oxidize organic molecules such as sugars, while others oxidize inorganic species such as H_2, Fe(II), or H_2S. Electrons derived from oxidation of organic or inorganic compounds can be passed to a variety of electron acceptors, including O_2, NO_3^-, and SO_4^{-2}. Such coupled oxidation and reduction reactions are termed "redox processes." During the thermodynamically downhill passage of

Frontiers of Astrobiology, ed. Chris Impey, Jonathan Lunine and José Funes.
Published by Cambridge University Press. © Cambridge University Press 2012.

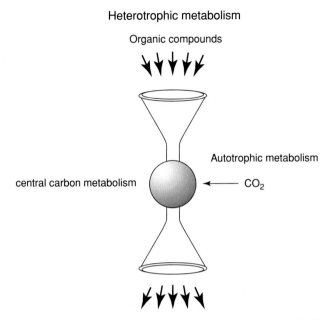

Figure 3.1 Organic compounds or CO_2 are fed into the central carbon metabolism. From this common core, pathways leading to amino acids, nucleotides, and various secondary metabolites emerge.

electrons from the initial electron donor to the final electron acceptor, energy is stored in the form of electrochemical gradients of ions (usually H^+, but sometimes Na^+) across lipid bilayers. These electrochemical gradients are subsequently used to drive energy-requiring processes such as movement, uptake of nutrients, and, most importantly, synthesis of ATP (see Figure 3.2). The chemical energy stored in ATP is used to drive hundreds of other energy-requiring processes.

Despite the variety of approaches microbes use to obtain carbon and energy from the environment, there is a strikingly conserved core of metabolic reactions in all forms of life. Pathways for synthesis of amino acids, ribose, and nucleotides are identical, or nearly so, in Bacteria, Archaea, and Eukarya, which constitute the three domains of life we recognize on Earth. The tricarboxylic acid cycle (also known as the TCA cycle) and glycolysis (the pathway for degradation of glucose) are also very similar in the three domains of life. The conservation of these pathways suggests that they are ancient and that they evolved in the last universal common ancestor (the LUCA) of the three domains of life. Since the LUCA, the ancestral metabolic network has been elaborated in certain lineages by addition of pathways for degrading carbon sources specific to certain environmental niches and for synthesizing natural products that allow organisms to manipulate other

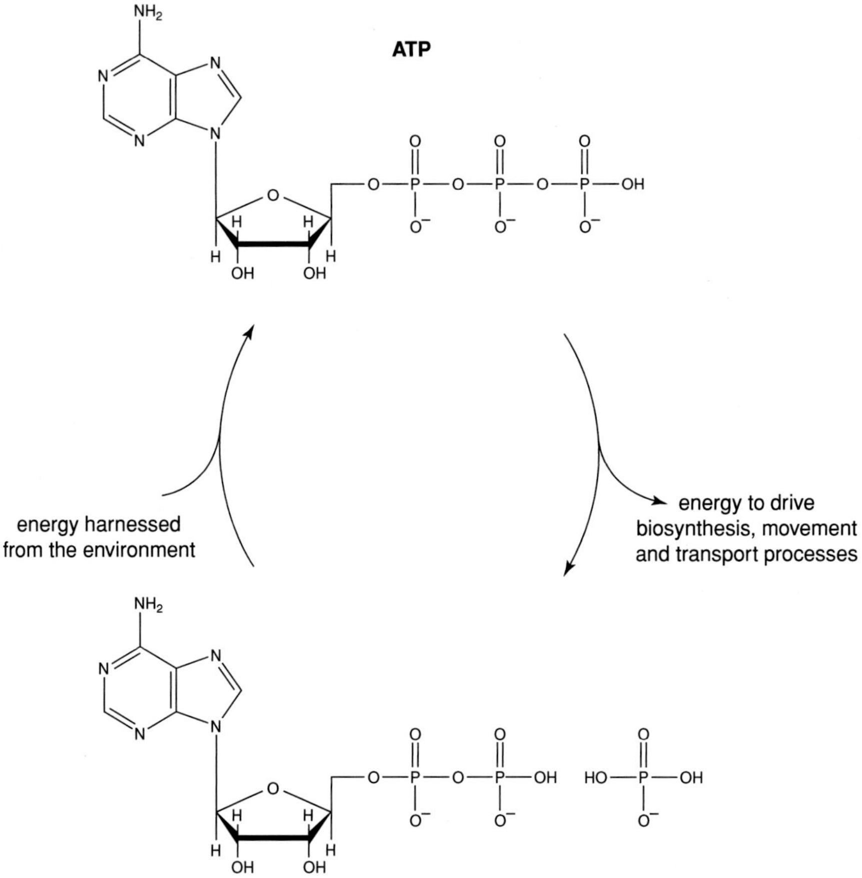

Figure 3.2 Hydrolysis of ATP (adenosine triphosphate) releases energy that can be used to drive processes that require energy. ATP is re-formed using energy harnessed from the environment.

organisms in their environment. It has also been pruned, again in certain lineages, to eliminate pathways that are unnecessary because the end-products are available in the environment. Thus, the metabolic networks of extant microbes are often quite different, but vestiges of the ancestral metabolism can still be seen.

The function of metabolic networks depends upon protein enzymes that catalyze nearly all of the chemical reactions in cells. Without catalysis, life would be impossible because many important reactions would be far too slow if left to run by themselves. Modern enzymes are marvelous catalysts, accelerating the rates of reactions by up to 20 orders of magnitude (Miller and Wolfenden 2002). Generally, each enzyme catalyzes a specific reaction, or sometimes a small number of related reactions, in the metabolic network. Fluxes through various parts of the metabolic network are controlled in response to environmental conditions by adjustment of enzyme activities by small molecules that either inhibit or stimulate activity, and by alterations in the levels of enzymes. These regulatory interactions allow efficient use of available resources by shutting down unnecessary pathways and diverting carbon toward metabolites that are needed most critically. Both the prodigious catalytic abilities of enzymes and the exquisite regulation seen in extant metabolic networks have been honed over billions of years as a result of competition between species for resources.

This chapter explores how the sedimentary rock record and the genomes of extant organisms provide information about the evolution of sophisticated metabolic networks over geological time. This is a difficult endeavor, as much of the evidence is both indirect and imperfect and there are divergent opinions over its interpretation. We begin with a discussion of evidence from the rock record and bioinformatics and how, together, these provide insights into the metabolism of early microbes. Sedimentary rocks record aspects of iron, carbon, sulfur, and nitrogen metabolism by microbes as early as 3.5 billion years ago and perhaps longer. Using bioinformatic methods, we can extrapolate back even further to the LUCA, which lived approximately 3.8 billion years ago. However, as we will see, the LUCA already had a fairly sophisticated metabolism. We have no direct information about the emergence of the metabolic network of the LUCA, and consequently we must resort to informed speculation based upon geochemical boundary conditions and experimental demonstrations of plausibility, which are limited to showing that certain things could have happened, but can never prove that they did.

Evidence from the rock record

Sedimentary rocks form when pre-existing rocks are broken down by the actions of water, ice, and microbes and the components transported to a new place by water or wind. Water, as it moves in rivers, lakes, and the oceans, is capable of relocating vast amounts of material, and this, when it comes to rest, forms a new deposit that captures some of that transport history, including the influences of water depth and chemistry, wave or tidal action, and, most importantly, evidence of the metabolisms it encountered along the way. Thick sequences of sedimentary rock accumulate over millions of years, especially on the margins of continents and in lakes. Some of these comprise relocated fragments of rock and mineral grains, while others are overwhelmingly dominated by direct biological input. Examples include organic matter from plants preserved in the form of coal beds, the skeletons of microscopic plankton forming the White Cliffs of Dover, and massive carbonate reef systems that appear in the geological record at about the same time as the first animal fossils.

Sediment transport and deposition also require that there be an uneven surface topography. Highs and lows in the crust are generated continuously through tectonism, that is, volcanism and the deformation of Earth's surface that accompanies heat loss from its interior mantle and core. Subsiding basins are topographic lows formed during the rifting of continents and other processes that accompany the movement of buoyant slabs of continental and ocean crust that are in constant motion across Earth's surface. Complementary to the process of rifting is the collision of landmasses, leading to formation of mountain ranges, or topographic highs, such as the modern-day Tibetan Plateau and Himalayas. Another important process, subduction, occurs when seafloor sediments are pushed under another piece of crust to be heated and metamorphosed. Thus, the movement of tectonic plates is forever resurfacing our planet. So, while new sediment piles are being formed, others are being lost to weathering, deep burial, or subduction. On average, the half-life of a sedimentary rock is about 300 million years, meaning that, as we go back in time, the amount of rock recording past events progressively decreases. Further, those rocks that remain are subject to various forms of confounding alteration and it becomes progressively more difficult to identify life's signals with confidence. We rely on just a few remnants of stable, preserved crust, called cratons, for information about Earth's early atmosphere, hydrosphere, and the metabolism of its inhabitants.

Unfortunately, no intact rocks remain to inform us about the first 600 million years of Earth's history. Models of planetary accretion and the cratering history evident on our Moon speak to hot and turbulent beginnings and geologists call this the Hadean Eon (Figure 3.3). As discussed in Chapter 6 of this book, compositions of volcanic gases and the fact that H_2 and Fe(II) are produced copiously when minerals such as olivine and pyroxene react with water suggest that the oceans and atmosphere were reducing during this interval. Chemical analyses of zircon mineral grains, nature's time capsules, dated to between 4.1 and 4.35 billion years ago (Gyr ago) and recovered from younger 3.5 Gyr old sediments, suggest that liquid water was present and, therefore, that the climate was equitable. However, the occurrence of these zircons also tells us that crustal differentiation and sediment recycling processes must have been in place sometime before 4 Gyr ago (Wilde et al. 2001, Watson and Harrison 2005).

The oldest known intact rocks, the ~3.96 Gyr old Acasta Gneiss of Canada (Bowring and Housh 1995) and 3.9–3.6 Gyr old Itsaq Gneiss Complex of SW Greenland (Kamber 2007), have seen such high temperatures that any remnants of metabolic activity, were they ever present, have been erased by recrystallization. The evolution of Earth's crust, from its original and relatively dense basaltic composition (like the Moon and Mars) to one with lower-density continental rocks, and especially a growing inventory of granites, is a consequence of the presence of oceans and active tectonic and hydrological cycles (Campbell and Taylor 1983). Earth is unique among inner Solar System bodies with its oceans, granites,

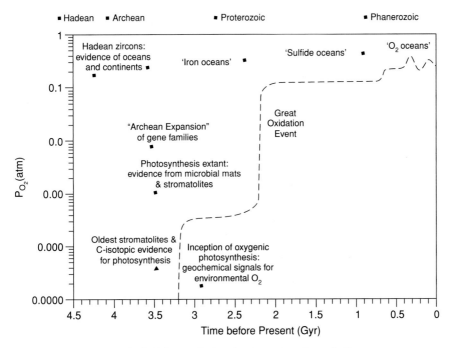

Figure 3.3 Geological time-scale showing inception of metabolic processes, hypothetical course of atmospheric oxygenation, and critical transitions in ocean chemistry.

continents, and life. Rosing *et al.* (2006) have gone so far as to propose that the presence of stable continents by 3.5 Gyr ago, if not enabled, was at least hastened by the advent of photosynthesis.

The oldest preserved sedimentary rock sequences, laid down between 3.5 and 3.0 Gyr ago, encode more tangible evidence for past biological activity. Microscopic objects, likely fossils of microbes, are found in several rock units of the Warrawoona Group of the Pilbara Craton that are dated at just under 3.5 Gyr ago (Awramik *et al.* 1983, Schopf and Packer 1987), and there is a compelling record of microbial activity in the form of stromatolites in sediments of the Strelley Pool Formation, also within the Warawoona sequences (Allwood *et al.* 2006, 2009). Of comparable age, within the sequences of the Barberton Greenstone Belt of Southern Africa, are many examples of microbially induced sedimentary structures (MISS) that record the presence of microbial mats in tidal settings with hydraulic characteristics (tidal range, average current velocities, sand compositon, etc.) equivalent to modern counterparts (Noffke *et al.* 2008, Noffke 2010). These are discussed in more detail in Chapter 5.

Early reports on these discoveries proposed that they constituted evidence for photosynthesis and, possibly, oxygenic photosynthetic metabolisms (Schopf and

Packer 1987, Awramik 1992, Des Marais 2000), and multiple lines of evidence are certainly consistent with this (Awramik 1992, Bosak *et al.* 2009). However, researchers have recently been more cautions about asserting which energy-harvesting metabolisms these communities are recording (Schopf 2006, Schopf *et al.* 2007, Allwood *et al.* 2009).

While the energy sources for early microbial activity are actively and vigorously debated, organic matter and the ratios of carbon isotopes in organic matter and in carbonate rocks attest to processes of biological carbon fixation. Organic carbon in the form of the complex macromolecular substance known as kerogen, that is preserved between layers of carbonate in the Strelley Pool Formation stromatolites and other sediments for the early part of the Archean Eon (3.8–2.5 Gyr ago) (Schidlowski *et al.* 1983), is depleted in carbon-13 to the same degree as modern biologically formed organic carbon (Marshall *et al.* 2007). The isotopic compositions of Strelley Pool Formation kerogens, expressed using the standard $\delta^{13}C$ notation, range from -28.3 to $-35.8‰$ relative to the international Vienna Pee Dee Belemnite (VPDB) standard. This reference point has a nominal value of 0‰ that is very close to that of dissolved inorganic carbon (DIC) in modern seawater. Similar isotopic depletions, or fractionations, also occur in the absence of biological processes, for example, the formation of gases such as methane in deep-sea hydrothermal systems (McCollom and Seewald 2006, Proskurowski *et al.* 2008). However, the trace element patterns evident in the carbonates of the Strelley Pool Formation stromatolites are close to what we find in ancient and modern seawater and different from what would be expected for hydrothermal fluids (Allwood *et al.* 2010).

Accordingly, biological carbon fixation best accounts for the origin of the organic matter in the stromatolitic reef member of the Strelley Pool Formation. And, as we look forward in geological time, it is a well-established fact that organic matter (i.e. kerogen), which is ubiquitously present in sediments and sedimentary rocks to the present day, is isotopically depleted relative to coeval carbonate. The range of depletion from 10 to 60‰ is wide and varies around a mean of $\sim30‰$. While many geological, environmental, biological, and biochemical parameters factor into this wide range of values (Des Marais 1997, 2001), the near constancy of the average separation is seen by researchers as evidence for the continued operation of a biogeochemical carbon cycle for at least 3.5 billion years and possibly longer (Schidlowski 1988). Similar arguments have been applied to the interpretation of the isotopic compositions of the fully reduced and oxidized forms of sulfur, sulfide, and sulfate, respectively, and can be used to place the advent of bacterial sulfate reduction to Early Archean times based on data acquired from the Warrawoona sediments of the Pilbara Craton in Australia (Canfield *et al.* 2000, Shen *et al.* 2001).

While the isotopic compositions of carbon, sulfur, and nitrogen preserved in sedimentary materials are interpreted to carry signatures of metabolism, the signals are subtle and it is generally difficult or impossible to infer specific reaction pathways. But the fact that we find morphological evidence for microbial communities and compelling isotopic data for the biological cycling of these elements in similarly aged sequences from Australia, Southern Africa, and North America suggests that microbes had developed complex syntrophic associations as are observed today in microbial mat ecosystems (Nisbet and Fowler 1999). Sedimentological evidence for their occurrence in shallow-water marine environments (Allwood *et al.* 2006), and far from the sources of electron donors such as reduced sulfur, iron, and H_2, further suggests they could have obtained carbon and energy by harvesting sunlight (Awramik 1992, Tice and Lowe 2004) with water-splitting oxygenic photosynthesis. The accumulation, concentration, and preservation of organic carbon in well-laminated stromatolites (Marshall *et al.* 2007, Allwood *et al.* 2009) forming in shallow-water, high-energy environments is especially notable because it requires that this photosynthesis was persistent and not interrupted by a lack of electron donors. This has led some scientists to propose an early appearance of photosynthesis where the freely available water was the electron donor and O_2 the waste product (Cloud 1972, Nisbet and Sleep 2001).

Arguably, the most overt signals for microbial metabolism operating on a global scale are those phenomena, beginning in the Archean Eon, which account for the oxidation of Earth's surface and interior following the advent of the first light-harvesting, O_2-producing metabolism. It is thought that iron-oxidizing bacteria prevalent in the early oceans could have harvested sunlight in an early stage of the evolution of photosynthesis. Electrons stripped from dissolved iron (II) released by sub-seafloor volcanic activity were then used to reduce carbon dioxide and generate organic compounds. The direct byproduct, insoluble iron (III) hydroxide, formed precipitates that sank to produce the first sedimentary iron deposits. Later, O_2-yielding photosynthesis provided molecular oxygen that could directly oxidize the iron (II) in the deep ocean to form vast deposits of iron oxides that, along with co-precipitated silica, are known as banded iron formations (BIF). BIF deposition began about 3.5 Gyr ago, peaked at about 2.5 Gyr ago, and had virtually ended by 1.8 Gyr ago; this pattern is observed globally.

The O_2 produced by photosynthesis in the surface ocean would have been closely balanced by numerous biological and abiotic sinks. A primary sink would have been O_2-respiring microbes consuming it for the energy premium it provides over anaerobic metabolisms (Fenchel and Finlay 1995). As the supply of iron (II) and other reductants in the ocean waned, together with a possible change in the composition of volcanic gases (Holland 2002), O_2 could escape the oceans and begin to accumulate in the atmosphere in what has become known as The Great

Oxidation Event (GOE). The timing appears to be relatively sudden at around 2.3–2.4 Gyr ago and the signs include subtle variations in isotopic patterns of elements such as sulfur, the weathering mobilization of redox-sensitive elements such as Mo, Re, and Ce, and the appearance of paleosols, that is, oxidized soils forming where basalts were exposed to air and water. It appears to have been an irreversible event in which the atmospheric O_2 content rose above a threshold of 10^{-5} of its present atmospheric level (PAL). This entire phenomenon is widely viewed as a signal for the power and persistence of microbial photosynthetic metabolisms and their ability to chemically alter a planetary surface (Cloud 1976).

Sulfur, another element with a central role in metabolism, is widely present as insoluble sulfide minerals in igneous and sedimentary rocks. With free O_2 present in the atmosphere and hydrosphere, these sulfides could be readily oxidized to soluble sulfate, transported to the ocean where they could be re-reduced to sulfide by bacterial sulfate reduction, and then trapped again in the form of pyrite (FeS_2), reflecting the affinity of sulfide for iron. In the absence of O_2, sulfides could be transported in water, but the grains were abraded to rounded shapes, thereby carrying the imprint of their transport. Rounded grains of transported pyrite disappeared from the geological record at the time of the GOE. Sulfides produced *in situ* as result of bacterial sulfide reduction, on the other hand, have a distinctive appearance that makes them recognizable as being "authigenic," i.e. formed in place. The decline of BIF after 2.5 Gyr ago is followed by a rise in the prevalence of sedimentary sulfide minerals. This, among other secular patterns in sedimentary mineral deposits, is further evidence of the role of microbial metabolism in processing redox-sensitive elements on a global scale (Lambert and Donnelly 1991).

To summarize what the geological record tells us, microbes had developed the ability to fix inorganic carbon into organic matter by 3.4 Gyr ago or before. The association of organic carbon with stromatolites, forming in shallow-water environments by this time, suggests that some form of photosynthesis was possible. This photosynthesis required a continuous supply of electron donors being transported into the shallow ocean. While such electron donors were liberated by volcanic activity in the deep ocean, there would have been rate limitations on their supply and transport that, in turn, would have placed severe constraints on where life could proliferate. Diverse forms of evidence suggest that microbes living near the ocean surface developed the ability to use water (which was not limiting) as an electron donor for photosynthesis by 2.8 Gyr ago. However, it could have been much earlier. Once this had been achieved, respiration and the products of volcanism kept a tight lid on aquatic O_2 concentrations until about 2.4 Gyr ago. By this time it is highly likely that all metabolisms that microbes utilize to harvest carbon and energy had been invented.

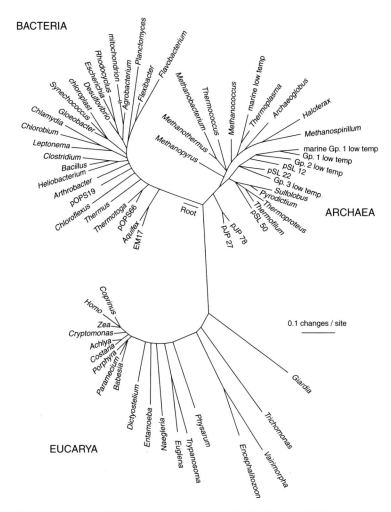

Figure 3.4 A tree of life based upon sequences of 16S ribosomal RNA.

The molecular "fossil" record

Figure 3.4 shows a tree of life that diagrams the relationships of all known life forms on earth. The tree is based upon the sequences of 16S ribosomal RNA (rRNA), a component of the ribosome, which is the molecular machine that synthesizes proteins in all recognized forms of life on Earth. Trees such as this one are built by comparing the sequences of 16S rRNA in different organisms and then drawing a map in which the lengths of branches on the tree are proportional to the number of differences observed in the RNA sequences. The root of the tree is the LUCA. The LUCA arose well before the first traces of life in the fossil record. Although all physical traces of the LUCA have vanished, we can reconstruct a remarkable amount of information about it by comparing characteristics, and in particular the genome sequences, of extant organisms in the three domains of life; traits shared by all three domains of life that are likely to have been inherited from the LUCA. This molecular "fossil" record allows us to extend our picture of early microbial life further back in time. For example, all organisms in the three domains of life have ribosomes, the sites of protein assembly in cells, so we can be certain that ribosomes were present in the LUCA. This in turn implies that template-encoded protein synthesis had emerged by the LUCA. The near-universality of the genetic code implies that the 20 amino acids commonly found in proteins were already being incorporated into proteins by the LUCA. The existence of proteins in the LUCA implies that proteins had already taken over many of the catalytic, structural, and regulatory functions that may have been provided by nucleic acids in earlier forms of life. Thus, many of the strategies used by extant microbes for growth and reproduction had already developed by this remarkably early stage.

When phenotypic characteristics vary in the three domains of life, it is more difficult to infer characteristics of the LUCA. For example, bacteria and eukaryotes synthesize membrane lipids by attaching fatty acids to sn-glycerol 3-phosphate via ester linkages, while Archaea synthesize membrane lipids by attaching isoprenoid chains to sn-glycerol 1-phosphate via ether linkages. This disparity in strategies makes it difficult to know what kind of membrane was present in the LUCA. One possibility is that the LUCA didn't have a lipid membrane at all, but existed within the porous walls of hydrothermal vents, and that the progenitors of bacteria and Archaea escaped separately from these inorganic compartments as independent pathways for lipid synthesis were invented (Koonin and Martin 2005). A competing theory suggests that the LUCA synthesized both types of lipids, and that one type of lipid was lost in the lineages leading to bacteria and eukaryotes, and the other type in the lineage leading to Archaea (Peretó et al. 2004). This possibility seems more likely; reconstruction of the probable gene content of the LUCA suggests that it had a number of membrane proteins, including transporters and ATP synthase, so it seems likely that the LUCA had some sort of lipid membrane. However, inorganic compartments may have been important in compartmentalization of an earlier form of life (see further below).

Confusion also arises when macromolecules that serve similar functions may have arisen independently in different lineages. For example, DNA is found in all recognized forms of cellular life, suggesting, by the logic explained above, that it arose in the LUCA. However, the core replicative DNA polymerases (enzymes that polymerize new DNA) of bacteria and Archaea are not related (Koonin 2006). Furthermore, the process of transcription of DNA into RNA is quite different in bacteria, Archaea, and eukaryotes. These discrepancies suggest that the LUCA may have had RNA but not DNA, and that the ability to synthesize DNA may have arisen independently after the divergence of the lineages leading to the three domains of life.

Metabolism in the LUCA

The core metabolic network of the LUCA was in many ways strikingly similar to those found in extant life. This assertion is based upon two lines of evidence. First, as mentioned above, pathways for synthesis of amino acids and nucleotides are identical or nearly identical in Archaea, bacteria, and eukaryotes. The TCA, or citric acid, cycle is present in representatives of all domains of life, although in some organisms only parts of the cycle are present, and in others the cycle runs in reverse. It is extremely unlikely that the same sequences of reactions would have been invented independently in different lineages.

The second line of evidence is based upon the predicted gene content of the LUCA. Assessing the metabolic capabilities of the LUCA is challenging because the content of metabolic enzymes in extant organisms has been shaped by multiple processes. Genes encoding metabolic enzymes can be inherited vertically from direct progenitors. They can also be gained by duplication of pre-existing genes and divergence of function between the resulting homologs, as well as by horizontal gene transfer between microbes, which has been rampant throughout the entire history of life. To further complicate the picture, genes encoding metabolic enzymes are often lost when they are not needed in particular environmental niches. As a consequence, few genes encoding metabolic enzymes are found in 100% of taxa in any domain of life. (Note the contrast between this situation and the universal conservation of ribosomes; protein synthesis is necessary in every environmental niche, so genes encoding ribosomal RNA are never lost.) A gene that is found in 40% of all taxa might have been present in the LUCA and subsequently lost in many lineages, or may have originated in one lineage and then appeared in other lineages via horizontal gene transfer. A decision between these alternative explanations requires assumptions about the relative frequencies of horizontal gene transfer and gene loss.

Careful analyses that take these uncertainties into account suggest that the LUCA contained a few hundred genes; Mirkin *et al.* (2003) estimated that the LUCA contained 572 genes. In a later study with a larger data set, Ouzounis *et al.* (2006) estimated that the LUCA contained 669 genes. Notably, genes encoding enzymes for synthesis of amino acids and nucleotides, for the TCA cycle, and for degradation of glucose, are predicted to have been present in the LUCA in both studies. Thus, we can be reasonably confident that, by the LUCA, many core metabolic pathways were firmly in place. Furthermore, a functional electron transport chain had emerged by the LUCA, as genes encoding subunits of the enzymes NADH dehydrogenase, succinate dehydrogenase, and the cytochrome b subunit of the cytochrome bc complex are predicted to have been present. The electron transport chain likely carried out its current function of generating an ion gradient across the membrane, since the reconstructed genome of the LUCA encodes several of the subunits of the F_0F_1 ATP synthase enzyme that uses the energy stored in an ion gradient to drive the synthesis of ATP. Clearly the LUCA captured energy from redox processes, but the specific types of electron donors and acceptors that were used cannot be determined.

Although the core of modern metabolism had emerged by the time of the LUCA, a considerable amount of innovation occurred as microbes proliferated on Earth, adapted to diverse habitats, and established ecosystems characterized by both syntrophic and competitive interactions between species. Existing free-living bacteria typically have 1000–2000 enzymes; *Streptomyces coelicolor* is predicted to have over 2800 enzymes (Freilich *et al.* 2005). This dramatic expansion of metabolic capabilities occurred by both acquisition of new protein folds and by gene duplication and divergence that allowed novel uses of existing protein folds.

Getting to the LUCA

It is remarkable that we can infer many of the characteristics of a life form that existed 3.8 billion years ago. The evidence discussed above suggests that the LUCA was a sophisticated life form, capable of replicating its genome (which may have consisted of RNA), synthesizing genetically encoded proteins using ribosomes, and synthesizing the nucleotide and amino acid building blocks of macromolecules. Unfortunately, the most intriguing part of the origin-of-life story, the processes that led to the origin of the complex metabolic and genetic systems of the LUCA, is beyond the reach of any hard physical or bioinformatic evidence. We can conjecture about these processes, but in this realm of speculation there is much more controversy.

An elegant hypothesis about the progenitor of the LUCA was put forward by Carl Woese (Woese 1998). Woese defined the "progenote" as an entity in which translation had just emerged but "had not developed to the point that proteins of the modern type could arise." The progenote, as the progenitor of the LUCA, is also a universal ancestor of life, but we distinguish it from the LUCA in this discussion because it lacked the accurate translation process required for reliable production of large proteins that are needed for faithful genome replication and efficient catalysis of metabolic reactions. Woese proposed that the progenote was a communal organism in which genetic information was extensively shared. Strategies for replication, transcription, translation, and metabolism would have been explored, refined, and shared within the entire community. The proposal that the progenote was a communal organism dovetails nicely with the proposition mentioned above that compartmentalization at the earliest stages of life was provided by the porous walls of hydrothermal vents, rather than by lipid membranes that would have provided a substantial barrier to transfer of genetic information.

Woese suggested that genome replication and translation in the progenote were rudimentary and inaccurate, a reasonable hypothesis since replication and translation could not have sprung into existence in their present sophisticated forms. Inaccurate replication would likely have limited the size of the progenote genome due to the risk of "error catastrophe," the accumulation of so many genetic mistakes that the organism is no longer viable. To illustrate this point, consider the problem of replicating a genome of one million bases, which is sufficient to encode a few hundred RNAs and proteins. (The smallest known genome for an extant free-living bacterium is that of *Pelagibacter ubique*, which consists of 1.3 million bases.) If replication were even modestly faithful, with an error frequency of 0.1%, every replication of a genome consisting of 1 million bases would result in 1000 errors, approximately one or two in every gene. Some of those errors would have been harmless, and a few might have been beneficial, but many would have been detrimental, leading to macromolecules with impaired functions. Inaccurate replication may have required the progenote to maintain multiple copies of each chromosome to provide genetic redundancy.

An important consequence – and a favorable one – of high error rates coupled with high genetic redundancy would have been an enhanced ability to explore sequence space and consequently a high rate of evolutionary innovation. The redundancy provided by multiple copies of chromosomes would mean that a functional copy of a gene was always available as a backup, even as the high mutation rate allowed a great deal of experimentation. When a gene encoding an improved macromolecule was discovered, it could replace the previous version and serve as the starting point for subsequent experimentation. This process would likely have resulted in fairly efficient testing of a variety of solutions to the problems of replicating the genome, producing proteins, harnessing energy from the environment, and synthesizing the building blocks of macromolecules.

The metabolic pathways that were present in the LUCA evolved as the progenote learned to synthesize the building blocks of macromolecules that initially were provided by the environment, either by delivery from the atmosphere or space or by geochemical processes. Early metabolic pathways probably relied on "generalist" enzymes that catalyzed certain generic reactions (for example, reduction of a carbonyl or phosphorylation of a carboxylate), probably with relatively low efficiency and low substrate specificity. Thus, the number of reactions that were included in an early metabolic network would have been much higher than the actual number of catalysts. As the progenote evolved toward the LUCA, both the number of enzymes and their specificity would likely have increased. More specific enzymes are advantageous for several reasons. First, an increase in specificity can increase catalytic power because a substrate can be oriented more precisely with respect to the active site machinery. Increased specificity also decreases the potential for catalysis of undesirable side reactions when certain substrates cannot be excluded from the active site. Finally, catalysis of reactions in more than one pathway by a generalist enzyme makes it difficult to optimize fluxes toward products that may be needed in very different quantities.

Metabolism in an "RNA" world

The progenote described by Woese is defined as an organism in which translation had just emerged. An even more primitive form of life must have existed before the progenote. This life form may have had an RNA genome, a metabolic network, and possibly small peptides, but no protein enzymes since translation had yet been invented. (The term "protein" is typically reserved for genetically encoded polypeptides that are produced by the ribosome.) In this life form, RNA molecules (or a similar nucleic acid) may have served as both the genetic material and as catalysts for genome replication as well as metabolic reactions. Numerous examples of RNA viruses that have RNA genomes and copy RNA directly into RNA (without a DNA intermediate) support the idea that a genome need not consist of DNA. The idea that nucleic acids were the earliest macromolecular catalysts was first proposed in the 1960s, and gained credence from the demonstration of catalysis by RNA molecules in the labs of Cech and Altman in the early 1980s (Kruger *et al.* 1982, Guerrier-Takada *et al.* 1983). Since then, ribozymes (catalytic RNA molecules) generated by *in vitro* evolution methods have been shown to catalyze a wide range of reactions involved in metabolism, including amino acid activation (Kumar and Yarus 2001), formation of coenzyme A (CoA), nicotinamide adenine dinucleotide (NAD), and flavin adenine dinucleotide (FAD) from 4′-phosphopantetheine, nicotinamide mononucleotide (NMN), and flavin mononucleotide (FMN), respectively (Huang *et al.* 2000), peptide bond synthesis (Illangasekare and Yarus 1999), and aldol condensation (Fusz *et al.* 2005). Thus, it appears that RNA molecules could have catalyzed all of the reactions needed to sustain life.

The first nucleic acid catalysts may not have actually been ribonucleic acid, which has a backbone based on the sugar ribose. Nucleic acids with intriguing alternative backbone structures include threonucleic acid (TNA) (Eschenmoser 1999, Schoning *et al.* 2000), peptide nucleic acid (PNA) (Nielsen 2007), and glycol nucleic acid (GNA) (Zhang *et al.* 2005). (This uncertainty is the reason for the inclusion of RNA in quotation marks in the title of this section.) Further, early nucleic acids may well have contained bases other than the four canonical bases found in RNA today. Although catalysis by nucleic acids has only been demonstrated for RNA and DNA, it is likely that alternative kinds of nucleic acids can catalyze chemical reactions, as well.

Although ribozymes have been shown to catalyze an impressive range of reactions, catalysis in the "RNA" world was likely not the exclusive purview of nucleic acids. Mineral surfaces, soluble metal ions, and small molecules, including amino acids and peptides, can also catalyze chemical reactions. These components presumably catalyzed reactions before the advent of macromolecular nucleic acids. Even after nucleic acid catalysts arose, metal ions and small molecules might have enhanced their catalytic abilities, either by stabilizing structures required for catalysis or by directly participating in catalysis (Copley *et al.* 2007, Cech 2009).

Prebiotic chemistry before the "RNA" world?

If the "RNA" world laid the foundation for the emergence of the progenote, then we must conclude that the emergence of life required a supply of nucleotides and probably simple amino acids, as well. The source of these critical components on the prebiotic Earth is the subject of considerable controversy.

One prevailing theory posits that life arose in ponds or lagoons supplied with organic compounds by in-fall from the atmosphere, interstellar dust particles, meteors, and comets. This theory is based on the early ideas of Oparin (1938) and Haldane (1929), and gained momentum from the famous Miller–Urey experiment, which demonstrated that electric discharges in a mixture of CH_4, NH_3, H_2O, and H_2 generate a complex mixture of organic compounds, including 22 amino acids (Miller 1953, Johnson *et al.* 2008). Although this mixture of gases was intended to model the early atmosphere, it is now believed that the early atmosphere was more likely a mixture of N_2, CO_2, CO, H_2O, and perhaps CH_4, and probably contained little NH_3 (Kasting 2008). However, amino acid precursors are generated when electrical discharges are passed through mixtures of these gases, as well (Cleaves *et al.* 2008). In addition to delivery of organics formed by atmospheric processes, an enormous amount of material was delivered to the early Earth from space. Chyba and Sagan (1992) estimated that $>10^9$ kg of carbon was delivered per year from interstellar dust particles, $>10^6$ kg by comets, and $>10^4$ kg by meteors.

A second important theory proposes that life originated at hydrothermal vents (Baross and Hoffman 1985, Martin *et al.* 2008) using organic compounds synthesized *in situ* from reactive small molecules in vent fluids. At mid-ocean spreading centers, hot hydrothermal fluids containing CO_2, H_2, and H_2S, and low levels of CH_4, are vented through porous minerals, including transition metal sulfides such as pyrrhotine (FeS), pyrite (FeS_2), chalcopyrite ($CuFeS_2$), and sphalerite (Zn,Fe)S (Kelley *et al.* 2002). Clay minerals (Murnane and Clague 1983) and phyllosilicates (Dekov *et al.* 2008) are also found in these hydrothermal systems. Off the main spreading axis, a fundamentally different type of vent is found (e.g. the Lost City) (Kelley *et al.* 2001, 2005). At these sites, reactions between seawater and newly exposed mantle olivine (($Mg,Fe)_2SiO_4$) generate serpentine ($Mg_3Si_2O_5(OH)_4$), magnetite (Fe_3O_4), and alkaline fluids that, upon mixing with cool seawater, generate towers of aragonite ($CaCO_3$) and brucite ($Mg(OH)_2$). Serpentine undergoes further reaction with CO_2 to give talc ($Mg_3Si_4O_{10}(OH)_2$). Vent fluids are cooler in these systems ($< 90\,°C$), and are rich in H_2 and CH_4. In both types of vents, the walls are porous, allowing chemical reactions between constituents of vent fluids to occur in pores lined with catalytic minerals at a range of temperatures varying from that of the inner conduit to that of the seawater bathing the external surface. Porous vent structures, as well as the surrounding highly fractured crust, would have provided an ideal environment for catalysis of prebiotic chemical reactions.

These two competing theories differ not only in the location at which life is believed to have emerged, but also in the mechanism by which metabolism emerged. Proponents of the first idea support the idea that the building blocks of macromolecules were initially supplied by in-fall from the atmosphere and from space, and that metabolism was invented only after the emergence of macromolecular nucleic acids capable of catalyzing reactions in biosynthetic pathways. This is referred to as the "genes first" hypothesis. In contrast, proponents of the second theory suggest that a proto-metabolic network consisting of chemical reactions catalyzed by minerals and small molecules led to the accumulation of the building blocks of macromolecules and enabled the emergence of macromolecules. This is referred to as the "metabolism first" hypothesis.

A conundrum posed by the first theory is that there is imperfect overlap between the set of compounds delivered from exogenous sources and the set of compounds that constitutes the core of metabolism. For example, over 100 amino acids have been found in meteorites (Pizzarello 2007). Of these, only eight (alanine, aspartate, isoleucine, leucine, glutamate, glycine, proline, and valine) are used in proteins. Conversely, 12 of the proteinogenic amino acids are not found in meteors. In spark discharge experiments, the proteinogenic amino acids aspartate, glutamate, serine, glycine, and alanine are formed, along with γ-aminobutyrate, β-alanine, and α-aminoisobutyrate, which are not found in proteins (Cleaves *et al.* 2008). If life emerged in surface waters, the presence of many components not found in the central metabolic networks of extant life would have required either substantial pruning of a complex network of abiotic compounds or some process of self-organization that selected only certain molecules for inclusion in the proto-metabolic network but excluded others that had similar properties.

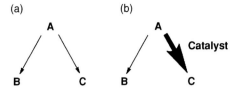

Figure 3.5 Catalysts influence the levels of products formed from a given reactant. (a) If the rates of conversion of A to B and to C are the same, B and C will be formed in equal amounts. (b) In the presence of a catalyst that increases the rate of conversion of A to C, the amount of C formed will be increased and the amount of B formed will be decreased.

The challenge for the theory that life originated at hydrothermal vents is how the extremely sparse network of small molecules found in vent fluids could have been elaborated to generate the 200–300 molecules necessary to support synthesis of nucleotides and amino acids. There is no doubt that prebiotically relevant reactions can be catalyzed by minerals found at hydrothermal vents (Cody 2004). For example, CO is reduced to methane thiol and other alkane thiols by phyrrhotite in the presence of H_2S at 100 °C (Heinen and Lauwers 1996), and carboxylic acids (Cody *et al.* 2004) and α-keto acids such as pyruvate (Cody *et al.* 2000) are formed from alkane thiols and CO by several transition metal sulfides. However, we are far from understanding how the building blocks of macromolecules could have been produced in a self-sustaining proto-metabolic network under geochemical conditions in vents.

Does extant metabolism resemble early proto-metabolic networks?

The structure of the metabolic network in extant life is dictated by the availability of catalysts, and this would have been true during all of the stages of the emergence of life, as well. Catalysts define the structure of metabolic networks by accelerating flux through certain reactions at the expense of slower competing reactions. For example, a catalyst that accelerates the rate of one of the possible reactions of a molecule by a relatively modest 50-fold alters the distribution of products to the point at which one product predominates and the others are formed in only very small quantities (see Figure 3.5). Figure 3.6 extends this idea to the effect of multiple catalysts on a more complicated chemical network, and illustrates how the availability of different sets of catalysts results in different network topologies and formation of different final products.

As discussed above, biosynthetic pathways in extant organisms clearly resemble those in the LUCA. A more difficult question is whether metabolism in the LUCA reflected the structure of a pre-existing proto-metabolic reaction network,

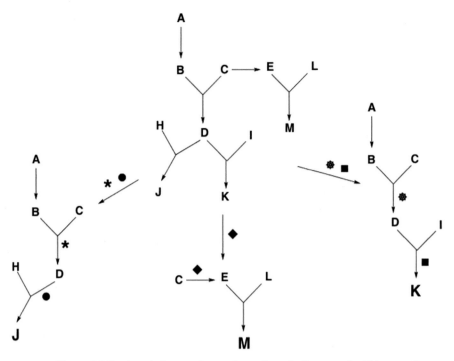

Figure 3.6 Catalysts influence the topology of metabolic networks. This example shows how the availability of catalysts (depicted by various symbols next to arrows) determines the topology of the network by directing flux primarily to certain products.

or replaced a pre-existing proto-metabolic reaction network. In the first scenario, metabolic pathways might have remained largely the same while ever more efficient catalysts were recruited to facilitate individual reactions, leading to a smooth transition from the earliest stages of mineral and small-molecule catalysis, through an intermediate stage involving proto-RNA and RNA catalysts (likely with catalytic auxiliaries provided by amino acids, peptides, and cofactors), and finally to protein enzymes. A point in favor of this argument is that it is undoubtedly easier to patch a single catalyst into a functioning pathway than to invent *de novo* an entirely different pathway whose efficiency surpasses that of a previously existing pathway.

A second hypothesis is that modern metabolic pathways have completely replaced primordial pathways due to the advent of more effective catalysts, probably at the stage of the RNA world (Benner *et al.* 1989). This viewpoint is based upon the assumption that a large number of highly effective catalysts arose in the RNA world, or at least by the LUCA, which together allowed flux through pathways that had never before been accessible. In the context of Figure 3.6, this would correspond to a switching between sets of catalysts with consequent reconstruction of the topology of the network. The answer to this question most likely lies somewhere between these two opposing theories. The idea that modern metabolism runs along pathways that were laid down before the emergence of the LUCA is appealing from the standpoint of continuity between pre-life and life, and because recruitment of catalysts one at a time is more likely than recruitment of several catalysts simultaneously to enable an entirely new pathway. However, recruitment of several catalysts simultaneously to enable a novel pathway can certainly occur once there is a sufficient collection of catalysts.

Conclusions

Isotopic analyses of kerogen preserved in ancient sediments indicate that biological carbon fixation and carbon cycling on a global scale were in place by 3.5 billion years ago. The striking resemblances between fossil and extant stromatolites and MISS suggest that the remarkably complex process of photosynthesis had emerged by that time. These phenomena are consistent with the idea that substantial metabolic complexity developed during the few hundred million years between the time of the LUCA (approximately 3.8 billion years ago) and the first fossil evidence of microbial mat communities. Biological sulfate reduction and Fe(II) oxidation likely played important roles in sustaining early life and in altering the geochemistry of the oceans over a period spanning over a billion years. We can extend our analysis further back in time using bioinformatic reconstructions of the LUCA genome that suggest that mechanisms for capturing energy and carbon from the environment, synthesis of amino acids and nucleotides, and use of genetic information stored in amino acids to direct synthesis of long proteins already resembled their modern counterparts. The path by which the LUCA emerged within a few hundred million years after the Earth cooled to the point where liquid water was possible is the most fascinating part of the story, but is also the most elusive. Experimental investigations of the planetary conditions that existed during this time will help establish the geochemical boundary conditions for the emergence of life. However, the environment of the early Earth was heterogeneous. The composition of, and fluxes through, prebiotic chemical networks would have been strongly influenced by local conditions of pH and temperature, the nature of mineral surfaces, and the concentrations of organic and inorganic compounds delivered from either the atmosphere and space at the surface, or by vent fluids at hydrothermal vents. Experimental and theoretical efforts to understand how metabolism might have emerged must take this complexity into account, while still imposing reasonable constraints defined by geochemistry.

References

Allwood, A. C., Walter, M. R., Kamber, B. S. *et al.* (2006). "Stromatolite reef from the early Archaean era of Australia," *Nature*, Vol. 441, pp. 714–718.

Allwood, A. C., Grotzinger, J. P., Knoll, A. H. *et al.* (2009). "Controls on development and diversity of early Archean stromatolites," *Proceedings of the National Academy of Sciences of the USA*, Vol. 106, pp. 9548–9555.

Allwood, A. C., Kamber, B. S., Walter, M. R. *et al.* (2010). "Trace elements record depositional history of an early Archean stromatolitic carbonate platform," *Chemical Geology*, Vol. 270, pp. 148–163.

Awramik, S. M. (1992). "The oldest records of photosynthesis," *Photosynthesis Research*, Vol. 33, pp. 75–89.

Awramik, S. M., Schopf, J. W., and Walter, M. R. (1983). "Filamentous fossil bacteria from the Archean of Western Australia," *Precambrian Research*, Vol. 20, pp. 357–374.

Baross, J. A. and Hoffman, S. E. (1985). "Submarine hydrothermal vents and associated gradient environments as sites for the origin and evolution of life," *Origins of Life and Evolution of Biospheres*, Vol. 15, pp. 327–345.

Benner, S. A., Ellington, A. D., and Tauer, A. (1989). "Modern metabolism as a palimpsest of the RNA world," *Proceedings of the National Academy of Sciences of the USA*, Vol. 86, pp. 7054–7058.

Bosak, T., Liang, B., Sim, M. S., and Petroff, A. P. (2009). "Morphological record of oxygenic photosynthesis in conical stromatolites," *Proceedings of the National Academy of Sciences of the USA*, Vol. 106, pp. 10939–10943.

Bowring, S. and Housh, T. (1995). "The Earth's early evolution," *Science*, Vol. 269, pp. 1535–1540.

Campbell, I. H. and Taylor, S. R. (1983). "No water, no granites – no oceans, no continents," *Geophysical Research Letters*, Vol. 10, pp. 1061–1064.

Canfield, D. E., Habicht, K. S., and Thamdrup, B. (2000). "The Archean sulfur cycle and the early history of atmospheric oxygen," *Science*, Vol. 288, pp. 658–661.

Cech, T. R. (2009). "Evolution of biological catalysis: ribozyme to RNP enzyme," *Cold Spring Harbor Symposium on Quantum Biology*, Vol. 74, pp. 11–16.

Chyba, C. and Sagan, C. (1992). "Endogenous production, exogenous delivery and impact-shock synthesis of organic molecules: an inventory for the origins of life," *Nature*, Vol. 355, pp. 125–132.

Cleaves, H. J., Chalmers, J. H., Lazcano, A. *et al.* (2008). "A reassessment of prebiotic organic synthesis in neutral planetary atmospheres," *Origins of Life and Evolution of Biospheres*, Vol. 38, pp. 105–115.

Cloud, P. E. (1972). "A working model of the primitive Earth," *American Journal of Science*, Vol. 272, pp. 537–548.

Cloud, P. E. (1976). "Beginnings of biospheric evolution and their biogeochemical consequences," *Paleobiology*, Vol. 2, pp. 351–387.

Cody, G. (2004). "Transition metal sulfides and the origin of metabolism," *Annual Review of Earth and Planetary Sciences*, Vol. 32, pp. 569–599.

Cody, G. D., Boctor, N. Z., Filley, T. R. *et al.* (2000). "Primordial carbonylated iron-sulfur compounds and the synthesis of pyruvate," *Science*, Vol. 289, pp. 1337–1340.

Cody, G. D., Boctor, N. Z., Bnades, J. A. *et al.* (2004). "Assaying the catalytic potential of transition metal sulfides for abiotic carbon fixation," *Geochimica et Cosmochimica Acta*, Vol. 68, pp. 2185–2196.

Copley, S. D., Smith, E., and Morowitz, H. J. (2007). "The origin of the RNA world: coevolution of genes and metabolism," *Bioorganic Chemistry*, Vol. 35, pp. 430–443.

Dekov, V. M., Cuadros, J., Shanks, W. C., and Koski, R. A. (2008). "Deposition of talc-kerolite-smectite at seafloor hydrothermal vent fields: evidence from mineralogical, geochemical and oxygen isotope studies," *Chemical Geology*, Vol. 247, pp. 171–194.

Des Marais, D. J. (1997). "Isotopic evolution of the biogeochemical carbon cycle during the proterozoic eon," *Organic Geochemistry*, Vol. 27, pp. 185–193.

Des Marais, D. J. (2000). "When did photosynthesis emerge on Earth?," *Science*, Vol. 289, pp. 1703–1705.

Des Marais, D. J. (2001). "Isotopic evolution of the biogeochemical carbon cycle during the Precambrian," *Reviews in Mineralogy and Geochemistry*, Vol. 43, pp. 555–578.

Eschenmoser, A. (1999). "Chemical etiology of nucleic acid structure," *Science*, Vol. 284, pp. 2118–2123.

Fenchel, T. and Finlay, B. J. (1995). *Ecology and Evolution in Anoxic Worlds*. Oxford: Oxford University Press.

Freilich, S., Spriggs, R. V., George, R. A. *et al.* (2005). "The complement of enzymatic sets in different species," *Journal of Molecular Biology*, Vol. 349, pp. 745–763.

Fusz, S., Eisenfuhr, A., Srivatsan, S. G. *et al.* (2005). "A ribozyme for the aldol reaction," *Chemical Biology*, Vol. 12, pp. 941–950.

Guerrier-Takada, C., Gardiner, K., Marsh, T. *et al.* (1983). "The RNA moeity of ribonuclease P is the catalytic subunit of the enzyme," *Cell*, Vol. 35, pp. 849–857.

Haldane, J. B. S. (1929). "The origin of life," *Rationalist Annual*, pp. 148–169.

Heinen, W. and Lauwers, A. M. (1996). "Organic sulfur compounds resulting from the interaction of iron sulfide, hydrogen sulfide and carbon dioxide in an anaerobic aqueous environment," *Origins of Life and Evolution of Biospheres*, Vol. 26, pp. 131–150.

Holland, H. D. (2002). "Volcanic gases, black smokers, and the great oxidation event," *Geochimica et Cosmochimica Acta*, Vol. 66, pp. 3811–3826.

Huang, F., Bugg, C. W., and Yarus, M. (2000). "RNA-catalyzed CoA, NAD, and FAD synthesis from phosphopantetheine, NMN, and FMN†," *Biochemistry*, Vol. 39, pp. 15548–15555.

Illangasekare, M. and Yarus, M. (1999). "A tiny RNA that catalyses both aminoacyl-RNA and peptidyl-RNA synthesis," *RNA*, Vol. 5, pp. 1482–1489.

Johnson, A. P., Cleaves, H. J., Dworkin, J. P., Glavin, D. P., Lazcano, A., and Bada, J. L. (2008). "The Miller volcanic spark discharge experiment," *Science*, Vol. 322, p. 404.

Kamber, B. S. (2007). "The enigma of the terrestrial protocrust: evidence for its former existence and the importance of its complete disappearance," in *Developments in Precambrian Geology*, R. H. S. Martin, J. van Kranendonk, and C. B. Vickie (eds.). Amsterdam: Elsevier, pp. 75–89.

Kasting, J. F. (2008). "The primitive earth," in *Prebiotic Evolution and Astrobiology*, J. T.-F. W. Wong and A. Lazcano (eds.). Austin, TX: Landes Biosciences.

Kelley, D. S., Karson, J. A., Blackman, D. K. *et al.* (2001). "An off-axis hydrothermal vent field near the Mid-Atlantic Ridge at 30 degrees north," *Nature*, Vol. 412, pp. 145–149.

Kelley, D. S., Baross, J. A., and Delaney, J. R. (2002). "Volcanoes, fluids, and life at mid-ocean ridge spreading centers," *Annual Review of Earth and Planetary Sciences*, Vol. 30, pp. 385–491.

Kelley, D. S., Karson, J. A., Fruh-Green, G. L. *et al.* (2005). "A serpentinite-hosted ecosystem: the Lost City hydrothermal field," *Science*, Vol. 307, pp. 1428–1434.

Koonin, E. V. (2006). "Temporal order of evolution of DNA replication systems inferred by comparison of cellular and viral DNA polymerases," *Biology Direct*, Vol. 1, p. 39.

Koonin, E. V. and Martin, W. (2005). "On the origin of genomes and cells within inorganic compartments," *Trends in Genetics*, Vol. 21, pp. 647–654.

Kruger, K., Grabowski, P. J., Zaug, A. J. *et al.* (1982). "Self-splicing RNA: autoexcision and autocyclization of the ribosomal RNA intervening sequence of tetrahymena," *Cell*, Vol. 31, pp. 147–157.

Kumar, R. K. and Yarus, M. (2001). "RNA-catalyzed amino acid activation," *Biochemistry*, Vol. 40, pp. 6998–7004.

Lambert, I. B. and Donnelly, T. H. (1991). "Atmospheric oxygen levels in the Precambrian: a review of isotopic and geological evidence," *Palaeogeography, Palaeoclimatology, Palaeoecology*, Vol. 97, pp. 83–91.

Marshall, C. P., Love, G. D., Snape, C. E. *et al.* (2007). "Structural characterization of kerogen in 3.4 Ga Archaean cherts from the Pilbara Craton, Western Australia," *Precambrian Research*, Vol. 155, pp. 1–23.

Martin, W., Baross, J., Kelley, D., and Russell, M. J. (2008). "Hydrothermal vents and the origin of life," *National Review of Microbiology*, Vol. 6, pp. 805–814.

McCollom, T. M. and Seewald, J. S. (2006). "Carbon isotope composition of organic compounds produced by abiotic synthesis under hydrothermal conditions," *Earth and Planetary Science Letters*, Vol. 243, pp. 74–84.

Miller, B. G. and Wolfenden, R. (2002). "Catalytic proficiency: the unusual case of OMP decarboxylase," *Annual Review of Biochemistry*, Vol. 71, pp. 847–885.

Miller, S. L. (1953). "A production of amino acids under possible primitive Earth conditions," *Science*, Vol. 117, pp. 528–529.

Mirkin, B. G., Fenner, T. I., Galperin, M. Y., and Koonin, E. V. (2003). "Algorithms for computing parsimonious evolutionary scenarios for genome evolution, the last universal common ancestor and dominance of horizontal gene transfer in the evolution of prokaryotes," *BMC Evolutionary Biology*, Vol. 3, p. 2.

Murnane, R. and Clague, D. A. (1983). "Nontronite from a low temperature hydrothermal system on the Juan de Fuca Ridge," *Earth Planetary Science Letters*, Vol. 65, pp. 343–352.

Nielsen, P. E. (2007). "Peptide nucleic acids and the origin of life," *Chemistry and Biodiversity*, Vol. 4(9), pp. 1996–2002.

Nisbet, E. G. and Fowler, C. M. R. (1999). "Archaean metabolic evolution of microbial mats," *Proceedings of the Royal Society of London, Series B: Biological Sciences*, Vol. 266, pp. 2375–2382.

Nisbet, E. G. and Sleep, N. H. (2001). "The habitat and nature of early life," *Nature*, Vol. 409, pp. 1083–1091.

Noffke, N. (2010). *Geobiology – Microbial Mats in Sandy Deposits from the Archean Era to Today*. Heidelberg: Springer.

Noffke, N., Beukes, N., Bower, D. *et al.* (2008). "An actualistic perspective into Archean worlds – (Cyano-)bacterially induced sedimentary structures in the siliciclastic nhlazatshe section, 2.9 Ga Pongola Supergroup, South Africa," *Geobiology*, Vol. 6, pp. 5–12.

Oparin, A. I. (1938). *The Origin of Life*. New York: Macmillan.

Ouzounis, C. A., Kunin, V., Darzentas, N., and Goldovsky, L. (2006). "A minimal estimate for the gene content of the last universal common ancestor – exobiology from a terrestrial perspective," *Research in Microbiology*, Vol. 157, pp. 57–68.

Peretó, J., López-García, P., and Moreira, D. (2004). "Ancestral lipid biosynthesis and early membrane evolution," *Trends in Biochemical Sciences*, Vol. 29, pp. 469–477.

Pizzarello, S. (2007). "The chemistry that preceded life's origin: a study guide from meteorites," *Chemistry and Biodiversity*, Vol. 4, pp. 680–693.

Proskurowski, G., Lilley, M. D., Seewald, J. S. *et al.* (2008). "Abiogenic hydrocarbon production at Lost City hydrothermal field," *Science*, Vol. 319, pp. 604–607.

Rosing, M. T., Bird, D. K., Sleep, N. H. *et al.* (2006). "The rise of continents – an essay on the geologic consequences of photosynthesis," *Palaeogeography, Palaeoclimatology, Palaeoecology*, Vol. 232, pp. 99–113.

Schidlowski, M. (1988). "A 3,800-million-year isotopic record of life from carbon in sedimentary rocks," *Nature*, Vol. 333, pp. 313–318.

Schidlowski, M., Hayes, J. M., and Kaplan, I. R. (1983). "Isotopic inferences of ancient biochemistries – carbon, sulfur, hydrogen, and nitrogen," in *Earth's Earliest Biosphere: Its Origin and Evolution*, J. W. Schopf (ed.). Princeton, NJ: Princeton University Press, pp. 149–186.

Schoning, K., Scholz, P., Guntha, S. *et al.* (2000). "Chemical etiology of nucleic acid structure: the alpha-threofuranosyl-(3′→2′) oligonucleotide system," *Science*, Vol. 290, pp. 1347–1351.

Schopf, J. W. (2006). "Fossil evidence of Archaean life," *Philosophical Transactions of the Royal Society, Series B: Biological Sciences*, Vol. 361, pp. 869–885.

Schopf, J. W. and Packer, B. M. (1987). "Early Archean (3.3-billion to 3.5-billion-year-old) microfossils from Warrawoona Group, Australia," *Science*, Vol. 237, pp. 70–73.

Schopf, J. W., Kudryavtsev, A. B., Czaja, A. D., and Tripath, A. B. (2007). "Evidence of Archean life: stromatolites and microfossils," *Precambrian Research*, Vol. 158, pp. 141–155.

Shen, Y., Buick, R., and Canfield, D. E. (2001). "Isotopic evidence for microbial sulphate reduction in the early Archaean era," *Nature*, Vol. 410, pp. 77–81.

Tice, M. M. and Lowe, D. R. (2004). "Photosynthetic microbial mats in the 3,416-Myr-old ocean," *Nature*, Vol. 431, pp. 549–552.

Watson, E. B. and Harrison, T. M. (2005). "Zircon thermometer reveals minimum melting conditions on earliest Earth," *Science*, Vol. 308, pp. 841–844.

Wilde, S. A., Valley, J. W., Peck, W. H., and Graham, C. M. (2001). "Evidence from detrital zircons for the existence of continental crust and oceans on the Earth 4.4 Gyr ago," *Nature*, Vol. 409, pp. 175–178.

Woese, C. R. (1998). "The universal ancestor," *Proceedings of the National Academy of Sciences of the USA*, Vol. 95, pp. 6854–6859.

Zhang, L., Peritz, A., and Meggers, E. (2005). "A simple glycol nucleic acid," *Journal of the American Chemical Society*, Vol. 127, pp. 4174–4175.

4

Planet Formation

SEAN N. RAYMOND AND WILLY BENZ

Introduction

Statistically speaking, one new star is born in our Milky Way Galaxy each year. This star will typically form in the densest parts of the Galaxy, either close to the central bulge or in the spiral arms. This is where giant clouds of cold molecular gas are concentrated, and these are the birthplaces of stars. Within such a cloud, the star forms when a small parcel of gas starts to "feel" its own gravity and begins a slow but accelerating inward collapse. It is not known whether an external trigger – such as shock waves from a nearby supernova – is needed to start this collapse, but once the collapse begins it follows a well-charted path. The angular momentum of the initial parcel of gas causes a disk to spin out as the gas contracts, and it is through this disk that gas is funneled onto the growing proto-star at the center. On a time-scale of 10^5 years, the central object reaches a large enough mass to increase its internal pressure and temperature above the critical value for nuclear fusion of hydrogen into helium and a star is truly born. An evolving disk of gas and dust orbits the star. This is where planets form.

The Hubble Space Telescope has taken exquisite images of nearby star-forming regions, some of which are, in cosmic terms, right next door. The Orion Nebula – found by Orion's sword in the constellation – is one of the most famous. Although it is more than a thousand light-years away, it is large and bright enough to be visible by eye. Within the Orion Nebula and in other star-forming regions, Hubble has imaged young stars and their dusty proto-planetary disks (O'Dell and Wen 1994). The disks themselves are far too small to be resolved in detail, but they

Frontiers of Astrobiology, ed. Chris Impey, Jonathan Lunine and José Funes.
Published by Cambridge University Press. © Cambridge University Press 2012.

confirm our basic picture: stars form in molecular clouds, and planets form in disks that are ubiquitous around young stars.

Like most of the universe, proto-planetary disks are made up almost entirely of hydrogen (H) and helium (He). Only about 1% of the mass in these disks (or anywhere in the nearby universe for that matter) is made up of heavier elements. However, it is from those heavy elements – and not even all of them but just a subset – that planets like Earth, Venus, and Mars must form, because the Earth is made up almost entirely of so-called condensable material, meaning elements that are able to exist as solids. By contrast, H and He are almost always found in gaseous form (except in high-pressure situations such as the interiors of stars or giant planets, where they can exist in more exotic states such as a liquid, plasma, or even "metallic" forms). The situation is somewhat different for gaseous planets like Jupiter and Saturn that are much closer to the mean Galactic composition, although we think that they probably needed to grow from solid "cores" that were themselves made from condensable material.

The question addressed in this chapter is the following: Within proto-planetary disks, how do planets form? We summarize our current, albeit incomplete, understanding that has been built up by hundreds of astronomers – both theorists and observers – over the past few hundred years (for recent theory, see Armitage 2007). Theories are constrained by a plethora of new observations: information on the masses, sizes, and distributions of proto-planetary disks, and how they change depending on the type of star, measurements of the dust content and evolution of disks, and of course the masses and orbital distribution of planets around other stars, which now number more than 700 (see http://www.exoplanets.org).

From disks to embryos

The naïve answer to the question framed by this chapter is that planets form by successive collisions between smaller bodies. In other words, starting with tiny grains, they simply bash together to form larger grains, which bash together to form even larger grains, and so on until full-sized planets are formed. However, it has been shown that this simple picture of collisional growth fails at remarkably small sizes (Blum and Wurm 2008). Micron-sized grains do indeed grow in low-speed collisions, and the resulting dust aggregates continue to grow until they reach a few millimeters in size. At this point, collisions simply do not result in further growth, as any pebble-throwing youngster playing on a sandy beach already knows. How then, are larger objects created?

Our current answer to this quandary is an interesting offshoot of what just a few years ago was called the "meter-sized barrier." Any object that does grow to 1 centimeter up to 1 meter in size in a proto-planetary disk should fall into the central star in just a century or two, thousands of times faster than the planetary-formation time-scale. In the collisional growth scenario this is a disaster, because if objects need to pass through this stage to reach the planetary scale, then everything should just grow to one meter then fall into the star. The result is no planets, case closed. But clearly, a huge body of evidence (including the existence of our own planet Earth) tells us that this thinking is flawed.

Let's look in more detail at what happens in a proto-planetary disk. The gas orbits the star slightly slower than the rocks (which have orbital speed given by Kepler's law) because the gas pressure offsets a fraction of the gravitational force. Dust grains are simply blown along with the gas, but larger rocky bodies feel the gas as a headwind (because the gas is moving more slowly) that slows down the objects and makes them spiral inward. Massive objects have enough inertia (in this case, what matters is the ratio of their volume to surface area) that they spiral in slowly. Smaller objects – those that are just big enough not to be blown around with the gas – are the unlucky ones that feel the gas most strongly and spiral in very quickly. These are the centimeter- to meter-sized objects.

The concept of the meter-sized barrier makes a key assumption that is its undoing. It assumes that proto-planetary disks are well behaved and that the properties of the gas disk are well regulated and smooth throughout. This is important because the meter-sized bodies that are thought to spiral inward actually just seek out maxima in the gas pressure (they follow the local pressure gradient). Real proto-planetary disks are decidedly lumpy and unsmooth. They are very turbulent. In fact, small-scale turbulence due to magnetic effects is thought to be the driving force behind their evolution. Even the simple settling of dust grains to the mid-plane of the disk creates turbulence. Turbulent disks create uneven structures at a variety of size-scales, and it is into these structures that centimeter- to meter-sized objects are drawn, not into the center of the star. And by the same arguments as before, objects should converge to these regions very quickly.

Once particles begin to accumulate in turbulent zones, there are feedback mechanisms that keep the process going. Even though they make up such a small fraction of the total mass, when sufficiently concentrated, solid bodies can have an effect on the gas and in fact increase the local pressure to attract ever more boulders. Once particles become so concentrated that they dominate the local mass distribution, they begin to feel their collective gravity and become gravitationally bound, much larger bodies (Johansen et al. 2007). This is a far gentler type of gravitational collapse than the parcel of gas that ended up as the star, but the general physical mechanism is the same: when the conditions are right, gravity trumps all. In this case, the outcome is objects with a spectrum of sizes – current simulations suggest that many of these objects are 100–1000 kilometers in size, although the details of this process are an active area of study. The most important point for our story is that these objects are on the right side of the meter-sized barrier. They are called planetesimals and are the building blocks of planets.

From embryos to planets

Once a sufficient number of planetesimals exist in the proto-planetary disk, they start to collide and grow. Once a planetesimal grows larger than its neighbors, its rate of growth also grows. The big planetesimal's increased gravity acts to focus the orbits of smaller planetesimals and, like the Death Star's tractor beam in *Star Wars*, it draws them in. The more massive the planetesimal gets, the stronger the focusing, leading to a rapid phase of runaway growth. Runaway growth slows down when the big planetesimal's gravity is so large that it stirs up nearby smaller planetesimals to the point that their relative speeds are high enough to shut down the gravitational focusing. At this point, the large bodies are called "planetary embryos."

Let's rewind the clock a little to consider the composition of the disk. The material that is available to form solid planets is not the same everywhere in the disk. The inner disk is hot, so only compounds that remain solid at high temperatures – namely, rock and iron – can be incorporated into planetesimals and planetary embryos. These are called "refractory" elements. Moving away from the star the disk cools off such that compounds that condense at lower temperatures, i.e. more "volatile" species, can exist as solids. For any given compound there is a condensation temperature that, in the context of a proto-planetary disk, corresponds to a radial transition (Lodders 2003). Interior to this transition, the compound will be in gas phase simply because it is too hot to condense, but exterior to the transition it will be a solid (note that liquids are virtually non-existent in proto-planetary disks because the pressure is too low). For particularly abundant compounds like water, these transitions are thought to be of vital importance for planet formation because exterior to the transition the abundance of solids increases substantially. In fact, the water transition front even has a special name: the snow line. Interior to the snow line, small planets should form from rock and iron but exterior they should contain a large fraction of ice and be much larger.

Back to our story. Planetary embryos take about a million years to form (Figure 4.1). In the terrestrial planet-forming part of the disk, embryos are Moon-sized to Mars-sized, but in the giant-planet-forming region embryos can grow much larger, to several Earth masses, because they are beyond the snow line (Kobuko and Ida 2002). Wherever they are, embryos will interact gravitationally with the gaseous disk in which they are embedded (Armitage 2011). While discovered much earlier, the importance of these interactions was really only realized once the first extrasolar planets were discovered in 1995.

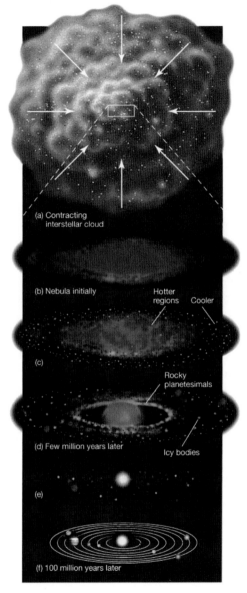

Figure 4.1 An illustration of the main phases of planet formation. First, a parcel of gas within a molecular cloud starts to collapse under its own gravity (panel a). As the parcel collapses, it inevitably creates a disk (panel b), and a star begins to form at the center (panel c). The inner disk is of course hotter than the outer disk, so rocky and metallic bodies condense in the inner disk while the outer disk becomes dominated by volatile, icy bodies (panel d). On time-scales of millions of year, these solid bodies accumulate and grow into both the cores of gas giant planets, the smaller and less gas-dominated ice giants (like Uranus and Neptune), and rocky, terrestrial planets (panel e). Within 100 million years the system's formation is complete (panel f); the system in this figure represents of course our own Solar System. *Image credit: Chaisson and McMillan, Astronomy Today, Pearson, 2011.* A colour version of this illustration appears in the colour plate section between pp. 148–149.

Planet migration

The gravitational interactions of a growing embryo with the gas disk will result in the launching of density waves in the disk. From resonant locations between these waves and the embryo, torques are exerted on the embryo's orbit. The first order of these so-called Lindblad resonances is located roughly a disk scale-height inside and outside the orbit of the planet while the higher orders are located further away. Gas located near outside resonances orbits the star slower than the embryo while gas located near inner resonances orbit the star faster. The zeroth-order resonance, or corotation, lies exactly at the orbit of the planet, and gas located near this corotation region follows so-called horseshoe orbits. The differences between all these different torques results in a net change of the angular momentum of the embryo and therefore to a net inward or outward motion depending upon whether angular momentum has been gained or lost. This motion of a body embedded in a gaseous proto-planetary gaseous disk has been called type I migration.

In a highly simplified proto-planetary disk, the net torque is such that the embryo loses angular momentum thus giving rise to *inward* type I migration. The rate at which it spirals inwards has been found to be directly proportional to the mass of the embryo (e.g. a 10 Earth-mass planet spirals inward 10 times faster than an Earth-mass planet). When the embryo reaches a few tenths of an Earth-mass, type I migration becomes so rapid that the body literally "falls" onto the star within ten to a hundred thousand years, a blink of an eye in this context. In other words, this simple analysis tells us that the more an embryo grows, the faster it will fall into its star and therefore no planet more massive than this should be able to survive very long (Lyra *et al.* 2010). As before, we know that this cannot be true since we observe countless planets (both inside and outside the Solar System) that are larger than a few tenths of an Earth-mass!

This problem with the short inward type I migration time-scale is still not fully resolved, although new studies are encouraging. The solution appears to lie once again in the reality that disks are not as simple and smooth as initially thought. The calculations suggesting very rapid migration made simplifying assumptions about the temperature structure of the disk, which is important because it controls the behavior of the density waves launched by the embryo. Complex new simulations that account for how the disk actually cools itself off by radiating away energy show that in reality type I migration probably does not behave in the predicted, simple fashion. In fact, for some conditions the corotation torques can largely dominate the dynamics and can be such that the embryo gains angular momentum and migrates outwards (Paardekooper *et al.* 2011). As the disk evolves or the embryo gains in mass, the corotation torques eventually saturate and the embryo migrates inwards again. To make matters even more complicated, in a given disk there can be several regions where the conditions necessary for outward migration are met. So we can imagine a disk in which embryos are moving inwards in some regions and outwards in others. Of particular interest are the locations where the two types of motion meet, since these define regions in which embryos migrate in a convergent manner and may efficiently accumulate into larger bodies (Lyra *et al.* 2010).

Although the implications of these new findings have not been fully fleshed out, this result is good news. Embryos are probably not overly susceptible to falling into their stars, although, like a toddler on a sugar rush, they have trouble staying in one place.

Planetary embryos can serve as the seeds of gas giant planets if they form fast enough and are massive enough. Once its gravity becomes substantial, an embryo gravitationally traps gas from the disk into an extended envelope that is much larger than any present-day atmospheres of Solar System bodies. This gas is gravitationally bound to the embryo but only loosely, and the amount of gas that is initially trapped is small simply because the gas is not very dense. The gaseous envelope extends all the way to the edge of the embryo's gravitational reach (called the "Hill sphere"), so in order to increase the mass in this gaseous envelope, the envelope itself must contract to make room for the new gas. For this to happen, the envelope needs to cool down by radiating its thermal energy away, which it does mainly at infrared wavelengths. However, this is easier said than done since it is also heated up by the simple act of falling deeper into the embryo's gravitational potential well and also by the dissipation of the kinetic energy of planetesimals falling into it. If one does the calculations, one finds that the net cooling rate remains quite small until the envelope becomes relatively massive. This slow cooling rate, in turn, implies a small mass accretion rate and since the overall lifetime of the gaseous disk is rather limited (less than 10 million years), this slow contraction represents a real bottleneck on the path of becoming a gas giant planet.

If the gas in the disk dissipates during this protracted stage then the embryo will end up as something like Uranus and Neptune, both of which contain some gas but are still dominated by solids (ices, mainly) (Mordasini *et al.* 2009). However, if the contraction of the embryo's gaseous envelope continues for long enough the mass in the envelope can become substantial. When the envelope mass is roughly equal to the mass of the solid core, the process becomes runaway. When this happens, the planet's mass increases very quickly as it accretes a large fraction of the gas in its local neighborhood. As more gas falls onto the growing planet, its Hill sphere grows and more distant gas falls within its reach. While technically gas accretion proceeds at a rate increasing with increasing mass, there comes a time when the gas supply is limited by the disk itself. This is essentially due to two factors. The first is that the disk cannot transport gas radially faster than the transport mechanism (usually assumed to be a form of turbulent viscosity) actually

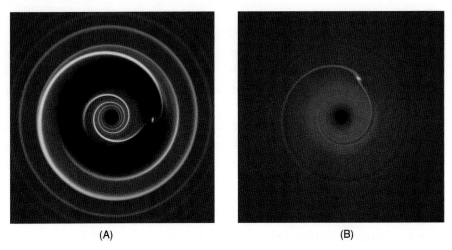

(A)   (B)

Figure 4.2 Snapshots from hydrodynamical simulations of low-mass (right) and
high-mass (left) planets embedded in gaseous proto-planetary disks. Here, the colors
represent the magnitude of the perturbations to the gas density caused by the planet's
gravity: yellow represents an increase in density while purple or black represents a
decrease. Note that the high-mass planet has completely cleared out an annular gap in
the disk, while the low-mass planet does not have enough gravity to create a gap. *Image
credit: Philip J. Armitage.* A colour version of this illustration appears in the colour plate
section between pp. 148–149.

allows. The second is that as the planet grows it eventually clears an annular gap
thereby quenching its own further growth.

As the planet clears this gap, type I migration becomes irrelevant. In fact, the
planet now migrates at a rate that is given by the transport of the gas itself. This can
be understood by considering that the planet simply receives angular momentum
from the inner edge region and transfers it to the outer edge region of the gap.
If the inner edge of the gap moves inwards and therefore further away from the
planet, the planet will receive less angular momentum while still transferring
as much to the outer edge. This will lead to an inward migration of the planet.
However, the outer region now receiving less angular momentum from the planet
will move inwards as well and the process repeats itself effectively capturing the
planet in the gap. This new type of migration is called type II migration and its rate
is equivalent to the rate at which the gas is transported radially in the disk thereby
linking the migration of the planet to the evolution of the disk itself (Figure 4.2).

The fact that planets migrate was a surprise. Even though theorists had an
inkling of the processes just described, simulations were quite primitive in the
1980s and nobody was willing to predict that planets change their orbits substantially, especially when the current orbits in our Solar System seem stable.

Growing gas giants

How does the disk evolve? Disks around other stars are observed to be slowly trickling onto their host stars at a rate that depends on their age. The simplest model that can explain the observations is that the disk has some internal viscosity and as such is simply spreading out over time. But, since it is in orbit around a star, the gas that spreads inward ends up falling onto the star. Over its lifetime, it is thought that three-quarters or more of the total gas disk falls onto the star. The rest of the disk is evaporated by absorbing high-energy (ultraviolet) photons from both the central star and from nearby massive stars. So, a given piece of disk gas is most likely to stream inward to either fall on the star or evaporate, but has about a one in four chance of streaming outward and eventually evaporating. Thus, as they are "bullied" by the disk most giant planets should type II migrate inward, but a fraction should go outward. Migration stops when the disk's dwindling mass drops to a fraction of the giant planet's mass (Papaloizou and Terquem 2006).

Gas giants' bulk – for example, Jupiter is 318 times Earth's mass and has a radius 11 times larger than Earth's – means they are the easiest planets to detect around other stars. The planets that have been found were not at all expected. What was expected were planetary systems more or less like our own, with small rocky planets close to their stars and gas giants farther out, with maybe an ice giant or two sprinkled amongst them. The very first planet confirmed around a Sun-like star was completely un-Solar System-like. It was a "hot Jupiter," a gas giant planet orbiting its star every 4.2 days (as opposed to Jupiter's own 12 *year* orbital period). The discovery was quickly followed by the realization that it could not have formed at the distance from its star where it was observed (Mayor and Queloz 1995, Lin *et al.* 1996). Another notable surprise in the giant extrasolar planet population is the abundance of very stretched-out or elliptical (rather than circular) orbits. The typical extrasolar planet has an eccentricity of 25%, five times larger than the eccentricity of Jupiter and Saturn. Extrasolar planet systems are extremely diverse. Based on the current sample, the architecture of our Solar System seems to be unusual. However, given the 10–15 years needed to detect true Jupiter analogs and the fact that many are now being found, only time will tell if we live in a typical or an unusual planetary system.

Where do hot Jupiters come from? Until very recently the answer was thought to lie with type II migration. Giant planets would form at Jupiter-like distances, then migrate in and stall at the inner edge of the disk, around 0.05 AU, right where hot Jupiters are seen to pile up. However, new observations have succeeded in measuring the orbital inclination of hot Jupiters with respect to the equatorial plane of their host stars. If we assume that proto-planetary disks are lined with the stellar equator, then hot Jupiters should have very small inclinations. However, what has been measured is an astounding range of inclinations, from coplanar to highly inclined to coplanar but retrograde! In retrograde systems, the planet orbits the star in the opposite direction than it started, like a minute hand of a clock going counter-clockwise. Neither retrograde nor highly inclined planets are easily explained with type II migration.

It turns out that both hot Jupiters and the eccentric orbits of more distant planets are telling us that the typical planet-forming environment is extremely violent. The story goes as follows. Giant planets form in groups of at least two, probably of three or more. After a time, their orbits become unstable; the instability probably starts as the gas disk evaporates because, like a TV for a pack of kids, the disk has a calming effect on the orbits of adjacent planets. The orbits of the giant planets cross, leading to close encounters and planet–planet scattering events (Chatterjee et al. 2008). During an encounter between two giant planets, each planet can be accelerated by up to the escape speed of the other (the escape speed is a measure of how fast something must be thrown from a planet's surface for it never to come back). Jupiter's escape speed is 60 kilometers per second but its orbital speed is only 13, so a single encounter with Jupiter can toss a planet onto any possible orbit including ejection from the star on a hyperbolic trajectory. The outcome of planet–planet scattering in a system of multiple giant planets is that one or more planets are removed from the system via either ejection or a collision with either the central star or another planet. The planets that survive this dynamical instability bear its mark with their eccentric and inclined orbits (Nagasawa et al. 2008).

The surviving planets explain the exoplanet eccentricities, but what about the hot Jupiters? During the instability, some scattered planets have extremely high eccentricities and therefore perihelion distances so small that tidal forces between the star and planet become important. Tidal dissipation within both the star and planet act to recircularize and shrink the planet's orbit and on the time span of a billion years or more the planet can become a hot Jupiter. The orbital inclination of the hot Jupiter will depend on both its starting value, which can take a very wide range of initial values after scattering, including retrograde configurations, and the tidal dissipation. Around stars less massive than roughly 1.2 solar masses, hot Jupiters are measured to have very low inclinations, but around higher-mass stars they have a huge range in inclination including retrograde values (Winn et al. 2010).

This is explained by a fundamental change in the internal structure of stars of different masses: stars less massive than 1.2 solar masses have substantial outer convective zones (comprising a few percent of their total mass) but stars more massive than 1.2 solar masses have virtually no convective zone. Since the convective zone is the outermost layer of the star, it makes sense that this layer is of vital importance in the dissipation of internal energy (tides). So, the observed inclinations of hot Jupiters are explained if tides are controlled by the width of the stellar convective zone. If this model is correct, hot Jupiters are really thrown into place rather than gently migrating.

Forming rocky planets

We have explored the formation and evolution of gas giants, but what about the rocky, terrestrial planets? Those are the most interesting planets of all in terms of the search for potentially habitable (or inhabited) worlds. So, once more, we will travel back to our proto-planetary disk. It is late in the gas disk lifetime, perhaps five million years after the parcel of gas began its collapse into a star–planet system. The gas is evaporating away and the gas giants are already formed. The terrestrial planet region contains perhaps 100 planetary embryos and a swarm of billions of planetesimals. The planetesimals continue to collide with the embryos and, once the mass in planetesimals and embryos is comparable, embryos begin to collide with other embryos. During this last stage of growth, terrestrial planets grow and settle onto stable orbits. Collisions between embryos can sometimes have dramatic consequences. High-speed, "hit and run" embryo–embryo impacts may bounce off one another and lead to net erosion rather than growth (Asphaug et al. 2006). Low-speed off-center impacts can spin a disk of material off from the larger body and lead to the formation of a large satellite like the Earth's Moon (Canup and Asphaug 2001). Within 50–100 million years terrestrial planet growth is finished.

For potentially habitable planets, water is a major concern. Embryos at 1 AU are thought to have been dry – the snow line (at least for the Sun) is thought to have been located somewhere between 2 and 4 AU. How, then, do Earth-like planets acquire water? The simplest answer (although this is a matter of debate) is that water must be delivered to these planets in the form of hydrated planetesimals that formed beyond the snow line. The isotopic signature of Earth's water is virtually identical to that of carbonaceous meteorites thought to have originated in the outer asteroid belt, and that is our best guess for the source of our water. This water-rich material from asteroids was probably "delivered" to the growing Earth in the form of millions of planetesimals and a few large embryos (Morbidelli et al. 2000, Raymond et al. 2004). The implication is that terrestrial planets form from material that comes from a range of orbital distances. In other words, the "feeding zones" of terrestrial planets are wide. This shouldn't be too surprising at this point, given that planets can migrate inward and/or outward at various times (Raymond et al. 2006); in fact, the inner Solar System may have been sculpted by the inward-then-outward migration of Jupiter (Walsh *et al.* 2011). Nonetheless, it is an important realization: the water we drink came from asteroids.

As we've seen before, giant planets form much faster than terrestrial planets and are therefore fully formed during the late stages of terrestrial planet growth. The giant planets' gravitational influence is a major influence during these stages by affecting the orbits of embryos and also by ejecting any bodies that stray too close. For example, a giant planet on a Jupiter-like orbit but with significant eccentricity clears out a large portion of the inner disk and excites embryos' eccentricities. The terrestrial planets that form in such a system are closer to the star (farther from the giant planet) and drier than for the case of a giant planet on a circular orbit, because water-rich material lies closer to the giant planet and is therefore preferentially ejected. In contrast, lower-mass giant planets and giant planets on circular orbits help rich systems of terrestrial planets to form with substantial water contents (Raymond *et al.* 2004).

Given our understanding of the influence of giant planets on terrestrial planet growth, what can we learn from the hundreds of known extrasolar giant planets? The majority are not good candidates for having formed a terrestrial planet, at least not in its star's habitable zone (Udry and Santos 2007). These systems either contain eccentric giant planets or hot Jupiters: bodies whose violent past probably destroyed all terrestrial building blocks. This still leaves over 90% of stars around which no close-in giant planet could be detected and which make potentially ideal hosts for terrestrial planets. As time goes on and with improving detection techniques, a growing population of truly Jupiter-like planets is slowly emerging as well as the recognition that planetary systems are probably extremely common. Our detection techniques are not quite yet capable of detecting truly terrestrial analogs, but serious efforts are underway and NASA's Kepler mission has the requisite sensitivity. It should not be too long before the first detection will be announced of a near-clone of the Earth.

To summarize, stars form in giant molecular clouds. Planets form in disks around those stars, from dust, then planetesimals and planetary embryos. This process is extremely dynamic and may involve large-scale orbital migration at several different phases. Giant planets form faster than terrestrial planets, and influence their growth. Given our evolving understanding of planet formation, there are sure to be more surprises to come.

References

Armitage, P. J. (2007). "Lecture notes on the formation and early evolution of planetary systems," published in extended form as *Astrophysics of Planet Formation*. Cambridge: Cambridge University Press.

Armitage, P. J. (2011). "Dynamics of protoplanetary disks," *Annual Reviews of Astronomy and Astrophysics*, Vol. 49, pp. 195–236.

Asphaug, E., Agnor, C. B., and Williams, Q. (2006). "Hit-and-run planetary collisions," *Nature*, Vol. 439, pp. 155–160.

Blum, J. and Wurm, G. (2008). "The growth mechanisms of macroscopic bodies in protoplanetary disks," *Annual Review of Astronomy and Astrophysics*, Vol. 46, pp. 21–56.

Canup, R. M. and Asphaug, E. (2001). "Origin of the Moon in a giant impact near the end of the Earth's formation," *Nature*, Vol. 412, pp. 708–712.

Chatterjee, S., Ford, E. B., Matsumura, S., and Rasio, F. A. (2008). "Dynamical outcomes of planet–planet scattering," *Astrophysical Journal*, Vol. 686, pp. 580–602.

Johansen, A., Oishi, J. S., Mac Low, M.-M. *et al.* (2007). "Rapid planetesimal formation in turbulent circumstellar disks," *Nature*, Vol. 448, pp. 1022–1025.

Kobuko, E. and Ida, S. (2002). "Formation of protoplanet systems and diversity of planetary systems," *Astrophysical Journal*, Vol. 581, pp. 666–680.

Lin, D. N. C., Bodenheimer, P., and Richardson, D. C. (1996). "Orbital migration of the planetary companion of 51 Pegasi to its present location," *Nature*, Vol. 380, pp. 606–607.

Lodders, K. (2003). "Solar system abundances and condensation temperatures of the elements," *Astrophysical Journal*, Vol. 591, pp. 1220–1247.

Lyra, W., Paardekooper, S.-J., and Mac Low, M.-M. (2010). "Orbital migration of low-mass planets in evolutionary radiative models: avoiding catastrophic infall," *Astrophysical Journal Letters*, Vol. 715, pp. 68–73.

Mayor, M. and Queloz, D. (1995). "A Jupiter-mass companion to a solar-type star," *Nature*, Vol. 378, pp. 355–359.

Morbidelli, A., Chambers, J., Lunine, J. I. *et al.* (2000). "Source regions and time scales for the delivery of water to Earth," *Meteoritics and Planetary Science*, Vol. 35, pp. 1309–1320.

Mordasini, C., Alibert, Y., and Benz, W. (2009). "Extrasolar planet population synthesis. I. Method, formation tracks, and mass-distance distribution," *Astronomy and Astrophysics*, Vol. 501, pp. 1139–1160.

Nagasawa, M., Ida, S., and Bessho, T. (2008). "Formation of hot planets by a combination of planet scattering, tidal circularization, and the Kozai mechanism," *Astrophysical Journal*, Vol. 678, pp. 498–508.

O'Dell, C. R. and Wen, Z. (1994). "Post-refurbishment mission Hubble Space Telescope images of the core of the Orion Nebula: proplyds, Herbig–Haro objects, and measurements of a circumstellar disk," *Astrophysical Journal*, Vol. 436, pp. 194–202.

Papaloizou, J. C. B. and Terquem, C. (2006). "Planet formation and migration," *Reports on Progress in Physics*, Vol. 69, pp. 119–180.

Paardekooper S.-J., Baruteau, C., and Kley, W. (2011). "A torque formula for non-isothermal Type 1 planetary migration – II. Effects of diffusion," *Monthly Notices of the Royal Astronomical Society*, Vol. 410, pp. 293–303.

Raymond, S. N., Quinn, T., and Lunine, J. I. (2004). "Making other Earths: dynamical simulations of terrestrial planet formation and water delivery," *Icarus*, Vol. 168, pp. 1–17.

Raymond, S. N., Mandell, A. M., and Sigurdsson, S. (2006). "Exotic Earths: forming habitable worlds with giant planet migration," *Science*, Vol. 313, pp. 1413–1416.

Raymond, S. N., Quinn, T., and Lunine, J. I. (2007). "High-resolution simulations of the final assembly of Earth-like planets. 2. Water delivery and planetary habitability," *Astrobiology*, Vol. 7, Issue 1, pp. 66–84.

Udry, S. and Santos, N. C. (2007). "Statistical properties of exoplanets," *Annual Review of Astronomy and Astrophysics*, Vol. 45, pp. 397–439.

Walsh, K. J., Morbidelli, A., Raymond, S. N. *et al.* (2011). "A low mass for Mars from Jupiter's early gas-driven migration," *Nature*, Vol. 475, pp 206–209.

Winn, J. N., Fabrycky, D., Albrecht, S., and Johnson, J. A. (2010). "Hot stars with hot Jupiters have high obliquities," *Astrophysical Journal Letters*, Vol. 718, pp. 145–149.

PART III HISTORY OF LIFE ON EARTH

5

The Early Earth

FRANCES WESTALL

In this chapter we will look at the environment of the early Earth as a habitat for life and at the primitive life forms that inhabited it. The early Earth was a very different planet from today's Earth. Hotter, much more volcanically active, with an oxygen-poor atmosphere and ocean waters that were probably slightly more acidic and more salty than today's ocean, at first glance the early Earth seems to have been an inhospitable planet. But this was the Earth upon which life first appeared. In fact, life could *not* have appeared on today's Earth because of the ubiquitous presence of oxygen – an active molecule that effectively destroys the organic ingredients of life by oxidation. Despite its apparent inhospitality, the early Earth was habitable because it had conditions that were conducive to the appearance of simple life forms: it had liquid water, carbon molecules, energy sources, and the elements necessary for both the building bricks of cells and for its metabolic processes (HNOPS, plus transition metals). And this early, different planet apparently teemed with primitive forms of life.

The environment of the early Earth

After consolidation of the planetesimals forming the proto-planet, early radiogenic heat from short-lived radiogenic species, such as ^{26}Al, fused the accreted planetesimals into a molten mass, producing a magma ocean, which allowed differentiation of the heavier elements, iron and nickel, into the core and the lighter elements, forming silicate minerals, into the mantle. Degassing of the early mantle expelled the lighter elements (volatile elements) that were originally contained in the planetesimals to create a weakly reducing atmosphere of N_2, CO_2, and water, with traces of other gases (Kasting and Brown 1998). About 40 My after

Frontiers of Astrobiology, ed. Chris Impey, Jonathan Lunine and José Funes.
Published by Cambridge University Press. © Cambridge University Press 2012.

the consolidation of the proto-Earth, it was impacted by another smaller planet having a composition not too different from that of the Earth. It is possible to estimate the timing of this impact from the age of differentiation of the cores of the Earth and the Moon. Using the ratio of the quantity of the radiogenic isotope ^{182}H and its daughter ^{182}W remaining in the mantle of the Earth and the Moon, the impact has been dated to approximately between 40 and 100 My after accretion (Yin *et al.* 2002, Kleine *et al.* 2009). The planetary material issuing from this glancing impact produced the Earth's satellite, the Moon. The existence of the Moon had a number of consequences. The attraction between the two planets reduced the orbital angle of the Earth in its progression around the Sun. This had the effect of stabilizing the obliquity of the planet with consequent effects on environmental conditions, an important aspect for life forms living on the planet. The other consequence was the resetting of Earth's evolutionary clock because of the loss of the primitive atmosphere and the formation of yet another magma ocean. These two processes meant that most of the early volatile inventory was lost and the terrestrial material became essentially "dry" (Martin *et al.* 2006), a problem for the appearance of life because liquid water is one of the essential ingredients of life.

The solution to this problem came from the continued importation of extrater-restrial materials in the form of carbonaceous meteorites, micrometeorites, and comets rich in volatiles and organics after the initial consolidation of the Earth. Although these materials would have been formed in the outer, cooler reaches of the solar nebular, recent modeling (Walsh *et al.* 2011) indicates that changes in the orbits of the very rapidly formed gas giants, Jupiter and Saturn, led first to their inward migration towards the proto-Sun and then outward migration to their present positions. The inward movement affected and entrained with it part of the population of volatile-rich asteroids that formed in the outer, cold reaches of the accretionary disk. Thus, the inner asteroid belt, the source region for the meteorites and micrometeorites raining down upon the Earth, was fed with volatile-rich materials, which continued to accrete onto the inner planets, forming a volatile-rich "late veneer" (Morbidelli *et al.* 2000, Dauphas *et al.* 2000).

There is a question about the timing of the formation of the "late veneer". Was it a continuous but gradually decreasing importation of left-over materials from the accretionary disk (Hartmann *et al.* 2000) or was most of the volatile material brought in during a cataclysmic peak centred around 4.0–3.85 Ga*, as suggested by lunar crater dates (Ryder *et al.* 2000)? Early calculations (Sleep *et al.* 1989) concluded that such a bombardment could have been catastrophic enough to have completely evaporated all the early Earth's oceans. In this case, it would have been impossible for life to have appeared until after about 3.85 Ga. Such catastrophic activity should have left its mark on the Earth, if the 1000 km-sized lunar craters are anything to

* Ga, billion years ago.

go by. There is, however, little trace of impact activity in rocks 3.8–3.7 Ga old from Greenland and Northern Canada (Anbar *et al.* 2001, Schoenberg *et al.* 2002), possibly because these rocks are too young. In fact, there are very few locations where rocks older than 3.8 Ga exist. The oldest known rocks are metamorphic gneisses dating between 4.0 and 4.2 Ga that occur in the northern provinces of Canada (Bowring *et al.* 1999, Papineau *et al.* 2011). Later tectonic recycling of the Earth's crust has effectively eliminated much of the first billion years of Earth's history.

The phenomenon of plate tectonics is of direct importance to habitability both on the short and the long term. On the short term, plate tectonics was one of the processes by which the interior of the Earth could be cooled, especially the base of the mantle at the core boundary. The differential temperatures between the base of the mantle and the core were fundamental for the development and continuation of convection in the core that led to the formation of a magnetic field (Breuer *et al.* 2010). And the magnetic field was of vital importance for protecting the delicate, ephemeral layers of surface volatiles, the oceans, and the atmosphere, from the highly erosive early solar wind. Plate tectonics was also important in terms of long-term habitability and the rise of oxygen in the atmosphere because of its implication in the burial of reduced carbon and carbonates in the mantle and thus the gradual removal of CO_2 from the atmosphere. The simple pumping of oxygen into the atmosphere by microbial activity was not sufficient. Thus, plate tectonics was essential for the build-up of oxygen in the environment and for indirectly helping environmental changes that promoted the further evolution of life. However, a very recent hypothesis also implicates the effect of thermochemical reduction of volcanic SO_4 in the oceans in aiding the rise of O_2 in the atmosphere (Gaillard *et al.* 2011).

The origin of plate tectonics is not well understood. Modelling suggests that planet size may be a factor in whether the planet does or does not develop plate tectonics. If it is too large or too small, a single crust or "stagnant lid" forms on top of the mantle (Plesa and Breuer 2010), such as on Mars or modelled "super-Earths", i.e. rocky planets that are larger than the Earth and that have been detected in other star systems from space (Charbonneau *et al.* 2009). The Earth appears to be just the right size. Initial mixing of the early crust with the mantle would have occurred through foundering of the crust into the mantle (Martin *et al.* 2006) but even then, it is still an enigma as to how this kind of plume tectonics involving rising masses of hot mantle material and convectional descent of cooler crust turned into a semi-controlled tectonic regime. Plate tectonics involves the extrusion of hot mantle materials along linear cracks in the crust and the descent of the crust under proto-continents, also along linear cracks.

The most immediate physical expression of these plate tectonics is the presence of continental masses consisting of rocks that are richer in silica and "lighter" minerals, such as feldspars, than the oceanic crust. These minerals are formed by the recycling of hydrated oceanic crust in the mantle through plate tectonics. The rock type granite is a typical product of transformations of hydrated crustal material and, under the influence of high temperature and pressures in the lower crust, it becomes metamorphosed into gneiss. Granites and gneisses are the cores of continents. The oldest surviving terrestrial rocks are metamorphosed granites called tonalite-trondjemite gneisses (TTGs). Older than 3.8 Ga, they occur as small enclaves in younger continental crustal terrains in Northern Canada and in Greenland. The oldest known gneisses have been dated to 4.2 Ga (Papineau *et al.* 2011). Interestingly, sediments deposited on the ancient seafloor were also preserved on Greenland (in the Isua greenstone belt) on the island of Akilia off the SW coast of Greenland (Mojzsis *et al.* 1996, Rosing 1999). These rocks are of particular interest because it has been claimed that they contain traces of ancient life in the form of specific isotopic compositions of carbon (op. cit.). We will return to this topic below.

Another product of the recycling of water-saturated crust into the mantle is the production of zircon crystals. These are extremely resistant minerals that can be eroded from their host rocks and redeposited many times over. The oldest known crystal on Earth is a zircon that dates back to 4.4 Ga (Wilde *et al.* 2001). It was found in sediments in Australia that are more than a billion years younger than the crystal itself. The particularity of this crystal is that its oxygen isotope ratios indicate that it originated from recycled, water-rich crust. This means that, already by 4.4 Ga, there was liquid water on the surface of the Earth and hydrothermal circulation within the crust, just as on today's Earth. However, the information contained in the zircon crystal does not tell us how much water there was, nor what its physico-chemical properties were.

We conclude this scene-setting section with a discussion of the concept of habitability on the early Earth (Figure 5.1). In this respect, the term habitability covers a number of concepts. The conditions for the origin and appearance of life are not necessarily those needed for supporting thriving life forms, or even for the survival of life. We do not know how much time was necessary for life to have appeared, or exactly where it appeared. The presently favoured locations are in the vicinity of hydrothermal vents (Figure 5.2) where there was heat and chemical energy and where there were minerals, such as sulphides (e.g. pyrite) or clays, with highly reactive surfaces that may have been implicated in the conformation and stabilization of the organic molecular building bricks of life (Deamer 2007). The littoral zone around volcanic islands where the swash could produce bubbles with organic membranes has also been suggested as another possibility (Deamer 2007). Whatever the location, the process must have been rapid because, in the turbulent and chemically active environment of the early Earth, the large, complex molecules forming the building bricks of life would have been easily broken up. If life appeared at hydrothermal vents, the temperatures needed to have been

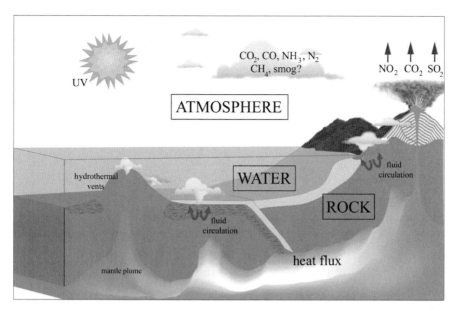

Figure 5.1 Sketch showing the kinds of habitable environments on a global scale on the early Earth – ranging from volcanic rocks and sediments in the ocean, to shallow-water environments around volcano peaks or early continents; from vent systems both in the deep ocean and in the littoral environment; and in the photon-bathed environments in the littoral (Westall/Bost).

Figure 5.2 Deep-sea hydrothermal vent. Life may have originated in this kind of beehive-shaped hydrothermal vent structure, which is very porous and has a steep temperature gradient between the hot fluids in the central pipe and the colder fluids in the outer pores of the structure. Photo IFREMER.

below about 80 °C otherwise the organic macromolecules would not have become dissociated and would not have survived. Also, reducing conditions were needed. It is not our intention here to discuss the origin of life; however, for the purposes of the discussion of the nature and duration of habitable conditions for the origin of life, we estimate that a period of hundreds of thousands to a few millions of years may have been sufficient for the first, very primitive cells to appear (Orgel 1998). This means relatively stable conditions in which liquid water with temperatures <80 °C was available in the presence of organic macromolecules, other essential elements (HNOPS and transition metals), and energy. Once life appeared, the process of cell formation would have been in competition with other physico-chemical processes but, being so much more efficient, would have "cannibalized" the available organic material in a runaway effect. It is possible that the very early cells appeared multiple times in multiple locations because of detrimental changes in the local environment, for example, drying up of a shallow hydrothermal system or destruction by an impactor. The successful early cells would have rapidly colonized all habitable locations, transported from one location to another by the currents on what was, at the time, basically an ocean planet (the early continents resembled submerged oceanic plateaus or emerged volcanic constructs).

Once it had appeared, life would either be destroyed or have to mutate to adapt to changing environmental conditions and to take advantage of even ephemerally habitable conditions. For example, on the modern Earth, photosynthesizing microorganisms can colonize the surface of a beach in between tidal cycles (Stahl 1994)! Thus, habitability for established life can be of much shorter duration than that necessary for the origin of life. As for the survival of life, the last 10 years has amply demonstrated the huge variety of strategies that life uses to survive in what seem to be the most hostile conditions. Although many of the conditions today considered to be hostile were actually the reality on the early Earth (lack of, or very limited, free oxygen in the environment, hot temperatures, pH extremes in hydrothermal systems, higher ocean salinity, and so on), it is clear that relatively simple life forms, such as the prokaryotes (Bacteria and Archaea), can resist for long periods of time in extreme conditions. Microbes have been discovered in 200 My-old salt deposits, apparently having survived there since the formation of the salt (Vreeland et al. 2000), although there is still debate as to whether these halophile microorganisms are really descendants of the original colonies and not later intruders. Microbes have also been resuscitated from 3-My-old permafrost (Gilichinsky et al. 2008).

The microbial habitat of the early Earth

Rocks are the library (or rather, in this case, lithotheque) of the Earth's history. Although rare, the oldest rocks and minerals can provide us with a wealth of information about the early Earth, its geological evolution, and the habitats available for microbial organisms. We have noted above the existence of very old zircon crystals dating back to the earliest days of Earth's history and the ancient gneisses that represented the cores of long destroyed proto-continents. For information regarding life, we need rocks that were formed in the presence of water, i.e. sediments (or, as we will see below, even the surfaces of volcanic lavas), because water is a primordial requirement of life. Where there is water, if the other ingredients of life are present, there is life, at least on Earth. The oldest sediments are found in ancient terranes dated at 3.7–3.8 Ga in south-west Greenland, the Akilia and Isua Greenstone Belts (Mojzsis *et al.* 1996, Rosing 1999). The rocks in both these terranes have been significantly altered by metamorphism and there is some discussion as to whether some of the metamorphosed sediments (metasediments) are really sediments and not some other rock type. Rosing (1999) describes turbiditic sediments, deposited at depth on the ocean floor, from Isua. Turbidites are avalanches of sediment that flow down the steep slopes at the edge of continents, forming characteristically stratified layers on the ocean floor at the bottom of the slopes. The significance of this interpretation is that there were continents and oceans on the early Earth and a sedimentation regime similar to that of today's Earth.

More information can be extracted from other terranes of ancient rocks that have not suffered such a high degree of metamorphism. The Barberton Greenstone Belt in eastern South Africa and the Pilbara in north-western Australia are enclaves having rocks dating between 3.5 and 3.2 Ga. The majority of these rocks are of volcanic origin, either lavas or intruded igneous rocks. Intercalated between the volcanic lavas are thin beds of sediments derived from volcanic detritus and chemically precipitated sediments. The seawater on the volcanically and hydrothermally active early Earth was rich in elements and minerals pumped out from the hydrothermal vents, especially silica and reduced iron oxides. At the interface between the crust and the ocean, both sediments and volcanic rocks were rapidly lithified and silicified by the silica-saturated seawaters.

Both the Pilbara and Barberton terranes consist of smaller geological blocks in which the layers of igneous and sedimentary rocks were intruded by granite cores. Each block represents a proto-continent and each terrane an assemblage of proto-continents, just as today's continents represent an amalgamation of previous continents sutured together by the forces of plate tectonics.

Apart from being among the earliest continental blocks preserved, the interest of the sediments from Barberton and the Pilbara is that they are exquisitely preserved. They have not undergone deep burial in the crust like the rocks from Isua or Akilia on Greenland. They were buried by a few kilometres of later deposited sediments and volcanic rocks that served to preserve them from erosion and, hence, disappearance. Relatively recent erosion of these covering rocks during the

Cenozoic epoch has exhumed them and, thus, gives us an unprecedented window into the past. Part of the reason for their excellent preservation, apart from having been covered until recently, is the fact that many of the formations have been silicified. Silica is almost inert and when it impregnates a sediment or a volcanic rock; it partially replaces the pre-existing lithology but, at the same time, the basic composition and textures are locked in and can thus be preserved over long periods of geological time – in this instance, for 3.5 Ga.

The volcanic lavas were mostly mafic and ultramafic in composition, i.e. rich in elements such as Fe and Mg. In fact, one type of lava, the komatiites, is so rich in Mg that it is believed that mantle temperatures must have been much higher than at present (by about 200 °C) in order to produce the hot melts that gave rise to such rocks (Arndt 1994). Ocean waters would have been enriched in the elements leached out of the igneous rocks by the more acidic seawater and by the abundant hydrothermal recycling in the upper crust. Early seawaters were believed to have been slightly more acidic (pH 6–7) (Grotzinger and Kasting 1993) than today's seawater that has a slightly alkaline pH of $\sim$8. This is because of the dissolution of atmospheric CO_2 in the seawater, the primary component of early atmosphere being CO_2. Mass fractionation of the sulphur isotopes by atmospheric processes indicates anoxic environmental conditions until about 2.4 Ga (Farquhar *et al.* 2000), although one hypothesis proposes possible O_2-rich oases on the early Earth (Ohmoto *et al.* 2006).

Seawater temperatures were higher, $\sim$50 °C and perhaps even 80 °C, judging from the silicon and oxygen isotopes (van den Boorn *et al.* 2007, Robert and Chaussidon 2006) contained in the chemical silica sediments, cherts, precipitated from the seawater and hydrothermal vents. The ocean waters were also apparently more salty than present seawater (de Ronde *et al.* 1997). Thus we have an environment that was volcanically and hydrothermally active, hotter, salty, rich in dissolved minerals and elements, and generally believed to be anoxic.

In terms of habitats for primitive, prokaryotic life forms, the early Earth provided abundant locations for colonization (Figures 5.1 and 5.3). The simplest of life forms would have been those based on a metabolic strategy whereby inorganic sources provided the carbon (CO_2 from seawater), the energy (reduction/oxidation reactions at the surfaces of reactive minerals), and the nutrients (elements in the seawater or leached from the minerals) (Figure 5.4). These are chemolithotrophic organisms. Other types of microorganisms would have obtained their carbon from previously existing organic sources and their energy from the oxidation of the same materials (the dead remains of other microorganisms, carbon produced by abiotic Fischer–Tropsch reactions in hydrothermal systems, or carbon from extraterrestrial materials). Their nutrients could be taken up from those liberated during the degradation of dead microbes dissolved in the seawater. These

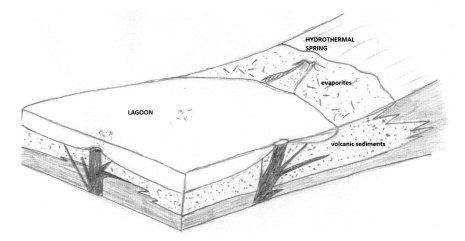

Figure 5.3 Sketch of habitable environments on a local scale showing a shallow-water basin/lagoon and the adjacent intertidal beach zone with hydrothermal vents and springs under water and on the beach surface. Evaporite minerals would have precipitated on the exposed beach surfaces from the more saline seawater and in the warmer environmental conditions. The volcanic lavas and sediments would have hosted chemolithotrophic life forms while the sediment/rock surfaces in the photic zone hosted phototrophic life forms.

are chemoorganotrophic microorganisms. On a microbial scale, habitats of the order of tens to hundreds of micrometres in size would have been sufficient for chemolitho- and chemoorganotrophic microorganisms. Photosynthetic micro-organisms are a more advanced microbial life form that obtains its energy from photons. These microorganisms are considered to be more advanced because the process by which they obtain their energy is complex but far more efficient than the other types of metabolisms (Des Marais 2000) and must have appeared by later mutation. In fact, this type of metabolism is so efficient that a second variant, oxygenic photosynthesis in which the water molecule H_2O is broken down and free oxygen is liberated, is believed to have contributed largely to the oxidation of the Earth's environment. This second variant appeared somewhat later in Earth's history, perhaps after about 3 Ga although, as noted above, there is some debate about the possibility of earlier oxygenic oases (Ohmoto *et al.* 2006). The characteristic of these microorganisms is that they are thought to have initially developed as colonies forming mat-like structures on surfaces in water depths shallow enough to receive sunlight. They tend to cover larger surface areas than the chemotrophic microorganisms. On the modern Earth, oxygenic microbial mats can be up to kilometres in areal extent. They can also cover much smaller areas. The remains of anoxygenic photosynthetic microorganisms in the Early Archaean rocks cover

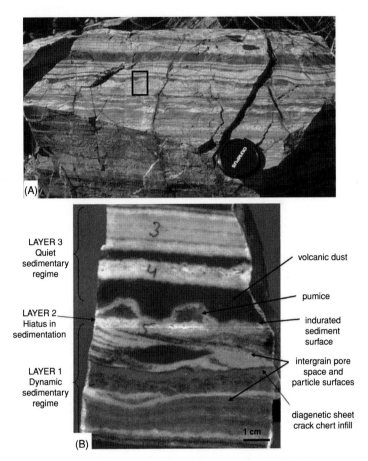

Figure 5.4 Microbial-scale habitats on the early Earth. (A) Field photograph of a
~3.5 Ga-old volcanic sediment (now silicified) from the Kitty's Gap Chert in the
Pilbara, Australia. The black and dark grey layers represent different layers of volcanic
sediment stacked one upon the other and viewed in vertical section. (B) The layers of
sediment tell the history of the rock, read from the bottom to the top. Layer 1 contains
sediments that were deposited in an infilling tidal channel. The filled-up channel was
exposed to the air and sedimentation stagnated for a short period of time (days/weeks)
(Layer 2). The littoral area was then submerged below the influence of waves and Layer
3 represents sediments that simply sedimented through the water column. The
surfaces of the volcanic particles and the pore spaces were host to chemolithotrophic
microorganisms, while a delicate biofilm formed on the resubmerged surface that also
included the transported and resedimented remains of photosynthetic microbial mats
(Westall *et al.* 2011a, b). Publication permission from Elsevier.

areas ranging from centimetres to tens or even hundreds of metres (Tice and Lowe 2004, Allwood *et al.* 2006, 2009, Noffke 2009). One important factor affecting life within reach of sunlight is the fact that this zone was bathed in what should have been lethal doses of UV radiation. There was insufficient free oxygen in the atmosphere to enable the formation of a protective ozone layer and harmful radiation doses would have been common on the early Earth. Cockell and Raven (2004) have estimated that the amount of DNA-weighted UV radiation in the time period in which we are interested was about 54 W/m^2, going up to 1000 W/m^2 in a worse case scenario, compared to 1 W/m^2 today. Volcanic dust or an organic haze from methane produced by methanogenic chemotrophs (Pavlov *et al.* 2001) could have reduced the amount of UV reaching the Earth's surface. On the other hand, there are well-known modern microorganisms which can survive very high doses of UV radiation, for example *Deinococcus radiodurans* (Cox and Battista 2005). They do so by having an efficient gene repair mechanism, a rapid reproduction rate, and unlimited access to the necessary ingredients of life. On the early Earth, the primitive microorganisms may have similar strategies for coping with high doses of UV radiation, including UV-blocking pigments. One of the characteristics of photosynthetic microbial mats is that they are embedded in great thicknesses of mucus, or extracellular substances that also served to protect them.

The important point here is that the size of microbial habitats is on a different scale from that of the regional and global geological processes influencing habitability (compare Figures 5.1 and 5.4). At this stage in the history of life on Earth, life was really surface specific, i.e. in direct contact with the underlying mineral, sediment, or rock surface. However, from this point of view, the whole of the Earth's surface was habitable, with the exceptions of hydrothermal vents and lava flows that were too hot. This means in and on the rocks, sediments and lower-temperature regions of hydothermal vents in the depths of the ocean, in the water-logged cracks and fractures in the upper crust, and in shallow water regions around the proto-continents and exposed volcanic edifices where volcanic lava flows, volcanically derived sediments, and hydrothermal vents and springs abounded.

Rocks dating between 3.5 and 3.3 Ga in the Pilbara and Barberton terranes were formed and deposited in mostly shallow-water environments around exposed landmasses, although some sediments deposited below wave base were also preserved. Purported traces of life have been described from many of these environments and we will examine these in the next section.

Early life

The earliest preserved life forms were microbial organisms using relatively simple metabologies, chemolitho- and chemoorganotrophs, and anoxygenic photosynthesizers (although, from a metabolic point of view, the latter are relatively advanced). Their remains occur in rocks 3.5–3.3 Ga old in the ancient terranes of Barberton in South Africa and the Pilbara in Australia. As noted above, these are the oldest known, well-preserved rocks on Earth, but they are not the oldest rocks known, having been formed more than a billion years after the formation of the Earth. The evidence for the presence of photosynthetic microorganisms in this epoch has important implications for the timing of the origin and early evolution of life because it means that life must have evolved much earlier. Unfortunately the rocks that would have preserved traces of the passage from inert components to the earliest cells, or even of the very early evolutionary steps of life, no longer exist at the Earth's surface. They have either been destroyed by crustal recycling or are buried by younger deposits. Nevertheless, the 3.5–3.2 Ga-old rocks from Australia and South Africa contain a wealth of information pertaining to the nature of the early inhabitants of the Earth and to their mode of life.

The first question is: How do we go about searching for fossilized traces of microbial life in very ancient rocks? These are organisms that are extremely small and that have no hard parts. They also have very simple structures. In terms of what can be preserved, the components of microorganisms can be broken down into three categories (Westall and Cavalazzi 2011): (1) the physical structure, i.e. cell, colony of cells (including microbially produced polymers or extracellular polymeric substances, EPS), and microbial constructions, such as mats, stromatolites, and microbially induced sedimentary structures (Noffke 2009); (2) the chemical components, i.e. the organic molecules making up the organisms and associated essential elements; and (3) evidence of the metabolic activity of the microorganisms, such as isotopic fractionation of elements like C and S; associated biominerals precipitated by microbially mediated processes; leaching of essential elements in rocks; influence of microbial materials on the formation of minerals, and so on. There is a huge number of subtle signatures that the presence of microbes and their metabolic activities can leave on rocks and minerals (Bandfield *et al.* 2002).

Not all of these types of traces of life are preserved in rocks. The highly altered, metamorphosed 3.8-Ga-old rocks from Greenland, for instance, contain no recognizable microbial structures but inclusions of carbon have an isotopic signature that is consistent with that produced by life (Mozjsis *et al.* 1996, Rosing *et al.* 1999). The ancient 3.5–3.3-Ga-old sediments from Barberton and the Pilbara contain the preserved traces of all three types of signature: the physical cells, colonies, microbial constructions in the form of stromatolites; the (highly) degraded remains of the carbon molecules that formed the cells (and their EPS) and associated essential elements; as well as evidence of metabolic activity in the form of isotopically fractionated carbon and microbially induced biominerals.

It is surprising that traces of such ephemeral forms of life can be preserved at all. Microbes degrade rapidly, broken down by a naturally produced enzyme upon death. Their remains are then used by chemoorganotrophic microorganisms as a source of carbon and energy. The degraded carbon molecules can be mixed in with the sediments formed or deposited in the environment in which the organisms lived. They may preserve something of the structure of the specific molecule. For instance, microbial cell walls are made of specific components, such as lipids, that aid protection of the precious genetic and metabolic machinery within the cell from the exterior environment. Hopanes are the degraded remains of one of the kinds of lipids found in cell membranes and they have been detected in carbon-rich sediments as old as 2.7 Ga (Summons *et al.* 1999, Brocks *et al.* 1999, although it has been suggested that these molecules are younger contaminants, e.g. Rasmussen *et al.* 2008). The fractionated isotopic signature of the carbon molecules may also be preserved, as in the purported cases of the carbon in the Greenland rocks (Mojzsis *et al.* 1996, Rosing 1999). The physical structures are more difficult to conserve because of the rapid degradation of microbial cells. Cellular remains can be preserved in very fine-grained, reducing sediments in the form of flattened carbonaceous structures. Pollen spores and larger algal remains, commonly termed achritarchs, are typically preserved in this manner (Javaux *et al.* 2001). These are structures having particularly robust external cell envelopes. Large (a few hundred microns in size) cellular structures of unknown affinity have even been detected in fine-grained, 3.2-Ga-old, deep-sea sediments from the Barberton region (Javaux *et al.* 2001). The kinds of microorganisms inhabiting the, by our standards, extreme conditions of the early Earth were small, of the order of a micron or less in size, producing colonies in the range of tens to hundreds of microns in diameter, or microbial biofilms and mats from centimetres to many metres in areal extent (for a review, see Westall 2011). Their cell envelopes were thinner and more delicate than those of the large, robust achritarchs.

Despite their relatively delicate forms, these kinds of microorganisms have been and are still being preserved by the precipitation and permeation of minerals on and in the cells (reviewed by Westall and Cavalazzi 2011). Experiments have shown that this can be a very rapid process, occurring within hours of insertion of the microorganisms in a mineral-saturated solution (Orange *et al.* 2009). Minerals in solution are chemically attracted to certain parts of molecules on the cell envelope called functional groups (Monty *et al.* 1991, Westall 1999, Orange *et al.* 2009). This is the microbially mediated part of the reaction in the sense that the microbes are simply passive supports for mineral fixation. The rest of the fossilization process is purely chemical and involves further precipitation of minerals onto those already fixed onto the organic support. In this way the microorganisms as individual cells or colonies of cells may be encased in a mineral coating that preserves the morphological structure. The next step involves the rapid lithification of the surrounding materials, sediments, or rock surfaces in order to preserve the still delicate, mineralized life forms. Serendipity also plays its part in the long-term preservation of the fossilized structures. Geological processes, such as metamorphism, can completely obliterate these biosignatures. Erosion can completely eliminate the rock formations in which they are entombed.

The Earth has always been geologically active and crustal recycling has been ongoing since at least the Hadean era (pre-4.0 Ga). It is thus perhaps surprising that rocks as old as ~3.5 Ga have survived at all and, moreover, intact and conserving delicate traces of life! One of the major reasons for survival of the microfossils and for the survival of the rocks encasing them is the aforementioned pervasive silicification of all materials at the crust–seawater interface during this Early Archaean era (4.0–3.3 Ga) (Hofmann and Bohlar 2007). Silica has a very small crystal lattice and thus faithfully preserves even nanometre-scale details in microorganisms (Westall *et al.* 1995, 2011a,b, Orange *et al.* 2009). It is also more or less impervious and extremely resistant to further change. Whereas carbonate, for example, is readily transformable, permeable, and easily broken up mechanically, silica is far more resistant. Thus, it is the silica-saturated seawater and silica-saturated hydrothermal fluids that caused the silicification of the colonies of microorganisms inhabiting the Early Archaean aqueous environments. As noted above, silicification was rapid because we find colonies containing turgid, living cells in division, as well as deflated cells that were already dead when fossilized, and even fragments of bent and contorted cells and microbial mats torn up from one environment and deposited in another nearby area.

Numerous studies have revealed carbon isotope signatures of life in the Early Archaean rocks, variously attributed to chemotrophic or phototrophic microorganisms (see the reviews by Westall and Southam 2006 and Westall 2011a,b). However, very detailed investigations are needed to reveal the fossilized cells and colonies *in situ* in their natural habitat. The microorganisms were silicified and embedded in a lithifying matrix of silica. Delicate chemical methods are necessary to extract them from the silica matrix. Etching using hydrofluoric acid is the only means of revealing them – the silica matrix is dissolved away but the silica fossilizing the microorganisms is bound to the organic molecules that act as an impurity, thus making the structures more resistant to the acid. The process is very delicate – too much etching and everything is lost; too little and nothing can be seen. Furthermore, artefacts produced during etching, for instance the precipitation of salts, can take on the simple morphologies of the microorganisms – spheres, rods, or filaments. There are many minerals that, at least superficially, produce "bacteriomorph" structures. Thus, other information is necessary to ensure that a particular structure is indeed a fossil microorganism and not an abiotic mineral feature (Westall and Cavalazzi 2011). Physical features include the association of microbial mucus, EPS, with cellular structures; evidence of cell division and lysed or deflated cells (life and death) together, as would be normal in a microbial colony consisting of a number of generations of cells; evidence of "hollowness" in dead cells from which the internal cytoplasmic material has escaped; mono-species colonies of cells having the same shape and size or multi-species colonies in which cells of the distinct species exhibit different sizes and shapes; different types of cells related to different micro-habitats; mat-forming behaviour with evidence of interaction with the directly adjacent environment, etc. Chemical features include the composition and structure of the organic molecules. Molecules about ~3.5 Ga old are extremely degraded but some remnants of the original composition and structure remain (Derenne *et al.* 2008, Westall *et al.* 2011a,b). Despite the great age of the organic molecules, it is still possible to measure concentrations of associated essential elements and to identify associated biominerals.

The caveat to these physical, organo-chemical, and mineralical biosignatures in the Early Archaean rocks is that, although perfectly preserved by silica, they are also extremely diluted by it, at least with respect to the chemical and mineralogical signatures. For *in situ* investigations, strong energy sources such as those associated with synchrotron radiation, proton probes, or various secondary ion mass spectrometers, are necessary for the analyses, and powerful, high-resolution electron microscopes are necessary to observe the structures.

Identification of a particular structure as a biosignature in these rocks is the first and necessary step in trying to extract information about ancient forms of life. The second step is making sure that these biosignatures were actually formed at the same time as the host rocks and are not a later intrusion. Fluids containing molecular remains can permeate through rocks and concentrate in formations that are older than those in which they originally formed (Papineau *et al.* 2011). Also, carbon-containing hydrothermal fluids can deposit carbon in shapes that resemble those of microorganisms (cf. Schopf 1993, Brasier *et al.* 2002). More importantly, microorganisms can infiltrate cracks and fissures in rocks, for protection from hostile environments and/or to extract carbon and nutrients from the rocks themselves. Epilithic (forming on rocks) and endolithic (within rocks) or chasmolithic (within cracks) organisms thrive in such habitats (Golubic *et al.* 1981). Given a suitable physico-chemical environment, these later intruders can also become fossilized within their older rocky matrix. Thus, silicified endolithic microorganisms <8 ka-old occur in rocks that are >3.7 Ga from Greenland (Westall and Folk 2003) (Figure 5.5). It is thus necessary to demonstrate co-formation of the biosignature under investigation with the host rock. Such an example could be, for instance, a biosignature distribution that demonstrates its emplacement before lithification of the rock, such as a fossilized microbial colony coating nutrient-rich

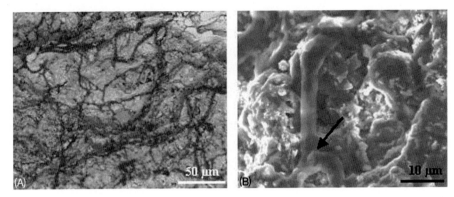

Figure 5.5 Recent endolithic fungal hyphae in a 3.8-Ga-old rock from Isua, Greenland (Westall and Folk 2003). (A) Scanning electron microscope (SEM-backscatter) image of the dark fungal hyphae forming a network on a fracture surface in the rock. (B) SEM (secondary) image showing branching hyphae. Publication permission from Elsevier.

volcanic particles in a silica-cemented matrix or microbial mats that formed on the surface of a sediment that were then overlain by another layer of sediment. The state of degradation of the organic molecules can also be an indication of their great age.

The search for ancient traces of life is full of pitfalls and false trails. This type of research is a steep learning curve and requires painstaking hours of study – but is very rewarding when the precious traces of very ancient life can finally be revealed. The *in situ* investigations, although laborious, have provided a wealth of information regarding the nature of primitive life and its metabolic strategies. The ancient volcanic substrates apparently supported thriving colonies of micro-organisms, probably chemolithotrophs that obtained their energy and nutrients from the volcanic raw materials given their direct contact with the volcanic rocks and detrital particles. Thus, traces of what is interpreted to be microbial corrosion has been observed on the vitreous surfaces of pillow lavas, extruded under water (Furnes *et al.* 2004, 2007) and in volcanic particle surfaces, both occurring in Barberton and the Pilbara (Westall *et al.* 2011b) (Figure 5.6). These traces are represented by tunnels lined with carbon and nitrogen, two of the essential elements of life. The carbon isotopic signatures of these materials could be interpreted as representing microbial fractionation. Support for such types of microorganisms comes from the *in situ* observation of silicified colonies of microorganisms directly on the surfaces of the volcanic particles deposited in a mud-flat environment, now at a location called the "Kitty's Gap Chert", 3.45 Ga in the Pilbara (Westall *et al.* 2006a, 2011b). These colonies consisted of two species, both coccoidal in shape but having different size modes, one ~0.5 µm and the other ~0.8–1 µm (Figure 5.7).

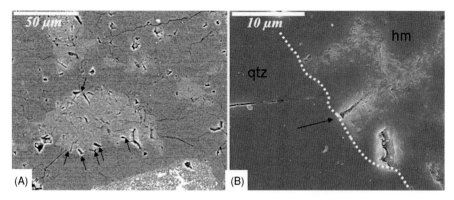

Figure 5.6 Fossil endolithic structures (i.e. syngenous with the formation of the rock itself) in a thin section of ~3.5-Ga-old sediments from the Kitty's Gap Chert, Pilbara, Australia. (A) SEM-backscatter view of a volcanic particle (light grey) showing cracks around its edges, embedded in a matrix of cementing silica (quartz, darker grey). (B) Detail showing that the cracks are actually corrosion tunnels produced by microbial processes in their search for nutrients (Westall *et al.* 2011b). Publication permission from Elsevier.

Microorganisms forming chains of coccoidal cells coexisted in the same micro-habitats. Different layers of sediments representing successive deposits in time were characterized by consortia of fossilized microorganisms having slightly different morphological characteristics. The pores of small 0.5–1-cm-sized pumice fragments, embedded in one of the sediment surfaces, were also filled and lined with the same kinds of silicified microorganisms. In this case, it is the microenvironment or microhabitat that is important as the source of ingredients of life – carbon from in CO_2 the seawater, nutrients and energy from the volcanic materials. Although the volcanic habitat suggests that the inhabiting microorganisms were probably chemolithotrophic, it would be difficult to distinguish them from microorganisms supporting themselves on a completely carbonaceous "diet", i.e. the chemoorganotrophs that could have colonized dead chemolithic microorganisms. Whatever the case, the chemolithotrophs had to have been there first in order to transform the primary resources into carbon cells.

The mud-flat environment in the Kitty's Gap Chert example was also an ideal environment for the formation of microbial mats of anoxygenic phototrophic microorganisms. Indeed, mat-fragments have been detected on one of the sedimentary horizons (Westall *et al.* 2006a, 2011b). From the Barberton region, a perfectly preserved photosynthetic microbial mat was conserved by hydrothermal silicification (Westall *et al.* 2006b, 2011a) (Figure 5.8). This mat actually formed on a beach and was episodically exposed. Exposure led to seawater evaporation

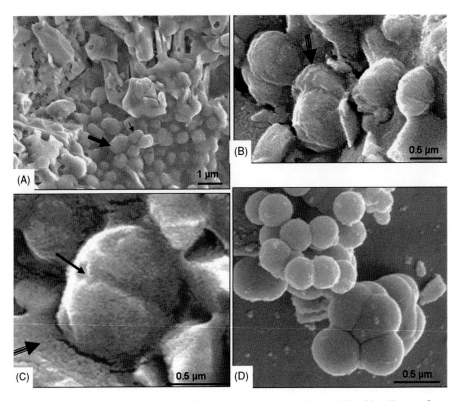

Figure 5.7 Details of the silicified microorganisms in the ~3.5-Ga-old sediments from the Kitty's Gap Chert, Pilbara, Australia. (A) SEM (secondary) view of a silicified colony of very small coccoidal microorganisms (<1 μm) colonizing the surface of a volcanic particle. (B) Detail of a colony containing two species of coccoids characterized by two different size modes (0.5 μm and 0.8 μm). Note that the coccoids were in the process of cell division when they were fossilized and one of the cells was dead and deflated (arrow). (C) Coccoidal microorganism in cell division showing details of the meniscus that forms as the cells separate. (D) Hydrothermal silica spheres (abiogenic) that have shapes similar to the microorganisms (bacteriomorphs). It is the fine details in (B) and (C) that help to determine the biogenicity of the fossil coccoids and that distinguish them from the abiogenic silica spheres. Publication permission from Elsevier.

and to the precipitation of layers of evaporite minerals in its surface, as well as to desiccation of the mat. This is the oldest known evidence of mat exposure – and also direct evidence that the microorganisms were well able to cope with high UV doses. It is also the oldest evidence of biomineral precipitation, the mat having been calcified *in situ*, as is common in photosynthetic microbial mats. This particular microbial mat contains morphological, chemical, and mineralogical evidence for the existence of a consortium of microorganisms inhabiting the photosynthetic mat. Like modern mats, the phototrophs at the surface of the mat

Figure 5.8 Photosynthetic microbial mat in 3.3-Ga-old volcanic sediments from Barberton, South Africa. This is an SEM view of a portion of a photosynthetic microbial mat formed in a tidal environment. It was formed by tiny (0.3 μm thick) microbial filaments. The arrows show a distinct orientation in the mat indicating that it was formed under flowing water (e.g. possibly a hydrothermal spring outflow channel). Publication permission from the Royal Society of London.

produced the primary biomass by using sunlight as energy. The lower, dead, and degraded layers of the mat were host to sulphur-reducing organisms that used the dead cells as a carbon and energy source. A byproduct of their metabolic activity was the *in situ* precipitation of carbonate within the mat.

That photosynthetic microorganisms were already common 3.5–3.3 Ga ago is testified by the many observations of mat remains in the silicified sediments (Walsh 1992, Tice and Lowe 2004). Photosynthetic microorganisms are known for their mat-forming strategies. Often, the mats form complicated three-dimensional structures called stromatolites whose form and size depend on many variables, including water depth, flow dynamics, nutrient availability, and more. Stromatolites were at their apogee in the Proterozoic era when oxygen had become readily available and these microbial structures had no predators. They decreased in frequency of occurrence during the Phanerozoic, probably due to predation, and today can be found only in isolated protected niches. Stromatolite-like structures were among the first potential biosignatures reported from the Early Archaean rocks of the Pilbara and Barberton (Lowe 1980, Schopf and Walter 1983). These were finely layered domical constructions that had been silicified. Later interpretations of their origin were more hesitant because of the possibility of such morphologies being produced by abiological processes (Lowe 1994), such as hydrothermal siliceous precipitations (Lindsay *et al.* 2005), or as in modern calcareous travertines (Pentecost 2005). However, detailed field mapping and microscopic examination have shown that, in one location in the Pilbara, the Strelley Pool stromatolite locality, a shallow-water carbonate platform existed about 3.45 Ga ago consisting of a variety of small localized microenvironments that were host to different types of stromatolites whose combined environmental, physical, and chemical characteristics could only have been produced by photosynthetic microbial mats (Allwood *et al.* 2006, 2009).

These oldest, well-preserved traces of life document life forms that were already relatively evolved. These are not the earliest, primitive cells that formed just after life appeared. They represent life forms that had already differentiated and that had exactly the same characteristics as chemotrophic life forms today. Moreover, life had already mutated and adapted to be able to exploit solar energy, therefore taking one of the most important steps in the evolution of life on Earth. The level of evolution demonstrated by these fossilized remains indicates that life must have evolved much earlier. The possible indications of life in the form of fractionated carbon isotopes in the 3.7–3.8-Ga-old rocks of Isua and Akilia, although contested (van Zuilen *et al.* 2002), are plausible. This also means that the Earth's oceans could not have been completely vaporized by a massive impact during the Later Heavy Bombardment period (4.0–3.8 Ga) during which, it has been hypothesized, life would have been exterminated. Life must have appeared much earlier in the pre-4.0 Ga Hadean era.

The time period 3.3–3.2 Ga may represent an evolutionary dichotomy in terms of both the geological evolution of the Earth and the evolution of life – or it may be an artefact of lack of preservation of other old rock formations. The type of sediments deposited after 3.2 Ga changes. Whereas previously the sediments were dominated by volcanic lithologies, afterwards more quartz-rich lithologies appeared, indicating the exposure and erosion of the granite and gneiss cores of continents. This does not mean that quartz grains of a similar provenance did not occur in older sediments – they were simply very rare. It is the massive appearance of such sediments, more similar to continental shelf sediments on today's Earth than the volcanic sediments of the early Earth, that indicates a change in the geology of the Earth. It is at this period that larger continental masses were being amalgamated from the dispersed small proto-continents of the early Earth. The formation of larger continental masses may have had the effect of a deepening of the global ocean and changes in circulation patterns in the oceans at different levels. Did such a phenomenon have an effect on microbial life? In the modern ocean, deep ocean circulation is important in distributing nutrients to surface waters and thus supporting a thriving planktonic population. There is little evidence for such a population, which would have had to have been photosynthetic, on the early Earth. If they were present, the relevant sediments may be lacking and/or their signatures diluted. However, sediments from the Pilbara and Barberton aged 3.2 Ga contain relatively large, complex carbonaceous structures of the order of 50 μm to several hundreds of microns in size that, if they are biogenic, may represent evolved mutants of more primitive life forms (Sugitani *et al.* 2007, Javaux *et al.* 2001). Their large size, platy form, and possible central gas bubble are somewhat suggestive of possible planktonic organisms – more work is needed.

Summary

The Earth was possibly already habitable by 4.4 Ga but the absence of well-preserved crust older than 3.5 Ga precludes study of the most important steps for life – the transformation of abiotic materials into simple, self-replicating, and (micro) environment-transforming cells and the early development of different metabolic strategies. The rare mineral and rock record between 4.4 and 3.5 Ga, however, provides us with much useful environmental information, although some important details, such as the exact composition of the early Earth's atmosphere, still escape us. The early Earth had a global ocean and proto-continents. It was basically a volcanic planet with an enormous amount of hydrothermal activity and hydrothermal circulation/alteration of the upper crustal materials. The ocean water was probably slightly acidic to neutral, saltier than today's ocean, and the atmosphere probably consisted mostly of CO_2 with the addition of some other gases, such as N_2, CH_4, water vapour, etc. The flux of extraterrestrial volatiles to the early Earth was high and there were probably many impacts, some rather large and regionally catastrophic. There is no evidence that any of the impacts totally vaporized the Earth's oceans during the Late Heavy Bombardment period (4.0–3.85 Ga).

The earliest traces of life on Earth represent simple, prokaryotic life forms that obtained their energy from inorganic and organic sources, as well as from the Sun. They were anoxygenic organisms; oxygenic photosynthesis had not yet developed. They were small (generally < 1 μm), chemolithotrophic and chemoorganotrophic cells that formed colonies on volcanic rock and particle surfaces, or phototrophs that formed microbial biofilms, mats, and stromatolites on the surfaces of shallow water sediments. Within the limits of our present knowledge, they appear to have colonized all the colonizable environments available on Earth except, perhaps, the open ocean.

From an evolutionary point of view, these earliest traces were already far removed from the first proto-cellular life. There are no rocks existing at the surface of the Earth today that might record the first appearance of such cells. There is, on the other hand, a large part of the surface of another planet, Mars, where rocks of a suitable age and recording a habitable environment could potentially hold this missing record. But would we be able to identify it? Recent experiments to fossilize viruses 50–100 nm in size (Orange *et al.* 2011) suggest that the task will be difficult.

References

Allwood, A. C., Walter, M. R., Kamber, B. S. *et al.* (2006). "Stromatolite reef from the Early Archaean Era of Australia," *Nature*, Vol. 441, 714–718.

Allwood, A. C., Grotzinger, J. P., Knoll, A. H. *et al.* (2009). "Controls on development and diversity of Early Archean stromatolites," *Proceedings of the National Academy of Sciences*, Vol. 106, 9548–9555.

Anbar, A. D., Zahnle, K. J., Arnold, G., and Mojzsis, S. J. (2001). "Extraterrestrial iridium, sediment accumulation and the habitability of the early Earth's surface," *Journal of Geophysical Research*, Vol. 106, 3219–3236.

Arndt, N. T. (1994). "Archean komatiites," in *Archean Crustal Evolution*, K. C. Condie (ed.). Amsterdam: Elsevier, pp. 11–44.

Bandfield, J., Moreau, J. W., Chan, C. S. *et al.* (2002). "Mineralogical biosignatures and the search for life on Mars," *Astrobiology*, Vol. 1, 447–465.

Baross, J. A., Hoffman Breuer, D., Labrosse, S., and Spohn, T. (2010). "Thermal evolution and magnetic field generation in terrestrial planets and satellites," *Space Science Review*, Vol. 152, 449–500.

Bowring, S. A. and Williams, I. S. (1999). "Priscoan (4.00–4.03 Ga) orthogneisses from northwestern Canada," *Contributions to Mineralogy and Petrology*, Vol. 134, 3–16.

Brasier, M. D., Green, O. R., Jephcoat, A. P. *et al.* (2002). "Questioning the evidence for Earth's oldest fossils," *Nature*, Vol. 416, 76–81.

Breuer, D., Labrosse, S., and Spohn, T. (2010). "Thermal evolution and magnetic field generation in terrestrial planets and satellites," *Space Science Review*, Vol. 152, 449–500.

Brocks, J. J., Logan, G. A., Buick, R., and Summons, R. E. (1999). "Archean molecular fossils and the early rise of eukaryotes," *Science*, Vol. 285, 1033–1036.

Charbonneau, D., Berta, Z. K., Irwin, J. *et al.* (2009). "A super-Earth transiting a nearby low-mass star," *Nature*, Vol. 462, 891–894.

Cockell, C. S. and Raven, J. A. (2004). "Zones of photosynthetic potential on Mars and the early Earth," *Icarus*, Vol. 169, 300–310.

Cox, M. M. and Battista, J. R. (2005). "Deinococcus radiodurans – the consummate survivor," *Nature Reviews: Microbiology*, Vol. 3, 882–892.

Dauphas, N., Robert, F., and Marty, B. (2000). "The late asteroidal and cometary bombardment of Earth as recorded in water deuterium to protium ratio," *Icarus*, Vol. 148, 508–512.

de Ronde, C. E. J., Channer, D. M. DeR., Faure, K. *et al.* (1997). "Fluid chemistry of Archean seafloor hydrothermal vents; implications for the composition of circa 3.2 Ga seawater," *Geochimica et Cosmochimica Acta*, Vol. 61, 4025–4042.

Deamer, D. W. (2007). "The origin of cellular life," in *Planets and Life: The Emerging Science of Astrobiology*, W. T. Sullivan and J. A. Baross (eds.). Cambridge: Cambridge University Press, pp. 187–209.

Derenne, S., Robert, F., Skryzpczak-Bonduelle, A. *et al.* (2008). "Molecular evidence for life in the 3.5-billion-year old Warreawoona chert," *Earth and Planetary Science Letters*, Vol. 272, 476–480.

Des Marais, D. J. (2000). "When did photosynthesis emerge on Earth?," *Science*, Vol. 289, 1703–1705.

Farquhar, J., Huiming Bao and Thiemens, M. (2000). "Atmospheric influence of the Earth's earliest sulfur cycle," *Science*, Vol. 289, 756–758.

Furnes, H., Banerjee, N. R., Muehlenbachs, K. *et al.* (2004). "Early life recorded in Archean pillow lavas," *Science*, Vol. 304, 578–581.

Furnes, H., Banerjee, N. R., Staudigel, H. *et al.* (2007). "Comparing petrographic signatures of bioalteration in recent to Mesoarchean pillow lavas: tracing subsurface life in oceanic igneous rocks," *Precambrian Research*, Vol. 158, 156–176.

Gaillard, F., Scaillet, B., and Arndt, N. (2011). "Atmospheric oxygenation caused by a change in volcanic degassing pressure," *Nature*, Vol. 478, 229–232.

Gilichinsky, D., Vishnivetskaya, T., Petrova, M. *et al.* (2008). "Bacteria in permafrost," in *Psychrophiles: From Biodiversity to Biotechnology*, R. Margesin, F. Schinner, J.-C. Marx and C. Gerday (eds.). Berlin: Springer, pp. 83–102.

Golubic, S., Friedmann, I., and Schneider, J. (1981). "The lithobiontic ecological niche, with special reference to microorganisms," *Journal of Sedimentary Petrology*, Vol. 51, 475–478.

Grotzinger, J. P. and Kasting, J. F. (1993). "New constraints on Precambrian ocean composition," *Journal of Geology*, Vol. 101, 235–243.

Hartmann, W. K., Ryder, G., Dones, L., and Grinspoon, D. (2000). "The time dependent intense bombardment of the primordial Earth/Moon system," in *Origin of the Earth and Moon*, K. Righter and R. Canup (eds.). Tucson: University of Arizona Press, pp. 493–512.

Hofmann, A. and Bolhar, R. (2007). "The origin of carbonaceous cherts in the Barberton greenstone belt and their significance for the study of early life in mid-Archaean rocks," *Astrobiology*, Vol. 7, 355–388.

Holm, N. G. (1992). "Why are hydrothermal systems proposed as plausible environ-ments for the origin of life?" *Origins of Life and Evolution of Biospheres*, Vol. 22, 5–14.

Javaux, E. J., Knoll, A. H., and Walter, M. R. (2001). "Morphological and ecological complexity in early eukaryotic ecosystems," *Nature*, Vol. 412, 66–69.

Kasting, J. F. and Brown, L. L. (1998). "The early atmosphere as a source of biogenic compounds?" in *The Molecular Origins of Life*, A. Brack (ed.). Cambridge: Cambridge University Press, 35–56.

Kleine, T. *et al.* (2009). "Hf-W chronology of the accretion and early evolution of asteroids and terrestrial planets," *Geochimica et Cosmochimica Acta*, Vol. 73, 5150–5188.

Knauth, L. P. (1998). "Salinity history of Earth's early ocean," *Nature*, Vol. 395, 554–555. *Geologie und Palaontologie-Monatshefte*, Vol. 4, 218–231.

Lindsay, J. F., Brasier, M. D., McLoughlin, N. *et al.* (2005). "The problem of deep carbon – an Archean paradox," *Precambrian Research*, Vol. 143, 1–22.

Lowe, D. R. (1980). "Stromatolites 3,400-Myr old from the Archean of Western Australia," *Nature*, Vol. 284, 441–443.

Lowe, D. R. (1994). "Abiological origin of described stromatolites older than 3.2 Ga," *Geology*, Vol. 22, 287–390.

Martin, H., Albarède, F., Claeys, P. *et al.* (2006). "Building of a habitable planet," *Earth, Moon and Planets*, Vol. 98, 97–151.

Maurette, M. (1998). "Carbonaceous micrometeorites and the origin of life," *Origins of Life and Evolution of Biospheres*, Vol. 28, 385–412.

Mojzsis, S. J., Arrhenius, G., Keegan, K. D. *et al.* (1996). "Evidence for life on Earth before 3,800 million years ago," *Nature*, Vol. 384, 55–59.

Monty, C. L. V., Westall, F., and Van Der Gaast, S. (1991). "The diagenesis of siliceous particles in Subantarctic sediments, ODP Leg 114, Hole 699: possible microbial mediation," in *Proc. ODP Sci. Results*, 114: P. F. Ciesielski, Y. Kristoffersen *et al.* (eds.). College Station, TX: Ocean Drilling Program, pp. 685–710.

Morbidelli, A., Chambers, J., Lunine, J. I. *et al.* (2000). "Source regions and timescales for the delivery of water to the Earth," *Meteoritics and Planetary Science*, Vol. 35, 1309–1320.

Noffke, N. (2009). "The criteria for the biogenicity of microbially induced sedimentary structures (MISS) in Archean and younger, sandy deposits," *Earth Science Reviews*, Vol. 96, 173–180.

Ohmoto, H., Watanabe, Y., Ikemi, H. *et al.* (2006). "Suphur isotope evidence for an oxic Archaean atmosphere," *Nature*, Vol. 442, p. 908.

Orange, F., Westall, F., Disnar, J.-R. *et al.* (2009). "Experimental silicification of the extremophilic Archaea Pyrocus abyssi and Methanocaldococcus jannaschii: applications in the search for evidence of life in early Earth and extraterrestrial rocks," *Geobiology*, Vol. 7, 403–418.

Orange, F., Chabin, A., Gorlas, A. *et al.* (2011). "Experimental fossilisation of viruses from extremophilic Archaea," *Biogeosciences*, Vol. 8, 1465–1475.

Orgel, L. E. (1998). "The origin of life – How long did it take?" *Origins of Life and Evolution of Biospheres*, Vol. 28, 91–96.

Papineau, D., De Gregorio, B. T., Cody, G. D. *et al.* (2011). "Young poorly crystalline graphite in the >3.8-Gyr-old Nuvvuagittuq banded iron formation," *Nature Geoscience*, Vol. 4, 376–379.

Pavlov, A. A., Kasting, J. F., Brown, L. L. *et al.* (2001). "Greenhouse warming by CH_4 in the atmosphere of early Earth," *Journal of Geophysical Research*, Vol. 105, 11981–11990.

Pentecost, A. L. (2005). *Travertines*. Berlin: Springer.

Plesa, A. C. and Breuer, D. (2010). "Influence of partial melting on mantle convection in a spherical shell: application to Mars," Earth and Planetary Science Congress, 24–29 September 2010. Rome, Italy.

Rasmussen, B., Fletcher, I. R., Brocks, J. J., and Kilburn, M. R. (2008). "Reassessing the first appearance of eukaryotes and cyanobacteria," *Nature*, Vol. 455, 1101–1104.

Robert, F. and Chaussidon, M. (2006). "A palaeotemperature curve for the Precambrian oceans based on silicon isotopes in cherts," *Nature*, Vol. 443, pp. 969–972.

Rosing M. T. (1999). "13C depleted carbon microparticles in >3700 Ma seafloor sedimentary rocks from West Greenland," *Science*, Vol. 283, 674–676.

Ryder, G., Koeberl, C., and Mojzsis, S. J. (2000). "Heavy bombardment on the Earth at ∼3.85 Ga: the search for petrographic and geochemical evidence," in *Origin of the Earth and Moon*, R. M. Canup and K. Righter (eds.). Tucson: University of Arizona Press, pp. 475–492.

Schoenberg, R., Kamber, B. S., Collerson, K. D., and Moorbath, S. (2002). "Tungsten isotope evidence from approximately 3.8-Gyr metamorphosed sediments for early meteorite bombardment of the Earth," *Nature*, Vol. 418, pp. 403–405.

Schopf, J. W. (1993). "Microfossils of the Early Archean Apex Chert: new evidence of the antiquity of life," *Science*, Vol. 260, pp. 640–646.

Schopf, J. W. and Walter, M. R. (1983). "Archean microfossils: new evidence of ancient microbes," in *Earth's Earliest Biosphere*, J. W. Schopf (ed.). Princeton, NJ: Princeton University Press, pp. 214–239.

Sleep, N. H., Zahnle, K. J., Kasting, J. F., and Morowitz, H. J. (1989). "Annihilation of ecosystems by large asteroid impacts on the early Earth," *Nature*, Vol. 342, pp. 139–142.

Stahl, L. J. (1994). "Microbial mats: ecophysiological interactions related to biogenic sediment stabilization," in *Biostabilization of Sediments*, W. E. Krumbein *et al.* (eds.). Oldenburg: Biblioteks und Informationssystem der Carl von Ossietzky Universität, pp. 41–54.

Sugitani, K., Grey, K., Allwood, A. *et al.* (2007). "Diverse microstructures from Archaean chert from the Mount Goldsworthy–Mount Grant area, Pilbara Craton, Western Australia: microfossils, dubiofossils or pseudofossils?" *Precambrian Research*, Vol. 158, 228–262.

Summons, R. E., Jahnke, L. L., Hope, J. M., and Logan, J. H. (1999). "2-methylhopanoids as biomarkers for cyanobacterial oxygenic photosynthesis," *Nature*, Vol. 400, 554–557.

Tice, M. and Lowe, D. R. (2004). "Photosynthetic microbial mats in the 3,416-Myr-old ocean," *Nature*, Vol. 431, 549–552.

van den Boorn, S., Van Bergen, M. J., Nijman, W., and Vroon, P. Z. (2007). "Dual role of seawater and hydrothermal fluids in Early Archean chert formation: evidence from silicon isotopes," *Geology*, Vol. 35, 939–942.

van Zuilen, M., Lepland, A., and Arrhenius, G. (2002). "Reassessing the evidence for the earliest traces of life," *Nature*, Vol. 418, 627–630.

Vreeland, R., Rosenzweig, W., and Powers, D. (2000). "Isolation of a 250 million year old halotolerant bacterium from a primary salt crystal," *Nature*, Vol. 407, 897–900.

Walsh, M. M. (1992). "Microfossils and possible microfossils from the Early Archean Onverwacht Group, Barberton Mountain Land, South Africa," *Precambrian Research*, Vol. 54, 271–293.

Walsh, K., Morbidelli, A., Raymond, S. N. *et al.* (2011). "A low mass for Mars from Jupiter's early gas-driven migration," *Nature*, Vol. 475, 206–209.

Walter, M. R. (1983). "Archean stromatolites: evidence of the Earth's earliest benthos," in *Earth's Earliest Biosphere*, J. W. Schopf (ed.). Princeton, NJ: Princeton University Press, pp. 187–213.

Westall, F. (1999). "The nature of fossil bacteria," *Journal of Geophysical Research*, Vol. 104, 16,437–16,451.

Westall, F. (2011). "Early life," in *Origins of Life, An Astrobiology Perspective*, M. Gargaud *et al.* (eds.). Cambridge: Cambridge University Press, pp. 391–413.

Westall, F. and Cavalazzi, B. (2011). "Biosignatures in rocks," in *Encyclopedia of Geobiology*, V. Thiel and J. Reitner (eds.). Berlin: Springer, pp. 189–201.

Westall, F. and Folk, R. L. (2003). "Exogenous carbonaceous microstructures in Early Archaean cherts and BIFs from the Isua greenstone belt: implications for the search for life in ancient rocks," *Precambrian Research*, Vol. 126, 313–330.

Westall, F. and Southam, G. (2006). "Early life on Earth," in *Archean Geodynamics and Environments*, K. Benn *et al.* (eds.). pp. 283–304. AGU Geophysics Monographs, 164.

Westall, F., Boni, L., and Guerzoni, M. E. (1995). "The experimental silicification of microbes," *Palaeontology*, Vol. 38, 495–528.

Westall, F., de Vries, S. T., Nijman, W. *et al.* (2006a). "The 3.466 Ga Kitty's Gap Chert, an Early Archaean microbial ecosystem," in *Processes on the Early Earth*, W. U. Reimold and R. Gibson (eds.). Geological Society of America Special Publication, Vol. 405, pp. 105–131.

Westall, F., de Ronde, C. E. J., Southam, G. *et al.* (2006b). "Implications of a 3,472–3,333-Ga-old subaerial microbial mat from the Barberton greenstone belt, South Africa, for the UV environmental conditions on the early Earth," *Philosophical Transactions of the Royal Society B*, Vol. 185, 1857–1875.

Westall, F. Foucher, F., and Cavalazzi, B. (2011a). "Early life on Earth and Mars: a case study from ~3.5-Ga-old rocks from the Pilbara, Australia," *Planetary and Space Science*, Vol. 59, 1093–1106.

Westall, F., Cavalazzi, B., Lemelle, L. *et al.* (2011b). "Implications of in situ calcification for photosynthesis in a ~3.3-Ga-old microbial biofilm from the Barberton greenstone belt, South Africa," *Earth and Planetary Science Letters*, Vol. 310, pp. 468–479.

Wilde, S. A., Valley, J. W., Peck, W. H., and Graham, C. M. (2001). "Evidence from detrital zircons for the existence of continental crust and oceans on the Earth 4.4 Gyr ago," *Nature*, Vol. 409, 175–178.

Yin, Q. Z., Jacobsen, S. B., Yamashita, K. *et al.* (2002). "A short timescale for terrestrial planet formation from Hf-W chronometry of meteorites," *Nature*, Vol. 418, 949–952.

6

Evolution of a Habitable Planet

JAMES KASTING AND JOSEPH KIRSCHVINK

Introduction

We saw in the previous chapter that Earth developed a climate that supported liquid water at its surface very early in its history, probably within the first few hundred million years. That was good for life, of course, because all life that we know about on Earth requires liquid water at least episodically. There are good chemical reasons for thinking that this requirement might be universal, some of which were discussed earlier in this volume.

Here we are concerned with a somewhat later stage in Earth's history, starting from when the rock record begins, around 3.8–4 billion years ago, or 3.8–4 Gyr ago, and continuing on until the rise of atmospheric oxygen, around 2.3 Gyr ago. This time interval overlaps almost precisely with the geologic time period called the Archean Eon. Although not formally defined as such, the beginning of the Archean corresponds with the beginning of the rock record, as marked by the oldest dated fragments of continental crust at Earth's surface. We know that Earth had older rocks, based on crystals of the mineral, zircon, which survive today as sand grains in mid-Archean quartzites from Western Australia, and can be dated as far back as 4.4 Gyr ago (Valley *et al.* 2002). Defining the end of Archean time to be precisely 2.5 Gyr ago was a somewhat arbitrary decision of the geological community: at about this time, the dominant type of sedimentary basin switches from the granite-greenstone belt configuration (dominated by ultramafic, magnesium-rich volcanics and chemically immature sedimentary rocks) to basins controlled by thermal subsidence along passive margins. This switch in rock types may or may not have influenced the rise of O_2, as discussed later in this chapter. We will

Frontiers of Astrobiology, ed. Chris Impey, Jonathan Lunine and José Funes.
Published by Cambridge University Press. © Cambridge University Press 2012.

henceforth use the term "Archean" to refer to the time period preceding the rise of O_2, recognizing that astrobiological and geological terminologies have slightly different meanings.

Earth's climate history

When we say that Earth's climate was clement after the first few hundred million years, we should acknowledge that there is large uncertainty in the mean surface temperature throughout much of Archean time. Liquid water is actually stable on Earth's surface at all temperatures below its critical temperature, $374\,°C$.[1] This is because the pressure of a fully vaporized ocean, ~300 bar, exceeds the critical pressure of 220.6 bar. So, the temperature constraint implied by the presence of liquid water is not as strong as is sometimes thought. Valley *et al.* (2002) argue that surface temperatures must have been lower than 200 °C to explain the oxygen isotope ratios in the ancient zircons that they have studied (see the discussion of these isotopes below). But this still does not tell us much about the potential for life.

Dubious constraints from oxygen isotope ratios in cherts

Somewhat tighter constraints are provided by oxygen isotope ratios in *cherts*. Chert is a particular mineral form of silica, SiO_2. The O atoms in chert can be either the normal isotope, ^{16}O, or one of the minor isotopes, ^{17}O or ^{18}O. We only care here about the ratio of the more abundant of these, ^{18}O, compared to ^{16}O. High temperatures in the water from which the chert is precipitated lead to lower $^{18}O/^{16}O$ ratios in the chert. Conversely, low temperatures lead to high $^{18}O/^{16}O$ ratios. If one examines cherts of different geologic ages, one finds that the older cherts have significantly lower $^{18}O/^{16}O$ ratios, suggesting that the early Earth was hotter than today (see e.g. Knauth 2005). Indeed, the mid-Archean appears to have been quite hot, ~70 °C, according to these data. By contrast, the present mean surface temperature is 15 °C.

Unfortunately, the isotopic data from cherts also suggest that Earth's surface remained impossibly warm until much more recently. For example, the same methods imply that temperatures at the beginning of the Cambrian Period, 542 Myr ago, were still around 55 °C, which is far above the temperature limits for growth of any known animal, and most eukaryotes. The molluscan fossils that appear around this time would have been cooked in their shells! Exactly which assumption is wrong remains unclear. Perhaps most or all cherts had their isotopic ratios reset during *diagenesis* (heating and chemical alteration of buried

---

1 All values given are for pure water. The behavior of saltwater is more complex, but the differences are not important for the arguments presented here.

sediments). Alternatively, the oxygen isotopic composition of seawater may have varied with time (Kasting *et al.* 2006), or the calculated temperatures may actually reflect those of widespread hydrothermal vent fluids, rather than that of the ocean. This last interpretation may be the most likely, as it might also explain systematic variations in silicon isotope ratios over time (van den Boorn *et al.* 2007). Whatever the case, there are good reasons to think that Archean surface temperatures were much cooler than 70 °C, some of which are discussed in the next section.

The long-term glacial record: firmer constraints on climate

Glaciers leave at least four very distinctive features in the rock record that do not form through any other known process. For example, rocks frozen into the ice at the base of a glacier are often scraped along the bedrock, gradually being ground down into glacial dust. If the rock is hard – such as a piece of chert or quartzite – it can become scratched and nicked with subparallel ridges across a surface. The ice, however, is dynamic and any individual stone will be tilted and tumbled numerous times and in different orientations before coming to rest. This produces a type of stone called a *multiply striated cobble*, which has only been found in glacial settings. Similarly, the surface of the bedrock over which the glacier moved also gets covered with long, parallel scratches as it is gradually sanded away, producing distinctive *glacial pavements* like those that grace Central Park in New York City. These pavements are often overlain by a pile of debris known as a *diamictite*, which is a rock that has two distinct grain sizes (typically fine rock flour and larger clasts). *Glacial till* is a type of diamictite that forms as the glaciers melt and drop the dust and stones they are carrying into a jumbled pile, usually with the stones "floating" in the finer-grained mud. Although debris flows can sometimes generate a diamictite, they tend to be local features. Glacial tills, in contrast, will form all along the edge of a melting ice sheet, leaving a much more extensive blanket of debris over the landscape. Finally, when a glacier reaches open water it will break up into icebergs that drift away, slowly melting and releasing the dust and rock that they hold. When a large rock slams into the mud at the bottom it deforms the sediment below and around it, and is then gradually buried by overlying sediment. Such distinctive *drop stones* can be as small as a pebble or as large as a house.

Firm evidence for glaciation exists both in the middle Archean at around 2.9–2.7 Gyr ago (Young *et al.* 1998), and near its end, between 2.45 and 2.22 Gyr ago (Young 1991), providing strong arguments that the early Earth was cool during the last half of Archean time (see Figure 6.1). The climate could, of course, have been hot in between these glaciations, much in the way that Earth has alternated between long (100 Myr scale) intervals of "ice-house" and "greenhouse" intervals for the past half billion years (Kump *et al.* 2010). The geological record supports

Geologic Time

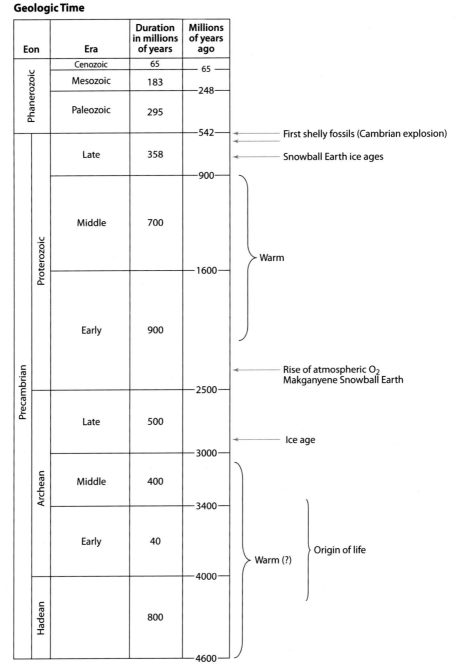

Figure 6.1 The geologic time-scale, showing major events in Earth's history back to its origin at 4.6 Gyr ago. Known periods of glaciation are indicated.

Figure 6.2 Example of a multiply striated cobble from the 2.22 Gyr old Makganyene diamictite of South Africa.

this concept. For example, the Pongola glaciation in Southern Africa at 2.9–2.7 Gyr ago has at least three distinct units of diamictite that can be traced laterally over hundreds of square kilometers and that contain occasional small drop stones. Similarly, early Proterozoic sediments (commonly known as the 'Huronian' interval from 2.45 to 2.22 Gyr ago) are globally characterized by all of these characteristic features, including multiple layers of laterally continuous diamictites, multiply striated cobbles (Figure 6.2), basal striated pavements (in both Canada and South Africa), and often abundant drop stones. Both the Huronian sequence in Canada and the Transvaal group in Southern Africa record several intervals of glacial advance and retreat. The total number of these glacial cycles is obviously unknown, but as discussed later, the last one – the Makganyene event in South Africa – reached down into low latitudes and most likely was a global 'Snowball Earth' event.

Because the geological record is incomplete, and because it becomes worse with increasing geological age, we have probably underestimated the extent and duration of these ancient glacial episodes. Indeed, most ice caps form at high latitudes, but the probability of a continental fragment being at the poles to record the presence of a glaciation is relatively small.

The faint young Sun problem

The central theoretical problem of Archean climate is not to explain how it could have been cool; rather, it is to explain why the Earth was not completely frozen. Solar evolution models (e.g. Gough 1981) predict that the Sun was ~30%

less luminous when it first formed. The reason lies at the heart of the Sun's mode of energy production: The Sun fuses hydrogen into helium in its core. This makes the core denser, causing it to contract and heat up, and this in turn makes the fusion reactions go faster. At the beginning of the Archean (3.8 Gyr ago), solar luminosity is thought to have been about 75% of its present value, increasing to roughly 83% at the end of that time (2.5 Gyr ago). Straightforward calculations (e.g. Kump et al. 2010) show that, because of this change in solar luminosity, Earth's surface would have been frozen prior to ~2.0 Gyr ago if its atmospheric greenhouse effect had remained unchanged over time.

Numerous hypotheses have been proposed to explain why the early Earth did not freeze. Sagan and Mullen (1972), who identified this problem originally, suggested that high concentrations of the greenhouse gases, methane (CH_4) and ammonia (NH_3), were what kept the Earth warm. Both CH_4 and NH_3 are reduced gases, meaning that they can react with O_2. Thus, when atmospheric O_2 concentrations were lower, it makes sense that the concentrations of CH_4 and NH_3 could have been higher. This expectation is borne out for CH_4, which is predicted to have had a concentration of 100–1000 parts per million (ppm) during the Archean, as compared to only 1.6 ppm today (Pavlov et al. 2001, Kharecha et al. 2005). NH_3, however, is photolyzed rapidly by solar ultraviolet (UV) radiation in the absence of shielding by atmospheric O_2 and O_3 (ozone) (Kuhn and Atreya 1979), and so until recently it was not thought to have played a significant role.

Sagan remained fond of this idea, however, and in a paper published posthumously (Sagan and Chyba 1997), he and his coauthor suggested that NH_3 was shielded from photolysis by an organic haze formed from CH_4 photolysis. Organic haze is observed in the CH_4-rich atmosphere of Saturn's moon, Titan, and is predicted to have been present at least some of the time on the Archean Earth (Domagal-Goldman et al. 2008). Simple models of the haze failed to provide enough UV shielding to protect NH_3 (Pavlov et al. 2001, Haqq-Misra et al. 2008). Just recently, however, Wolf and Toon (2010) argued that these models may be misleading. Actual hydrocarbon particles are more like carbonaceous "snowflakes" rather than the simple spheres used in most radiative transfer calculations. By treating these snowflakes as *fractals* – many tiny spheres packed inside of large ones – Wolf and Toon showed that the UV shielding is dramatically improved. Indeed, preliminary unpublished calculations by Tian et al. suggest that the shielding is good enough to allow NH_3 to again become an important greenhouse gas.

This said, another important piece of the climate puzzle involves carbon dioxide, CO_2. Along with water vapor, CO_2 is one of the two most important greenhouse gases in Earth's present atmosphere. Its concentration in the early Earth's atmosphere may have been higher, for a variety of reasons.[2] CO_2 is produced from volcanism, and volcanic activity is predicted to have been higher on the hot, young Earth. Furthermore, removal of CO_2 by silicate weathering, followed by deposition of carbonate sediments, depends strongly on temperature (Walker *et al.* 1981, Kump *et al.* 2010). All other things being equal, lower solar luminosity early in Earth's history should have led to lower surface temperatures, slower weathering rates, and higher atmospheric CO_2. This negative feedback loop could easily have generated CO_2 concentrations 100–1000 times higher than today's level of ~300 ppm.[3] If CO_2 was actually this high, then it could have kept Earth's surface temperature above freezing, even in the complete absence of reduced greenhouse gases (Owen *et al.* 1979, Kasting 1987, von Paris *et al.* 2008).

Theories about past atmospheric composition and greenhouse warming have begun to be tested by measurements made on ancient rocks. Rye *et al.* (1995) and Sheldon (2006) have published limits on atmospheric CO_2 concentrations based on *paleosols* (ancient soils). Sheldon's kinetic analysis is to be preferred, as the thermodynamic data used by Rye *et al.* are out of date and the analysis method itself is suspect. Sheldon's published pCO_2 estimate is equivalent to about 10–100 times present at 2.2 Gyr ago, decreasing to about 1–10 times present at 1 Gyr ago. This is less than would have been needed to keep the Earth warm at these times, and so it suggests that other warming mechanisms must have been present. More recently, Rosing *et al.* (2010) published a much lower limit on CO_2, about 3 times present, based on the mineralogy of *banded iron-formations*, or BIFs.[4] BIFs contain a variety of iron-bearing minerals, including magnetite, Fe_3O_4, and siderite, $FeCO_3$. The Rosing *et al.* limit is based on equilibrium reactions between these two minerals

$$Fe_3O_4 + 3CO_2 + H_2 \leftrightarrow 3FeCO_3 + H_2O. \tag{6.1}$$

A number of objections to this idea have been raised, including the fact that neither magnetite nor siderite is a primary mineral and that organic carbon, not hydrogen, was the probable reductant (Dauphas and Kasting 2011). It remains to be seen which side will prevail in this debate. For our purposes here, the outcome of this dispute does not matter too much. We know empirically that Earth did remain habitable during

---

2 In the following discussion, we will ignore the short-term *organic carbon cycle*, which affects atmospheric CO_2 concentrations on time-scales of months to thousands of years. The slower *inorganic carbon cycle*, or *carbonate-silicate cycle*, is the chief process controlling atmospheric CO_2 on time-scales of millions of years or longer.

3 Today's CO_2 concentration is changing year to year as a consequence of fossil fuel burning. The pre-industrial (pre-1800) CO_2 concentration was about 280 ppm. Today's level is closer to 390 ppm.

4 Rosing *et al.* make up for the lack of greenhouse warming by postulating that the early Earth had a lower albedo, or reflectivity, caused by the lack of biogenic sulfur gases which could act as cloud condensation nuclei.

the Archean eon, and we understand that there are processes, particularly the CO_2-weathering feedback, that promote habitability. This bodes well for the prospects of eventually finding other Earth-like planets with stable, liquid-water-supporting climates.

Atmospheric composition and redox state

Besides climate and the availability of liquid water, the other environmental variable that has most affected life throughout Earth's history is the atmospheric and oceanic redox state. Today, Earth's atmosphere is rich in O_2 – a powerful oxidant (electron acceptor) that drives the metabolism of aerobic organisms through *respiration*. Chemically, we can express respiration as

$$CH_2O + O_2 \rightarrow CO_2 + H_2O. \tag{6.2}$$

The symbol "CH_2O" is geochemists' shorthand for organic matter, which contains carbon, hydrogen, and oxygen in roughly a 1:2:1 ratio. To a geochemist, respiration is the opposite of *oxygenic photosynthesis*, which is just reaction (6.2) run backwards. All higher organisms, including plants, animals, and single-celled, *eukaryotic* organisms[5] such as algae, depend on respiration to drive their metabolism. Hence, the origin of oxygenic photosynthesis, and the subsequent rise in atmospheric O_2, played a major role in the evolution of advanced life on Earth.

Curiously, O_2 is actually an incredibly poisonous molecule. If it can grab an electron from a common biological cation like Fe^{+2} or Mn^{+2}, it will form a superoxide radical, O_2^*, which will then react with water and generate hydroxyl radicals, OH^*. Although they have an extremely short lifetime, hydroxyl radicals can attack almost any organic substance in their path, including the backbone of DNA. These and similar O_2^*-induced reactions had to be contained before Earth's biosphere could exploit the enormous chemical energy potential provided by the availability of O_2. The puzzle of how this might have evolved is addressed at the end of this chapter.

Atmospheric composition prior to the rise of O_2

Prior to the "Great Oxygenation Event", somewhere around 2.4 Gyr ago (Farquhar *et al.* 2000), the atmosphere must have had a very different composition. Reduced gases, such as molecular hydrogen (H_2) and methane, should have been more abundant, and free O_2 should have been virtually absent at ground level. Some O_2 would have been produced from photolysis of CO_2 up in the stratosphere, followed by recombination of O atoms to form O_2; however, the amount formed

---

5 *Eukaryotic* organisms are those that have cells with nuclei. *Prokaryotic* organisms, including Bacteria and Archea, lack cell nuclei.

was too small to block out short-wavelength solar UV radiation or to form an effective ozone screen.

Just because oxygenic photosynthesis had not yet been invented does not mean, however, that life had no effect on the early atmosphere. Prior to the origin of life, the main reduced gas in the atmosphere was probably H_2.[6] As first pointed out by Walker (1977), the H_2 concentration of the prebiotic atmosphere should have been determined largely by the balance between outgassing of H_2 and other reduced gases from volcanoes and loss of hydrogen to space. Volcanoes today release mostly H_2O and CO_2, but H_2 is a minor component, along with CO and various sulfur gases (SO_2 and H_2S). The present H_2 outgassing rate is of the order of 1×10^{12} mol/yr, or roughly 4×10^9 molecules cm^{-2} s^{-1} (Holland 2009). If hydrogen escapes to space at the *diffusion-limited rate* (Walker 1977), then its atmospheric mixing ratio, fH_2, can be estimated from the equation

$$\Phi_{esc}(H_2) \cong 2.5 \times 10^{13} \, f H_2 \text{ molecules cm}^{-2} \text{ s}^{-1}. \tag{6.3}$$

Here, we are for the moment neglecting the contribution of other hydrogen-bearing gases to the volcanic outgassing rate (which is probably a safe assumption for the prebiotic atmosphere). Using this formula, and assuming that the H_2 escape rate equals the present outgassing rate, yields an atmospheric H_2 mixing ratio of about 1.6×10^{-4}, or 160 ppm. Higher volcanic release rates during the Archean could yield H_2 mixing ratios of 10^{-3} or higher, or about 0.1%. This creates what is typically referred to as a *weakly reduced atmosphere* (Figure 6.3).

Effect of early life on the atmosphere

Although 0.1% H_2 may not sound like a lot, it is enough to power several different forms of *anaerobic* metabolism, or those that occur in the absence of oxygen. One of these is *methanogenesis*, which produces methane. Methanogenesis is carried out by a type of anaerobic Archaea called *methanogens*. Methanogens produce methane by a variety of different pathways. Most, or all, of them, however, can utilize the reaction

$$CO_2 + 4H_2 \rightarrow CH_4 + 2H_2O. \tag{6.4}$$

The H_2 threshold at which methanogenesis can occur depends on the amount of CO_2 that is present, but it is at least 10–100 times lower than the 0.1% H_2 that is thought to

---

6 Although we have not discussed it explicitly, we assume that the bulk of the atmosphere was composed of molecular nitrogen, N_2, as it is today. N_2 is relatively inert and has likely been partitioned mostly into the atmosphere since early in Earth's history. Indeed, Goldblatt (2009) has argued that N_2 partial pressures could have been 2–3 times higher than today, and that nitrogen has subsequently been sequestered in the mantle by subduction of N-bearing sediments. We acknowledge this possibility but point out that it probably has little bearing on Earth's habitability, other than providing a few degrees of warming by pressure broadening of other greenhouse gas absorption lines.

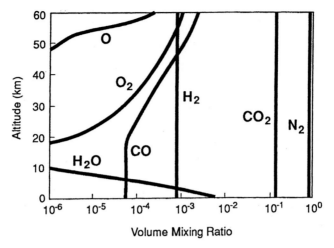

Figure 6.3 Vertical profiles of atmospheric species for a typical, weakly reduced, "prebiotic" atmosphere. The hydrogen abundance is determined by balancing volcanic outgassing with hydrogen escape. O_2 concentrations are predicted using a one-dimensional photochemical model (from Kasting 1993).

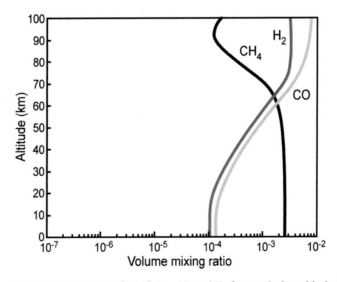

Figure 6.4 Vertical profiles of CH_4, CO, and H_2 for a typical postbiotic, but "pre-photosynthetic", atmosphere (from Kharecha *et al.* 2005, Fig. 7c).

have been available. Hence, as Walker (1977) suggested many years ago, methanogens, once they evolved, probably converted much of the existing H_2 into CH_4. This prediction has been quantified by Kharecha *et al.* (2005), who used a numerical model to estimate the rate at which methanogens could have metabolized, along with the coexisting concentrations of H_2 and CH_4 (Figure 6.4). In this model, the rate of methanogenesis is limited by the rate at which H_2 can diffuse downward through the atmosphere–ocean interface and by the rate at which CH_4 can diffuse upward. A surprising result is that the predicted rate of methane production is within a factor of 3 of the present value, 3.6×10^{13} mol/yr (Prather *et al.* 2001), despite the fact that the Archean and modern ecosystems bear little resemblance to each other.

Methanogenesis was probably an early metabolic invention, according to most phylogenetic analyses (e.g. Woese and Fox 1977), and there is direct evidence of biogenic methane trapped in "bubbles" from 3.5 Gyr old cherts in Australia (Ueno *et al.* 2006). But other forms of anaerobic metabolism likely evolved at about the same time, including various forms of *anoxygenic photosynthesis*. Unlike oxygenic photosynthesis, in which the electrons needed to reduce CO_2 to organic carbon are derived from water (reaction 2 in reverse), in anoxygenic photosynthesis the electrons are provided by various reduced species, such as ferrous iron (Fe^{+2}), sulfide (S^{2-}), or H_2. Fe^{+2} and H_2 were the most widely available reductants (Kharecha *et al.* 2005, Canfield *et al.* 2006), and they fueled marine biological productivity at rates that were of the order of 0.01 times today's value, $\sim 4 \times 10^{15}$ mol/yr (Prentice *et al.* 2001). As many other authors have pointed out previously, this illustrates the gigantic leap in productivity that became available once cyanobacteria learned to split the water molecule. The Archean Earth was probably teaming with life, but it was still only a fraction as "alive" as today.

What influence would such anoxygenic photosynthesizers have had on the atmosphere? Surprisingly, their predicted effect is not greatly different from that of methanogens (Kharecha *et al.* 2005). H_2-based photosynthesizers living in the oceans would have been limited by the downward flux of H_2 from the atmosphere, as before. Their productivity would have been roughly 10 times higher than that of methanogens, because they obtained their energy from sunlight rather than from chemical reactions. However, the organic matter that they produced would likely have decayed by *fermentation* and methanogenesis, and the rate of cycling of H_2 and CH_4 was still limited by diffusion. Hence, the flux of CH_4 into the atmosphere could not have been substantially higher than it was previously.

Finally, we should tie this discussion back to the earlier one concerning climate. Methane, as pointed out previously, is a reasonably good greenhouse gas on its own; furthermore, it photolyzes to produce organic haze, which can shield other greenhouse gases like ammonia from photolysis. Hence, the methane-producing Archean biosphere likely played a significant role in warming Earth's climate right up to the point when atmospheric O_2 concentrations began to rise. At that point, the photochemical lifetime of methane dropped, atmospheric CH_4 concentrations declined precipitously, and Earth was thrown into a series of glaciations – the Paleoproterozoic glaciations discussed above. So, the methane-rich Archean atmosphere discussed here is broadly consistent with Earth's inferred climate history. That does not mean it is necessarily right, but it provides at least a starting point for thinking about the Archean environment.

The great debate about the Great Oxygenation Event

As pointed out many years ago by Roscoe (1969), the Paleoproterozoic glaciations appear to have occurred right when atmospheric O_2 concentrations increased for the first time – the so-called "Great Oxidation Event" (Holland 2006). Naturally, this makes us suspect that the two events are causally connected, and indeed, two separate hypotheses have been proposed to explain how this might have occurred. In the first (Kasting *et al.* 1983, Pavlov *et al.* 2000), the rise of O_2 caused a decline in atmospheric CH_4 concentrations. CH_4 is a greenhouse gas that helps to warm Earth's surface; thus, when CH_4 levels fell, the climate cooled and Earth was thrown into glaciation.

Kopp *et al.* (2005) extended this hypothesis by suggesting that the sudden evolution of oxygenic photosynthesis, followed by the exponential growth of the first cyanobacteria, was rapid enough to destroy this methane-based greenhouse faster than other feedbacks involving CO_2 could respond. In this case the net result would be a global freeze-over producing a "Snowball Earth", consistent with paleomagnetic evidence for the low-latitude deposition of the Makganyene glaciation in South Africa between ~2.32 and 2.22 Gyr ago (Kirschvink *et al.* 2000, Evans *et al.* 1997). This later scenario is still consistent with the hard geological evidence for the onset of oxygenation of the atmosphere and oceans, including the massive Kalahari Manganese field that was deposited in the direct wake of the Makganyene Snowball (at 2.22 Gyr ago). This is the oldest unchallenged firm constraint on environmental oxidation (Kirschvink and Kopp 2008).

Alternative scenarios for global oxygenation hold that oxygenic photosynthesis might have evolved much earlier in Earth history, perhaps as far back as 3.8 Gyr ago (Rosing 1999), but that some environmental limitations on their growth kept them from oxidizing the planet until about 2.35 Gyr ago. Evidence in support of this scenario ranges from evidence of uranium mobility in highly metamorphosed black shales from the 3.8 Gyr old Isua providence in Greenland (Rosing and Frei 2003), fossil biomarkers thought to require molecular oxygen in their biosynthetic pathways between 2.7 and 2.5 Gyr ago (Brocks *et al.* 1999, Summons *et al.* 1999), and elevated levels of redox-sensitive trace elements like Mo, Re, and various iron phases found in ~2.5 Gyr old cores from various Archean and early Proterozoic scientific drilling programs (Anbar *et al.* 2007).

However, every one of these arguments has cogent counter-arguments. The redox potential required to mobilize uranium[7] is about the same as the ferric/ferrous couple, and both iron and uranium redox chemistry is done easily by anaerobic biochemistry. The biomarker constraints have been called into

---

7 Oxidized uranium, U^{+6}, is soluble, while reduced uranium, U^{+4}, is insoluble.

question both because of the ability of oils to migrate through porous sedimentary rock (Rasmussen *et al.* 2008), and the fact that many modern oxygen-dependent enzymatic steps in biochemical synthesis have replaced older, anaerobic versions (Kirschvink and Kopp 2008). Finally, all of the black shales that record a supposed "whiff of oxygen" signature (Anbar *et al.* 2007) have been completely remagnetized by subsequent hydrothermal events. As this remagnetization directly affects iron minerals through the growth of sulfide phases, and these in turn are known to scavenge these redox-sensitive trace elements, it is not surprising that the geochemistry could be misinterpreted as "oxygen whiffs."

Although this is not the place to try to decide which of these two hypotheses is correct, there are some puzzles that need to be resolved. Growth requirements for cyanobacteria are actually rather simple and uniform across all divisions within this bacterial group, and hence were most likely present in their last common ancestor. It is clear that depriving them of the metals needed for growth and nitrogen fixation can provide a powerful check on their ability to oxidize the atmosphere and upper oceans (Fennel *et al.* 2005). However, Kopp and Kirschvink (2005) noted that glacial dust is rich in all trace elements needed for explosive cyanobacterial growth. The quantities of these elements being released into Earth's oceans today (and during the ice ages for the past 2.5 million years) would have been more than sufficient to make the global environment jump over the nitrate limitation barrier of Fennel *et al.* into their oxic world. Estimates of the severity of the Pongola and Huronian glacial advances indicate that they were more intense than those of the Quaternary (Hambrey and Harland 1981). Hence, if cyanobacteria had evolved prior to either the Pongola (2.9–2.7 Ga) or the first two Huronian glaciations (2.45–2.35 Gyr ago), the "Great Oxygenation Event" should have happened sooner than the rock record suggests.

The final puzzle is a biochemical "chicken and the egg" problem. All oxygen in Earth's atmosphere today is a product of oxygenic photosynthesis, released by the oxygen evolving complex of Photosystem-II. That complex system depends upon the ability to control the toxic byproducts of oxygen production, particularly the superoxide (O_2^*) and hydoxyl (OH$*$) free radicals, and yet the enzymes that do this clearly required the presence of this poison (O_2) in the environment to evolve in the first place! Although the authors of this chapter are still arguing about the exact mechanism involved, it is most likely that some product of UV light hitting Earth's surface environment is responsible, either as trace levels of H_2O_2 frozen out onto the surface of ice sheets (Liang *et al.* 2006) or some nitrogen or sulfur-based compounds (Sagura *et al.* 2007). In either case it is worth noting that all known Precambrian glaciations are associated with mineralogical and/or geochemical indicators showing local oxidation increases (Raub and Kirschvink 2008), even though true vertical redox gradients in the sediments – like those needed to support populations of magnetotactic bacteria – do not appear until after the Makganyene glaciation (Kopp and Kirschvink 2008). It is interesting to speculate that oxygenic photosynthesis might not evolve on Earth-like planets that are too close to their parent stars to have glaciation.

Summary

Although some aspects of Earth's early climate history and the rise of atmospheric O_2 are by now understood, the details of what controlled climate in the distant past and what triggered the rise of oxygen are still being debated. The discussion necessarily involves biologists, geologists, and even astronomers, and hence is a prime research area for the developing field of astrobiology. Understanding how Earth's environment has evolved over time will help us understand biological evolution on our planet and may eventually provide insights into atmospheres that we observe on planets orbiting nearby stars.

References

Anbar, A. D., Duan, Y., Lyons, T. W. *et al.* (2007). "A whiff of oxygen before the great oxidation event?" *Science*, Vol. 317, pp. 1903–1906.

Brocks, J. J., Logan, G. A., Buick, R., and Summons, R. E. (1999). "Archean molecular fossils and the early rise of eukaryotes," *Science*, Vol. 285, pp. 1033–1036.

Canfield, D. E., Rosing, M. T., and Bjerrum, C. (2006). "Early anaerobic metabolisms," *Philosophical Transactions of the Royal Society, Series B: Biological Sciences*, Vol. 361, pp. 1819–1834.

Crowell, J. C. (1999). *Pre-Mesozoic Ice Ages: Their Bearing on Understanding the Climate System*. Boulder, CO: Geological Society of America.

Dauphas, N. and Kasting, J. F. (2011). "Low pCO$_2$ in the pore water, not in the Archean air" *Nature*, Vol. 474, pp. E2–E3.

Domagal-Goldman, S. D., Kasting, J. F., Johnston, D. T., and Farquhar, J. (2008). "Organic haze, glaciations and multiple sulfur isotopes in the Mid-Archean Era," *Earth and Planetary Science Letters*, Vol. 269, pp. 29–40.

Evans, D. A., Beukes, N. J., and Kirshvink, J. L. (1997). "Low-latitude glaciation in the Proterozoic Era," *Nature*, Vol. 386, pp. 262–266.

Farquhar, J., Bao, H., and Thiemans, M. (2000). "Atmospheric influence of Earth's earliest sulfur cycle," *Science*, Vol. 289, pp. 756–758.

Fennel, K., Follows, M., and Falkowski, P. G. (2005). "The co-evolution of the nitrogen, carbon and oxygen cycles in the Proterozoic ocean," *American Journal of Science*, Vol. 305, pp. 526–545.

Goldblatt, C., Claire, M. W., Lenton, T. M. *et al.* (2009). "Nitrogen-enhanced greenhouse warming on early Earth," *Nature Geoscience*, Vol. 2, pp. 891–896.

Gough, D. O. (1981). "Solar interior structure and luminosity variations," *Solar Physics*, Vol. 74, pp. 21–34.

Hambrey, M. J. and Harland, W. B. (1981). *Earth's Pre-Pleistocene Glacial Record*. Cambridge: Cambridge University Press.

Haqq-Misra, J. D., Domagal-Goldman, S. D., Kasting, P. J., and Kasting, J. F. (2008). "A revised, hazy methane greenhouse for the early Earth," *Astrobiology*, Vol. 8, pp. 1127–1137.

Hayes, J. M. and Bengtson, S. (1994). "Global methanotrophy at the Archean–Preoterozoic transition," in *Early Life on Earth*, S. Bengston (ed.). New York: Columbia University Press, pp. 220–236.

Hayes, J. M. and Schopf, J. W. (1983). "Geochemical evidence bearing on the origin of aerobiosis: a speculative hypothesis," in *Earth's Earliest Biosphere: Its Origin and Evolution*, J. W. Schopf (ed.). Princeton, NJ: Princeton University Press, pp. 291–301.

Holland, H. D. (2006). "The oxygenation of the atmosphere and oceans," *Philosophical Transactions of the Royal Society, Series B: Biological Sciences*, Vol. 361, pp. 903–915.

Holland, H. D. (2009). "Why the atmosphere became oxygenated: a proposal," *Geochimica Cosmochimica Acta*, Vol. 73, pp. 5241–5255.

Kasting, J. F. (1987). "Theoretical constraints on oxygen and carbon dioxide concentrations in the Precambrian atmosphere," *Precambrian Research*, Vol. 34, pp. 205–229.

Kasting, J. F. (1993). "Earth's early atmosphere," *Science*, Vol. 259, pp. 920–926.

Kasting, J. F., Zahnle, K. J., and Walker, J. C. G. (1983). "Photochemistry of methane in the Earth's early atmosphere," *Precambrian Research*, Vol. 20, pp. 121–148.

Kasting, J. F., Howard, M. T., Wallmann, K. *et al.* (2006). "Paleoclimates, ocean depth, and the oxygen isotopic composition of seawater," *Earth and Planetary Science Letters*, Vol. 252, pp. 82–93.

Kharecha, P., Kasting, J. F., and Siefert, J. L. (2005). "A coupled atmosphere–ecosystem model of the early Archean Earth," *Geobiology*, Vol. 3, pp. 53–76.

Kirschvink, J. L. and Kopp, R. E. (2008). "Paleoproterozic icehouses and the evolution of oxygen mediating enzymes: the case for a late origin of Photosystem-II," *Philosophical Transactions of the Royal Society, Series B*, Vol. 363, pp. 2755–2765.

Kirschvink, J. L., Gaidos, E. J., Bertani, L. E. *et al.* (2000). "Paleoproterozoic Snowball Earth: extreme climatic and geochemical global change and its biological consequences," *Proceedings of the National Academy of Sciences of the USA*, Vol. 97, pp. 1400–1405.

Kopp, R. E. and Kirschvink, J. L. (2008). "The identification and biogeochemical interpretation of fossil magnetotactic bacteria," *Earth-Science Reviews*, Vol. 86, pp. 42–61.

Kopp, R. E., Kirschvink, J. L., Hilburn, I. A., and Nash, C. Z. (2005). "Was the Paleoproterozoic Snowball Earth a biologically-triggered climate disaster?" *Proceedings of the National Academy of Sciences of the USA*, Vol. 102, pp. 11131–11136.

Knauth, L. P. (2005). "Temperature and salinity history of the Precambrian ocean: implications for the course of microbial evolution," *Palaeogeography, Palaeoclimatology, Palaeoecology*, Vol. 219, pp. 53–69.

Kuhn, W. R. and Atreya, S. K. (1979). "Ammonia photolysis and the greenhouse effect in the primordial atmosphere of the Earth," *Icarus*, Vol. 37, pp. 207–213.

Kump, L. R., Kasting, J. F., and Crane, R. G. (2010). *The Earth System*. Upper Saddle River, NJ: Pearson.

Liang, M. C., Hartman, H., Kopp, R. E., Kirschvink, J. L., and Yung, Y. L. (2006). "Production of hydrogen peroxide in the atmosphere of a Snowball Earth and the origin of oxygenic photosynthesis," *Proceedings of the National Academy of Sciences of the USA*, Vol. 103, pp. 18896–18899.

Owen, T., Cess, R. D., and Ramanathan, V. (1979). "Early Earth: an enhanced carbon dioxide greenhouse to compensate for reduced solar luminosity," *Nature*, Vol. 277, pp. 640–642.

Pavlov, A. A., Kasting, J. F., Brown, L. L. *et al.* (2000). "Greenhouse warming by CH_4 in the atmosphere of early Earth," *Journal of Geophysical Research*, Vol. 105, pp. 11981–11990.

Pavlov, A. A., Kasting, J. F., and Brown, L. L. (2001). "UV-shielding of NH_3 and O_2 by organic hazes in the Archean atmosphere," *Journal of Geophysical Research*, Vol. 106, pp. 23267–23287.

Prather, M. *et al.* (2001). "Atmospheric chemistry and greenhouse gases," in *Climate Change 2001: The Scientific Basis*, J. T. Houghton *et al.* (eds.). New York: Cambridge University Press, pp. 239–288.

Prentice, I. C., Farquhar, G. D., Fasham, M. J. R. *et al.* (2001). "The carbon cycle and atmospheric carbon dioxide," in *Climate Change 2001: The Scientific Basis*, J. T. Houghton *et al.* (eds.). Contribution of Working Group I to the Third Assessment Report of the Intergovernmental Panel on Climate Change. New York: Cambridge University Press, pp. 183–238.

Rasmussen, B., Fletcher, I. R., Brocks, J. J., and Kilburn, M. R. (2008). "Reassessing the first appearance of eukaryotes and cyanobacteria," *Nature*, Vol. 455, pp. 1101–1104.

Raub, T. D. and Kirschvink, J. L. (2008). "A pan-Precambrian link between deglaciation and environmental oxidation," in *Antarctica: A Keystone in a Changing World*, A. K. Cooper, P. Barrett, B. Story, E. Stump, and W. Wise (eds.). Washington, DC: National Academy Press, pp. 83–90.

Roscoe, S. M. (1969). *Huronian Rocks and Uraniferous Conglomerates in the Canadian Shield*, Geological Surveys of Canada Paper 68–40.

Rosing, M. T. (1999). "^{13}C-depleted carbon microparticles in >3700-Ma sea-floor sedimentary rocks from West Greenland," *Science*, Vol. 283, pp. 674–676.

Rosing, M. T. and Frei, R. (2003). "U-rich Archaean sea-floor sediments from Greenland – indications of >3700 Ma oxygenic photosynthesis," *Earth and Planetary Science Letters*, Vol. 6907, pp. 1–8.

Rosing, M. T., Bird, D. K., Sleep, N. H., and Bjerrum, C. J. (2010). "No climate paradox under the faint early Sun," *Nature*, Vol. 464, pp. 744–747.

Rye, R., Kuo, P. H., and Holland, H. D. (1995). "Atmospheric carbon dioxide concentrations before 2.2 billion years ago," *Nature*, Vol. 378, pp. 603–605.

Sagan, C. and Chyba, C. (1997). "The early faint Sun paradox: organic shielding of ultraviolet-labile greenhouse gases," *Science*, Vol. 276, pp. 1217–1221.

Sagan, C. and Mullen, G. (1972). "Earth and Mars: evolution of atmospheres and surface temperatures," *Science*, Vol. 177, pp. 52–56.

Sagura, A., Meadows, V. S., and Kasting, J. *et al.* (2007). "Abiotic production of O_2 and O_3 in high-CO_2 terrestrial atmospheres," *Astrobiology*, Vol. 7, pp. 494–495.

Sheldon, N. D. (2006). "Precambrian paleosols and atmospheric CO_2 levels," *Precambrian Research*, Vol. 147, pp. 148–155.

Summons, J. R., Jahnke, L. L., Hope, J. M., and Logan, G. A. (1999). "Methylhopanoids as biomarkers for cyanobacterial oxygenic photosynthesis," *Nature*, Vol. 400, pp. 554–557.

Ueno, Y., Yamada, K., Yoshida, N. *et al.* (2006). "Evidence from fluid inclusions for microbial methanogenesis in the Early Archaean Era," *Nature*, Vol. 440, pp. 516–519.

Valley, J. W., Peck, W. H., King, E. M., and Wilde, S. A. (2002). "A cool early Earth," *Geology*, Vol. 30, pp. 351–354.

van den Boorn, S. H. J. M., van Bergen, M. J., Nijman, W., and Vroon, P. Z. (2007). "Dual role of seawater and hydrothermal fluids in early Archean chert formation: evidence from silicon isotopes," *Geology*, Vol. 35, pp. 939–942.

von Paris, P., Rauer, H., Lee Grenfell, J. *et al.* (2008). "Warming the early Earth – CO_2 reconsidered," *Planetary and Space Science*, Vol. 56, pp. 1244–1259.

Walker, J. C. G. (1977). *Evolution of the Atmosphere*. New York: Macmillan.

Walker, J. C. G., Hays, P. B., and Kasting, J. F. (1981). "A negative feedback mechanism for the long-term stabilization of Earth's surface temperature," *Journal of Geophysical Research*, Vol. 86, pp. 9776–9782.

Wolf, E. T. and Toon, O. B. (2010). "Fractal organic hazes provided an ultraviolet shield for early Earth," *Science*, Vol. 328, pp. 1266–1268.

Woese, C. R. and Fox, G. E. (1977). "Phylogenetic structure of the prokaryotic domain: the primary kingdoms," *Proceedings of the National Academy of Science of the USA*, Vol. 74, pp. 5088–5090.

Young, G. M. (1991). *Stratigraphy, Sedimentology, and Tectonic Setting of the Huronian Supergroup*. Report of Annual Meeting, Toronto, 1991, Field Trip B5, Guidebook, Toronto: Geological Association of Canada.

Young, G. M., von Brunn, V., Gold, D. J. C., and Minter, W. E. L. (1998). "Earth's oldest reported glaciation; physical and chemical evidence from the Archean Mozaan Group (~2.9 Ga) of South Africa," *Journal of Geology*, Vol. 106, pp. 523–538.

7

Our Evolving Planet

From Dark Ages to Evolutionary Renaissance

ERIC GAIDOS AND ANDREW KNOLL

Introduction

Earth records its own history in the physical, chemical, and biological features of sedimentary rocks. In particular, the history of life is recorded by the remains of organisms buried and preserved in accumulating sediments, by physical traces of organisms' activity in sediments (e.g. burrowing), and by chemical changes wrought by organisms (e.g. oxygen produced by land plants, algae, and cyanobacteria). The process of sediment accumulation, so essential to preservation, has biased the fossil record: organisms that lived in environments where burial was likely are relatively well represented in the geologic record, whereas organisms that lived in habitats characterized by net erosion seldom become fossils.

There is a second bias to the fossil record. The organisms most likely to be preserved as fossils are those that produce "hard parts," mineralized skeletons or decay-resistant organic compounds such as the lignin in wood. In contrast, organisms with no readily preservable components fossilize only under exceptional circumstances, although some leave a record in the form of "trace" fossils such as tracks and burrows. Some microorganisms produce walls, spores, and extracellular envelopes that also preserve well in accumulating sediments; thus, we have a fossil record of bacteria and unicellular eukaryotes that predates the conventional record of animals and land plants. As in the case of animals and their skeletons, some microorganisms routinely produce preservable structures, whereas others never do. There are also microbial trace fossils, recorded by the influence of microbial mat communities on bedding and stromatolites, distinctive

Frontiers of Astrobiology, ed. Chris Impey, Jonathan Lunine and José Funes.
Published by Cambridge University Press. © Cambridge University Press 2012.

three-dimensional structures formed where large colonies of microbes influenced or controlled the formation, texture, and/or mechanical properties of sediments. In general, then, the fossil morphologies that document early life largely record microorganisms that (1) lived where burial facilitates preservation and (2) made decay-resistant organic walls or sheaths. Cyanobacteria are well represented in Proterozoic sedimentary rocks; Archaea are unknown as microfossils (e.g. Knoll 2003).

Less conventionally, life can leave a sedimentary signature in the form of decay-resistant biomolecules, especially lipids, and these can preserve a record of microorganisms that form conventional fossils only rarely, if at all. Some biological processes also result in isotopic fractionation, and these leave an important geologic record in the isotopic abundances of carbon, sulfur, and nitrogen found in both organic and inorganic constituents of sedimentary rocks. Finally, life leaves a signature through its physical and chemical interactions with accumulating sediments.

The Earth is a dynamic planet, and the tectonic processes that produce fossil-bearing sedimentary rocks at one time will ultimately work to diminish that record via erosion and metamorphism. Mapped abundances of sedimentary rocks deposited through Earth history can be modeled assuming they are destroyed according to an exponential decay law (Veizer and Jansen 1979). This means that, much like human history, early chapters in our planet's history are more poorly preserved than more recent ones. Remarkably, however, the known record of surface processes on the Earth extends backward some four billion years. Previous chapters introduced the fundamental events that established life on the early Earth, a history stitched together from fragmentary geological data, inferences from comparative biology, and laboratory research on prebiotic chemistry.

Other chapters in this book examine Earth's earliest geologic record. Here we pick up the narrative 2.4 billion years ago, as Earth's atmosphere and surface oceans began to accumulate oxygen. About this time, the sedimentary, geochemical, and paleontological records dramatically improve, enabling us to reconstruct evolutionary and environmental history with greater confidence. We carry the narrative forward through about 2 billion years, to the initial diversification of animal life in the oceans.

Out of the Dark Ages

We know of no sedimentary rocks that preserve a record of Earth's first 700 million years, and relatively few that document events older than about 2800 million years ago. By analogy to European history, then, we might think of this foundational interval as Earth's Dark Ages. With the expansion of large continents, or cratons, on which rocks can avoid destruction by subduction, the sedimentary record improves. Rare microfossils and biomarker molecules, as well as common isotopic signatures and stromatolites, indicate that the late Archean (2800–2500 million years ago) Earth supported thriving microbial ecosystems that cycled carbon, sulfur, and other essential elements through the biosphere (e.g. Buick 2008). Tantalizing but hotly debated evidence suggests that cyanobacteria – the only group of organisms ever to evolve the capacity to generate electrons needed for photosynthesis from water – may have been present through at least part of this interval. Cyanobacteria were certainly present by 2.3–2.4 billion years ago, because the chemistry of sedimentary rocks indicates that by this time oxygen had begun to accumulate in the atmosphere and surface oceans (Holland 2006). Oxygenic photosynthesis is the only known process capable of generating the required quantities of oxygen on the early Earth.

Other chapters have summarized the continuing debate on what combination of biological and physical processes led to this so-called Great Oxygenation Event. Here we consider its consequences for life. The most obvious consequence is that oxygen became widely available for aerobic respiration, dramatically increasing the yield of energy from the oxidation of glucose and other organic molecules. Other oxidants grew more abundant as well, especially sulfate (SO_4^{2-}) and, to a much lesser extent, nitrate (NO_3^-) and nitrite (NO_2^-). Thus, the diversity of respiratory pathways expanded, even in environments without measurable O_2. This increased supply of oxidants allowed for an expanded metabolic repertoire among microbes, enlarging the diversity of pathways by which biologically important elements are cycled through surface environments. Perhaps the most profound biological consequence of increasing oxygen levels was the symbiosis between a host cell (quite possibly an archaeon) and an aerobic bacterium (Embley and Martin 2006). From this emerged the eukaryotic cell, with its bacterially derived organelle for aerobic respiration, the mitochondrion. Eukaryotes established no fundamentally novel energy-harvesting pathways, but they added new tiers to microbial food webs. Unlike prokaryotes, many eukaryotic microorganisms can ingest particulate food, establishing scavenging and predation as part of the biosphere.

Earth's Middle Ages

The Great Oxidation Event

The Proterozoic Eon extends from 2500 to 542 million years ago, both beginning and ending with seminal biological and environmental events. The beginning of the eon is marked by the Great Oxidation Event (Holland 2006) introduced in the previous section. Several independent lines of geochemical evidence indicate that oxygen began to accumulate in the atmosphere and surface ocean about 2.4 billion years ago, but also that the environmental transition was protracted, and its end-product was not the modern world, but an intermediate state of the Earth system distinct from both the preceding Archean and the Phanerozoic Eon that would follow (Anbar and Knoll 2002, Scott *et al.* 2008, Frei *et al.* 2009). Earth's Middle Ages persisted until near the end of the Proterozoic Eon, when a series of tectonic, geochemical, climatic, and biological events coincident with, and possibly causally related to, the breakup of the supercontinent Rodinia (Bogdanova *et al.* 2009) culminated in the diversification of animals throughout the oceans (Knoll 1992). The intervening billion year interval was, *prima facie*, much less dramatic, so much so that Buick *et al.* (1995) described it as "the dullest time in Earth's history." Indeed, it has become popular to call this interval the "boring billion," yet sedimentary rocks deposited during this interval preserve evidence for subtle planetary change, and we are reasonably confident that many of the important biological innovations that laid the foundation for the evolution of animal life occurred as well (Knoll *et al.* 2006).

Geochemistry and biogeochemistry discriminate between stable isotopes of C, N, O, and S in a manner proportional to atomic weight (known as "mass dependent" fractionation). Some ultraviolet photochemistry, however, can produce fractionation patterns that deviate from those predicted on the basis of mass. The dearth of evidence for such "mass-independent" fractionation of sulfur isotopes in all but the earliest Proterozoic sediments indicates that the atmosphere was sufficiently oxidizing for the source of sulfur, volcanic SO_2, to have a relatively short chemical lifetime and not be dissociated by UV light in the stratosphere (Farquhar *et al.* 2000). The transition from mass-independent to mass-dependent fractionation of S is predicted to occur at an O_2 concentration of 10^{-5} present atmospheric level (PAL) (Guo *et al.* 2009), and this can be regarded as a lower limit for Proterozoic atmospheric oxygen.

How much higher could oxygen levels be in the Proterozoic atmosphere? The relative paucity of the redox-sensitive element molybdenum in Proterozoic sediments compared to modern values indicates that oxidative weathering of the continents may have been less vigorous and that widespread sulfidic water masses removed dissolved Mo from seawater (Scott *et al.* 2008). Also, isotopes of chromium (Cr) are fractionated during oxidation from insoluble Cr III to soluble Cr VI, but such fractionation is absent in 1880 million year old sediments (Frei *et al.* 2009). Model-based estimates suggest low Proterozoic oxygen but depend on unknown rates of ocean circulation and the global net rates of organic matter production during photosynthesis (Kasting 1987).

The presence of sedimentary banded iron formations in Earth's oldest sedimentary successions was first interpreted in terms of deep ocean anoxia, because reduced (ferrous, Fe^{2+}) iron is soluble – and, hence, transportable, in anoxic waters but essentially insoluble in oxic waters of moderate pH (e.g. Cloud 1972). By the same logic, the disappearance of BIFs after about 1800 million years ago was interpreted as indicating the oxygenation of the deep ocean and the "rusting" out of Fe before it could reach the continental shelf (Beukes and Klein 1992). In 1998, however, Donald Canfield proposed an alternative explanation for the BIF record. In Canfield's (1998) scenario, atmospheric oxygen reacts with pyrite in continental sedimentary rocks during weathering to form highly soluble sulfate. The sulfate washes into the deep oceans, where, in the absence of oxygen, it is reduced by anaerobic bacteria to sulfide. This sulfide, in turn, reacts with ferrous iron to form pyrite. Canfield hypothesized that after the appearance of oxygen in Earth's atmosphere, marine sulfate and sulfide levels began to rise until iron was titrated from the oceans, at least in the oxygen minimum zone just beneath the surface mixed layer.

The fact that iron formation disappeared some 600 million years after oxygen began to accumulate (Poulton *et al.* 2004, Wilson *et al.* 2010) indicates that this transition was protracted, or, as iron formations are not common in 2400–1900 million year old successions, that tectonic events transiently bolstered hydrothermal iron supply 1800–1900 million years ago. Evidence from iron chemistry, molybdenum chemistry, and organic geochemistry all support the hypothesis that by 1800 Ma or so, the oxygen minimum zone of the world's oceans tended toward euxinia (anoxic and sulfidic conditions; Shen *et al.* 2002, 2003, Brocks *et al.* 2005, Scott *et al.* 2008). The chemistry of deep oceans during Earth's Middle Ages is not well constrained but likely included water masses that were dysoxic (low oxygen), others that were sulfidic, and still others that were anoxic but not sulfidic (Slack *et al.* 2007, Johnston *et al.* 2009). Iron formation made a comparatively brief and modest reappearance near the end of the Proterozoic (Ilyin 2009), discussed further in the next section.

Oxygen stabilization and regulation

Once molecular oxygen appeared in Earth's atmosphere and surface waters, what regulated its level throughout the Proterozoic Eon? Molecular oxygen is produced during oxygenic photosynthesis, when H_2O molecules donate electrons for the production of the energy-yielding molecule adenosine triphosphate (ATP) and reducing power needed to reduce CO_2 to sugar. On short time-scales, there is a net gain in molecular oxygen when organic carbon produced by oxygenic photosynthesis exceeds that consumed by aerobic respiration. On longer time-scales, the major regulatory processes include the burial of organic carbon in sediments, sequestering it from aerobic metabolism, and the consumption of O_2 during the oxidative weathering of continental rocks. Changes in the relative strengths of this source and sink are usually considered to control atmospheric oxygen on geologic time-scales (Lasaga and Ohmoto 2002, Berner *et al.* 2003). This control consists of two negative feedbacks. If atmospheric oxygen increases, then accelerated oxidative weathering of organic matter and pyrite in sedimentary rocks is supposed to enhance consumption. Greater oxygenation of the deep oceans facilitates aerobic respiration of marine detritus by microbes, with concomitant consumption of oxygen (Betts and Holland 1991). As a result of these feedbacks, atmospheric oxygen is expected to be relatively stable through geologic time. On the other hand, some mechanism is required to enhance or suppress carbon burial (or suppress or enhance weathering) during episodes when atmospheric oxygen levels are thought to have changed (Logan *et al.* 1995, Lenton and Watson 2004).

In an alternative view (Gaidos *et al.* 2007, Gaidos 2010), oxygen stabilizes at a level where total global respiration plus abiotic oxidation approximately equals photosynthetic production, complicated when at least a fraction of primary production occurs via photosynthetic organisms that do not produce oxygen (Johnston *et al.* 2009). If photosynthetic production is approximately independent of oxygen concentration, then oxygen levels are entirely determined by the overall uptake kinetics of the aerobic (plus anaerobic) biosphere. The system has a stable point as long as respiration increases with oxygen concentration. If global respiration decreases, then organic carbon will accumulate and oxygen levels will rise until respiration increases to compensate. In this picture, oxygen levels are controlled not by the efficiency of organic matter burial or the weathering of continental sediments, but by the kinetics of total respiration. That depends on who is respiring oxygen, how they are doing it, and where they are.

Different aerobic taxa use oxygen down to differing partial pressures because of physiological differences in metabolism and the efficiency with which oxygen exchanges between their interiors and the surrounding environment. For example, in marine invertebrates, oxygen uptake falls at less than $\sim$50% PAL O_2, and often ceases entirely by $\sim$5% PAL (Mangum and Van Winkle 1973). Many vascular plants will die because of soil anoxia at atmospheric oxygen levels of a few percent PAL. In contrast, microaerophilic bacteria can use oxygen below 1% PAL (and in some cases, well below this level; Stolper *et al.* 2010). The relative contributions of different groups of organisms to the modern global respiration rate of around 2×10^{16} mol O_2 yr^{-1} is very poorly measured, but it seems to be dominated by the land biota ($\sim$70%) and is roughly equally split between terrestrial microbes/marine microplankton (which includes eukaryotic microalgae), and terrestrial plants and marine meso- and macrofauna (Raich and Schlesinger 1992, Hanson *et al.* 2000, del Giorgio and Duarte 2002, Raich *et al.* 2002, Hernandez-Leon and Ikeda 2005). In a Proterozoic world with little grazing and minimal food chains, bacteria would dominate respiration, and their total respiration might be able to balance oxygen production at very low oxygen levels.

The conditions under which respiration occurs will also affect the kinetic curve. The most notable of these is temperature. All else being equal, respiration will be more rapid in warmer waters (i.e. much of the ocean surface) than in colder water (i.e. deep water in the modern ocean) (Wohlers *et al.* 2009). Thus the steady-state concentration of oxygen in the surface waters of a warmer ocean will be less than that of a colder ocean (Rivkin and Legendre 2001). Moreover, the solubility of oxygen decreases in warmer water. A 6 °C increase in temperature, thus, increases bacterial respiration by a factor of 2–3 (Wohlers *et al.* 2009) while decreasing oxygen solubility by 10–15%. Another factor is the rate at which the ocean mixes. In a more sluggishly circulating ocean, deep water will contain less oxygen, leading to slower aerobic respiration of organic matter and, eventually, an increase in atmospheric oxygen until the balance is restored.

In the hypoxic, warmer Proterozoic Eon, then, much of the deep ocean may have been depleted of oxygen, in which case, microbial respiration would switch to anaerobic pathways. Sulfide produced by bacterial sulfate respiration is highly soluble and reacts readily with oxygen in surface waters. If Canfield's model of the Proterozoic ocean is correct, water masses below the mixing layer became sulfidic by 1800 million years ago. Sulfide, formed by reduction of sulfate by organic matter, will accumulate in the deep ocean to the extent that it is not re-oxidized by the oxygen originally liberated during photosynthesis or removed by reaction with ferrous iron, forming pyrite. Thus the sulfide in the ocean will have its stochiometric equivalent in the atmosphere, e.g. 1 millimolar of H_2S will have a stochiometric equivalent pO_2 of 7.5% PAL. Additional O_2 will be present to balance buried organic carbon that has escaped remineralization, and sulfur present in insoluble iron sulfides.

Life in the Proterozoic

There is biomarker evidence for the presence of phototrophic green and purple sulfur bacteria in Middle Age oceans, and thus for the presence of sulfide within the sunlit surface mixed layer (Brocks *et al.* 2005). It is plausible, therefore, to assume that oxygen and sulfide could coexist in the shallow Proterozoic ocean. An approximate value of the oxygen level at which both oxygen and sulfide can exist in the mixing zone of the ocean can be estimated by a stochiometric balance between the input of H_2S, $u[H_2S]$ from upwelling with velocity u, and the mixing of O_2, pO_2kK_H, where pO_2 is the oxygen partial pressure, k is a "piston" velocity, and K_H is Henry's constant. Then $pO_2 = 2u[H_2S]/(kK_H)$; for standard values (Broeker and Peng 1982, Kump *et al.* 2005) and a sulfide concentration of 1 millimolar (Hurtgen *et al.* 2005), the stochiometric pO_2 is about 3% PAL. At a higher level, sulfide would be removed from the photic zone; at a lower level, the photic zone would become anoxic.

We can also equate the reaction kinetics of sulfide oxidation with plausible rates of organic matter (carbon) production. The modern, globally averaged net primary production in surface waters is about 1 g C per cubic meter per year (Schlesinger 1997), which is equivalent to 200 micromolar d^{-1}. The abiotic (approximately second-order) reaction rate of sulfide oxidation is $\sim$5 mM d^{-1} (Millero 2005) and thus a steady-state oxygen level would be 40 mM l^{-1} (equivalent to $\sim$15% PAL). However, sulfur-oxidizing bacteria will accelerate the reaction rate many-fold, and so this provides only an upper limit. Microaerophilic sulfur-oxidizing bacteria can exploit oxygen at least as low as 0.5% of PAL (1 micromolar) and thus would "clamp" oxygen concentrations. A full marine sulfur cycle operating in the water column would therefore have allowed oxygen consumption to balance net primary production at an oxygen level only a few percent or less of the modern value.

Thus, we have arrived at a picture of an ocean of Earth's Middle Ages that is commonly sulfidic, at least in the oxygen minimum zone, with low oxygen in surface waters permitting aerobic microbes and some single-celled eukaryotes, but excluding motile animals larger than a mm or so (Runnegar 1982). Iron (from hydrothermal vents) was immobile in such an environment and many other biologically important transition metals such as molybdenum may have been depleted due to their efficient removal by reaction with sulfide (Anbar and Knoll 2002).

Climate in the Proterozoic

Another important aspect of the Proterozoic environment was the climate. Globally widespread ice ages occurred near the beginning and, again, near the end of the eon, but no widespread glacial deposits are known for the intervening "boring billon" years (Hoffman and Schrag 2002). Climate is modulated by the greenhouse gas CO_2, and we can make some inferences about this control using carbon isotopes. Carbon isotopes are fractionated by carbon-fixing organisms; ^{12}C is preferentially included in the organic product, leaving the inorganic remainder (CO_2 or bicarbonate HCO_3^-) enriched in ^{13}C. Inorganic carbon in the oceans is eventually included in carbonate rocks where the isotopic enrichment is recorded: the greater the enrichment (indicated as $\delta^{13}C$), the greater the proportional burial rate of organic carbon, a sink for CO_2. The Proterozoic carbon isotopic record begins with the pronounced positive Lomagundi excursion (Aharon 2005), continues with a billion years of comparatively stable values close to modern seawater (Knoll et al. 1995, Buick et al. 1995), and ends with strongly fluctuating values that exceed both the maximum and minimum C-isotopic values observed in the previous or subsequent 500 million years (Knoll et al. 1986, Halverson et al. 2005, see below).

The increasing magnitude of Proterozoic carbon isotope excursions may reflect a secular decrease in the marine inorganic carbon reservoir: a smaller reservoir perturbed by a fixed amount will result in a larger excursion (Bartley and

Kah 2004). Inorganic carbon (principally bicarbonate at the ocean's present pH) is removed from the ocean by the Urey reaction: the weathering of crustal rocks, the release of alkalinity into the ocean, and precipitation of carbonates as limestone. The Sun's luminosity is thought to have increased by about 16% over Proterozoic history, resulting in an increase in surface temperature, enhanced weathering, and, therefore, elevated formation of carbonates leading to a drawdown of the marine inorganic carbon reservoir. Corroborative evidence for increasing rates of weathering comes in the form of the ratios of the two isotopes of strontium, ^{86}Sr and ^{87}Sr. The ratio is a proxy for the relative contributions of weathering of continental crust and seafloor to ocean chemistry, and the increasing ratio through the Proterozoic is indicative of greater continental weathering (when corrected for crustal evolution; Shields 2007). Even with no change in weathering rates, atmospheric CO_2 is expected to have decreased with time as the Sun brightened and volcanism and metamorphism of crustal rocks declined: from a level of ~0.1 bar at the beginning of the Proterozoic to a hundredth of that value towards the end (Kasting 1992). If the ocean's pH did not vary dramatically, its inorganic carbon content would have tracked decreasing atmospheric CO_2. On the other hand, other greenhouse gases may have been important, namely CH_4 (Kasting 2005) and N_2O (Buick 2007). Sinks and sources of methane and nitrous oxide do not have the same temperature dependence as those of CO_2, but elevated temperatures could have accelerated microbial metabolism and methanogenesis. It is possible that the greenhouse effect of these two gases in combination could have allowed weathering to draw down CO_2 to a very low level. Below a certain CO_2 level, the stabilizing effect of its greenhouse contribution may have been overwhelmed by positive feedbacks, leading to climate instability and alternating intense greenhouse and ice house (glacial) conditions, i.e. "snowball Earth."

A second oxygen rise

Oxygen rose again at the end of the Proterozoic (see below). Mechanisms for this fall into two categories; those that invoke geologic or geochemical changes, and those that propose that biological evolution was itself responsible (Gaidos *et al.* 2007). The former include exceptional episodes of organic carbon burial on the shelves of continents spawned by a supercontinent breakup. The latter includes the advent of the metazoan gut and sinking of fecal pellets (Logan *et al.* 1995). Climate itself may have been important: weathering reactions are highly temperature dependent and thus the flux of sulfur (as sulfate) into the Proterozoic ocean from continental rocks would have been climate-controlled; greenhouse conditions would have corresponded to sulfidic oceans and low atmospheric oxygen; ice ages would have led to low sulfide and elevated pO_2. Thus, the first-order pattern of Proterozoic climate, global ice ages that bracket a long interval of warmth, may explain, at least in part, the dynamics of the carbon cycle and the long-term paleo-environmental pattern of low oxygen, before a major shift toward modern levels near the close of the eon.

What was the nature of biological communities in the oceans of Earth's Middle Ages? Biomarker molecules indicate that marine primary production was dominated by bacteria, at least until about 800 million years ago (Knoll *et al.* 2007). This observation is consistent with the ability of some photosynthetic bacteria, including some cyanobacteria, to use sulfide as a source of electrons for photosynthesis, as well as the widespread intolerance of eukaryotes for sulfide (Martin *et al.* 2003). It is also consistent with models that suggest low concentrations of fixed nitrogen in mid-Proterozoic oceans (Anbar and Knoll 2002, Fennel *et al.* 2005); many photosynthetic bacteria can fix nitrogen, but no photosynthetic eukaryotes can do so.

Fossils indicate that mid-Proterozoic oceans supported a modest diversity of eukaryotic microorganisms, mostly in coastal waters least susceptible to sulfide mixing (Javaux *et al.* 2001) (Figure 7.1A). Heterotrophic (organic-consuming) eukaryotes accommodated to sulfide toxicity by avoiding it, by harboring bacteria symbionts to detoxify ambient waters (Bernhard *et al.* 2006), by deploying an alternative oxidation pathway in mitochondria that avoids the steps vulnerable to sulfide inhibition (McDonald *et al.* 2009), or by permanently altering mitochondria for anaerobic metabolism (Hjort *et al.* 2010). Physiological adaptations to environmental sulfide would have resulted in lower ATP gain per mole of glucose oxidized. Thus, while present in Earth's Middle Ages, both photosynthetic and heterotrophic eukaryotes played a smaller role in marine ecosystems than they do today. Cyanobacteria and other photosynthetic bacteria remained the major primary producers in the oceans (Figure 7.1B,C). Freshwater environments are less well sampled by preserved rocks, but early eukaryotes might well have diversified early in lakes and streams (Gaidos 2010).

A biological Renaissance

How did the Earth escape from its protracted Middle Ages? Again, the precise mechanism of environmental transformation remains a subject for research, but increasingly geologic and geochemical data place key pieces of the puzzle in stratigraphic context. Beginning no later than about 800 million years ago, subsurface water masses rich in sulfide began to retreat, replaced not by oxic waters but by anoxic waters containing ferrous iron (Canfield *et al.* 2008, Johnston *et al.* 2010). The distribution of iron in mudstones indicates that anoxia remained common, even in water masses within a few tens of meters of the sea surface (Johnston *et al.* 2010); thus oxygen levels must have remained low. At about the same time,

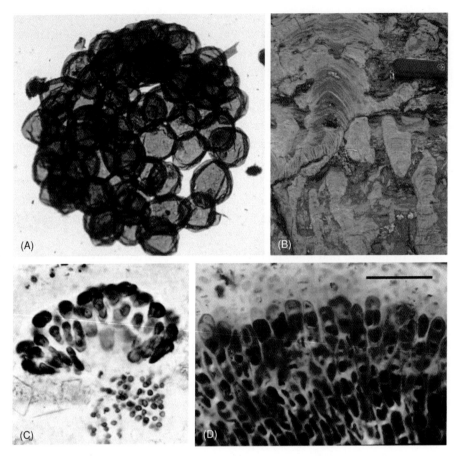

Figure 7.1 (A) Carbonaceous compression fossils of possible early eukaryotic organisms, ca. 1000 million years ago, Lakhanda Group, Siberia. (B) Stromatolites accreted by microbial mat communities, ca. 800 million years ago, Upper Eleonore Bay Group, Greenland. (C) *Eoentophysalis*, cellularly preserved cyanobacteria from the ca. 1200 million year old, Debengda Formation, Siberia. (D) Tghallophyca, a red algal thallus from 632–511 million year old beds of the Doushantuo Formation, China. Bat in $D = 150$ μm for A, $D = 5$ cm for B, $D = 75$ μm for C, and $D = 40$ μm in D.

however, the diversity and environmental distribution of eukaryotic microfossils also increased markedly (Figure 7.2) (Knoll *et al.* 2006), consistent with the hypothesis that sulfide toxicity constituted an environmental challenge to earlier eukaryotes in the oceans (Martin *et al.* 2003).

Several lines of evidence indicate that a second oxygenation event took place about 580–560 million years ago. As early as 1992, Derry *et al.* modeled organic carbon burial rates, based on the emerging stratigraphic patterns of carbon and strontium isotopes. Organic carbon burial is the geologic process most likely to

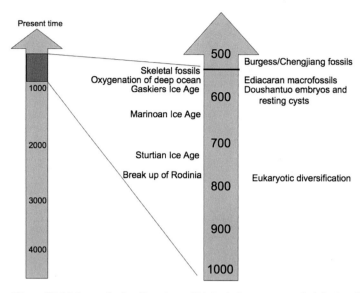

Figure 7.2 Major geologic, climatic, and biological events recorded during the interval of Earth's transformation to a state that supported large, metabolically active animals.

result in atmospheric oxygen accumulation, so the finding that organic burial rates spiked 580–560 million years ago points to a time of transformation in the biosphere (Figure 7.2). Several ways of looking at the sulfur isotopic record independently point toward 580–560 million years ago as a time of increasing oxygen levels (Canfield and Teske 1996, Fike *et al.* 2006), as do both the abundance and isotopic composition of molybdenum in marine sedimentary rocks (Scott *et al.* 2008, Dahl *et al.* 2010). Collectively, available geochemical data suggest during the latest Proterozoic transition atmospheric oxygen increased from a few percent to perhaps 20–50% of today's levels (Dahl *et al.* 2010).

A survey of modern oceans shows that macroscopic animals with high energy demands do not occur in waters containing less than about 10% of the oxygen levels in surface waters in direct contact with the atmosphere (Diaz and Rosenberg 1995, Vaquer-Sunyer and Duarte 2008). As geochemical data strongly suggest that this level of oxygen was exceeded in surface ocean waters for the first time 580–560 million years ago, we might ask whether the fossil record records a biological response in the form of animal diversification. Molecular "clocks" (chronologies based on an assumed constant rate of divergence in genome sequences in sister lineages) estimate that animals diverged from their unicellular ancestors some 800 million years ago, when the challenge of environmental sulfide began to abate (e.g. Peterson *et al.* 2008). The biomarker molecule 24-isopropylcholestane has been isolated from shales older than 635 million years; as the precursor sterol for this biomarker is synthesized mostly by one group of sponges, the record can be

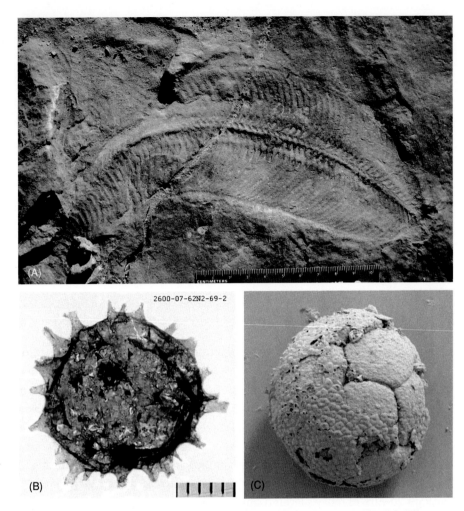

Figure 7.3 Early evidence of animal evolution. (A) *Pteridinium*, problematic Ediacaran animal impressions from ca. 543 million year old, beds of the Nama Group, Namibia. (B) Large, ornamented microfossil interpreted as the egg or diapause resting stage of an early animal, Ediacaran Vychegda Formation, northern Russia. Scale = 100 μm in B and C. (C) Multicellular structure contained within a cyst-like hull, interpreted as an early cleavage stage of an animal embryo; ca. 580–551 million year old, phosphatic rocks from the Doushantuo Formation, China.

interpreted as a minimum date for the diversification of sponges, the simplest of extant animals (Love *et al.* 2009). Sponges do not have high oxygen requirements, so it is not surprising that their earliest record begins before geochemistry suggests widespread oxygen renaissance. Remarkably preserved embryos (Xiao and Knoll 2000) (Figure 7.3C) provide further evidence of animal life in 580 million year old rocks, and large, highly ornamented microfossils interpreted as the egg and cysts of animals with a resting stage in their life cycles also occur in 635–560 million year old successions (Figure 7.3B) (Cohen *et al.* 2009, and references therein). Extant animals form resting stages when their eggs and embryos have a high probability of landing where they cannot grow, and at the time these cysts formed, seafloors on continental shelves remained unstable with respect to oxygen.

By 575 million years ago, macroscopic organisms populated oxygenated seafloors, as evidenced by remarkable fractal-like fossils in Newfoundland (Figure 7.3A) (Narbonne 2004). Fossils of large, motile bilaterian animals (with left–right symmetry) occur in 555–560 million year old sandstones (Martin *et al.* 2000), providing the earliest evidence of animals with high metabolic oxygen demand at just the point in Earth history when more oxygen became available. This is compelling evidence that levels of free molecular oxygen limited the appearance of animals. With motile bilaterian predators and denser benthic populations on the seafloor came the stimulus for enhanced neural activity, arguably the spark that culminated more than 500 million years later in technological intelligence.

Although only bilaterians (animals with left–right symmetry) have true brains and complex nervous systems, the cnidarians (which include jellyfish and corals) have simple neural networks and sensory cells, the "lowly" placozoans – tiny animals with only a few cell types likely to be the sister group to cnidarinas plus bilaterian animals (Degnan *et al.* 2009) – appear to have a vestigial or primitive system of nerve-like cells (Schuchert 1993), and even sponges have genes homologous to those involved in the formation of synaptic complexes (Sakarya *et al.* 2007). It seems likely, then, that the last common ancestor of all extant animals already contained cells that sensed the environment and transmitted information from one group of cells to another. As animals again upped the ante for ecological complexity, by evolving the capacity to capture and ingest macroscopic prey, natural selection must have favored populations with increased sensory perception and an improved capacity to respond behaviorally to external cues. With animal life, then, the biosphere transitioned from a biogeochemical ecology to a more modern community ecology in which interactions with other organisms, be they prey, predators, mates, or competitors, became principal determinants of fitness and survival.

Clearly, the empirical evidence for late Proterozoic biospheric transition provides a classic puzzle of linked tectonic, environmental, climatic, and biological change. The full spectrum of cause and effect remains uncertain, but very likely the relationships involve a tectonic driver (the breakup of Rodinia, and rapid subsequent continental reamalgamation), which both shifted the balance between sulfide and ferrous iron in favor of the latter and helped to modulate the carbon cycle in ways that increased the likelihood of ice ages. High Ediacaran rates of sediment accumulation would have increased rates of organic carbon burial, providing a mechanism for oxygen increase in the atmosphere and oceans (e.g. Campbell and Squire 2010, Tosca *et al.* 2010), and with more oxygen, large bilaterian animals became physiologically feasible (e.g. Knoll 1992, Johnston *et al.* 2010). And, as discussed above, with macroscopic animals would come positive feedbacks to oxygen regulation via rates of respiration. However, the interlocking puzzle pieces get resolved, it is clear that the late Proterozoic dawn of a modern biosphere reflects the integration of physical and biological processes.

Discussion

Caution is advised when drawing lessons from our experience on Earth to life as it may exist throughout the universe, but as we have no second example of a life-bearing planet, our most informative perspective on variations in planets and life comes from the record of Earth through time.

The diversity of rocky planets in our Solar System suggests how differences in mass and distance from the central star can lead to vastly different surface conditions. Other factors, not accessible in our Solar System, may be equally important, e.g. the mass of the star, the composition of the planet, and contingencies such as the satellite (moon)-forming impact. The surfaces of Earth-mass planets in a habitable zone of another star might, thus, have divergent evolutionary histories because of small differences in elemental composition, the amount of water on the surface, geologic activity, or stellar radiation.

The geologically rapid origin of life on Earth and the tenacity with which it has adapted to "extreme" environments supports the astrobiology optimist's view that microbial life may be common on distant planets. Less clear is the probability that complex multicellularity will evolve from simple ancestors. Simple, multicellularity, in which organisms have multiple cells but little differentiation, has evolved at least two dozen times among eukaryotes (Knoll 2011). In contrast, complex multicellularity, when multiple cell and tissue types differentiate during development and, notably, and in which processes of bulk transport free interior cells from the sharp constraints of diffusion, has evolved only six times (Knoll 2011). Only one of these origins resulted in the emergence of technological intelligence. The elapsed interval between the first definitive evidence for life and the appearance of bilaterian animals, about 3 billion years – two-thirds of Earth history and a quarter of the main-sequence lifetime of our Sun – suggests that the road to complex life and intelligence is not direct and may involve many contingencies. Among these is the presence of abundant atmospheric oxygen.

Molecular oxygen is considered the electron acceptor *nonpareil* because of its high energy yield, and copious oxygen is thought to have been required for the evolution of large and complex multicellular life (Runnegar 1982, Catling *et al.* 2005). If this is so, then the relevant questions are: On what fraction of planets is abundant O_2 inevitable, and on what time-scale(s) does it appear? Future space observatories might be able to answer this question because O_2 can be detected by absorption lines at either visible or infrared wavelengths (Heap 2010).

Oxygen might never appear on a planet because photosystem II (the water-splitting moiety of oxygenic photosynthesis) never evolved. Or, the atmosphere could be sufficiently dense as to scatter and absorb light at the relatively blue wavelengths that are energetically required. It is also conceivable that oxygen might be produced, but be restricted by efficient aerobic respiration to levels not permissive of animal-like complexity. Or, tectonic circumstances might allow oxygen to accumulate, but only on a time-scale longer than the main-sequence lifetime of the star.

The Middle Ages of an Earth-like planet around another star may have different durations depending on the parent star's mass and the planet's distance from the star. Lower-mass stars evolve more slowly and the Middle Ages of their planet might, in consequence, be proportionally longer: if oxygenation correlates with luminosity history, an "Earth" evolving around an M dwarf with a mass half that of the Sun might have a Proterozoic interval 16 billion years long! The location of the planet within the circumstellar habitable zone (HZ), where surface liquid water can persist on an Earth-like planet, is also important (Kasting *et al.* 1993). As the star brightens over time, the CHZ expands, and its width can traverse a habitable planet's orbit. Near the inner edge of the habitable zone the steady-state CO_2 concentration in the atmosphere is near zero (Caldeira and Kasting 1992). Thus, a planet whose orbit is closer to the inner edge of the HZ will experience climate instability, ice ages, and a rise in atmospheric oxygen to levels permissive of animal life after a shorter interval of time. Perhaps the most important lesson of Earth history for astrobiology is that speculation and extrapolation from a single planet will get us only so far. Exploration and discovery provides the only sure route to understanding life in the universe.

References

Aharon, P. (2005). "Redox stratification and anoxia of the early Precambrian oceans: implications for carbon isotope excursions and oxidation," *Precambrian Research*, Vol. 137, pp. 207–222.

Anbar, A. D. and Knoll, A. H. (2002). "Proterozoic ocean chemistry and evolution: a bioinorganic bridge?," *Science*, Vol. 297, pp. 1137–1142.

Bartley, J. K. and Kah, L. C. (2004). "Marine carbon reservoir, C_{org}–C_{carb} coupling, and the evolution of the Proterozoic carbon cycle," *Geology*, Vol. 32, pp. 129–132.

Berner, R. A., Beerling, D. J., Dudley, R. *et al.* (2003). "Phanerozoic atmospheric oxygen," *Annual Reviews of Earth and Planetary Sciences*, Vol. 31, pp. 105–134.

Bernhard, J. M., Habura, A., and Bowser, S. S. (2006). "An endosymbiont-bearing allogromid from the Santa Barbara Basin: implications for the early diversification of foraminifera," *Journal of Geophysical Research*, Vol. 111, Go3002.

Betts, J. N. and Holland, H. D. (1991). "The oxygen content of ocean bottom waters, the burial efficiency of organic carbon, and the regulation of atmospheric oxygen," *Global and Planetary Change*, Vol. 97, pp. 5–18.

Beukes, N. J., and Klein, C. (1992). "Models for iron-formation deposition," in *The Proterozoic Biosphere*. J. W. Schopf and C. Klein (eds.). Cambridge: Cambridge University Press, pp. 147–152.

Bogdanova, S., Pisarevsky, S., and Li, Z. (2009). "Assembly and breakup of Rodinia (some results of IGCP Project 440)," *Stratigraphy and Geological Correlation*, Vol. 17, pp. 259–274.

Brocks, J. J., Love, G. D., Summons, R. E. *et al.* (2005). "Biomarker evidence for green and purple sulphur bacteria in a stratified Palaeoproterozoic sea," *Nature*, Vol. 437, pp. 866–870.

Broecker, W. S. and Peng, T.-H. (1982). *Tracers in the Sea*. Palisades, NY: Eldigio Press.

Buick, R. (2007). "Did the Proterozoic 'Canfield Ocean' cause a laughing gas greenhouse?," *Geobiology*, Vol. 5, pp. 97–100.

Buick, R. (2008). "When did oxygenic photosynthesis evolve?," *Philosophical Transactions of the Royal Society B: Biological Sciences*, Vol. 363, pp. 2731–2743.

Buick, R., Des Marais, D., and Knoll, A. H. (1995). "Stable isotope compositions of carbonates from the Mesoproterozoic Bangemall Group, Australia: environmental variations, metamorphic effects and stratigraphic trends," *Chemical Geology*, Vol. 123, pp. 153–172.

Caldeira, K., and Kasting, J. F. (1992). "The life span of the biosphere revisited," *Nature*, Vol. 360, pp. 721–723.

Campbell, I. H. and Squire, R. J. (2010). "The mountains that triggered the late Neoproterozoic increase in oxygen: the second great oxidation event," *Geochimica et Cosmochimica Acta*, Vol. 74, pp. 4187–4206.

Canfield, D. E. (1998). "A new model for Proterozoic ocean chemistry," *Nature*, Vol. 396, pp. 450–453.

Canfield, D. E. and Teske, A. (1996). "Late Proterozoic rise in atmospheric oxygen concentration inferred from phylogenetic and sulphur-isotope studies," *Nature*, Vol. 382, pp. 27–132.

Canfield, D. E., Poulton, S. W., Knoll, A. H. *et al.* (2008). "Ferruginous conditions dominated later Neoproterozoic deep water chemistry," *Science*, Vol. 321, pp. 949–952.

Catling, D. C., Glein, C. R., Zahnle, K. J., and McKay, C. P. (2005). "Why O_2 is required by complex life on habitable planets and the concept of planetary 'Oxygenation Time'," *Astrobiology*, Vol. 5, pp. 415–438.

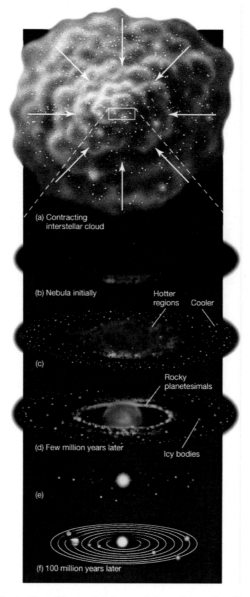

Figure 4.1 An illustration of the main phases of planet formation. First, a parcel of gas within a molecular cloud starts to collapse under its own gravity (panel a). As the parcel collapses, it inevitably creates a disk (panel b), and a star begins to form at the center (panel c). The inner disk is of course hotter than the outer disk, so rocky and metallic bodies condense in the inner disk while the outer disk becomes dominated by volatile, icy bodies (panel d). On time-scales of millions of year, these solid bodies accumulate and grow into both the cores of gas giant planets, the smaller and less gas-dominated ice giants (like Uranus and Neptune), and rocky, terrestrial planets (panel e). Within 100 million years the system's formation is complete (panel f); the system in this figure represents of course our own Solar System. *Image credit: Chaisson and McMillan, Astronomy Today, Pearson, 2011.*

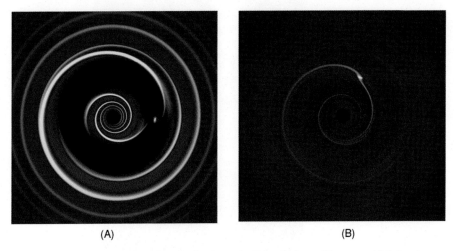

(A)   (B)

Figure 4.2 Snapshots from hydrodynamical simulations of low-mass (right) and high-mass (left) planets embedded in gaseous proto-planetary disks. Here, the colors represent the magnitude of the perturbations to the gas density caused by the planet's gravity: yellow represents an increase in density while purple or black represents a decrease. Note that the high-mass planet has completely cleared out an annular gap in the disk, while the low-mass planet does not have enough gravity to create a gap. *Image credit: Philip J. Armitage.*

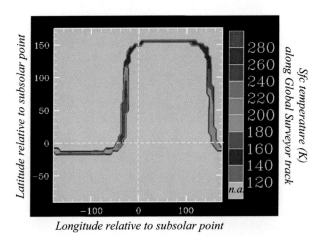

Longitude relative to subsolar point

Figure 8.2 Martian surface temperature along the track of the TES instrument on Mars Global Surveyor. Position is given as latitude/longitude relative to the subsolar point.

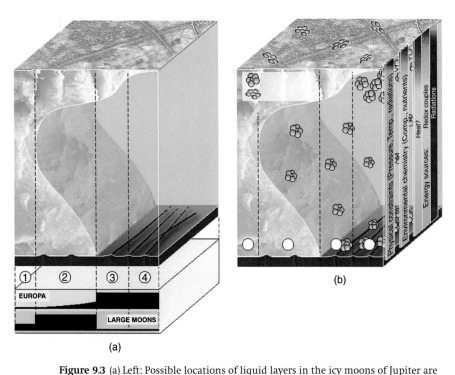

Figure 9.3 (a) Left: Possible locations of liquid layers in the icy moons of Jupiter are plotted here as a function of depth: (1) completely frozen; (2) three-layered structures impeding any contact between the liquid layer and the silicate floor; (3) thick upper icy layer (>10 km) and a thick ocean; (4) very thin upper icy layer (3–4 km). Structures 3–4 are the most probable for Europa. The larger moons Ganymede and Callisto are located in the left region (1 or 2) where internal pressures are sufficient to allow for the formation of high-pressure ice-phases. Oceans in Ganymede and Callisto, if they exist, should be enclosed between thick ice layers (from Lammer *et al.* 2009). (b) Right: Present habitability of Europa. Possible locations of present life and biosignatures have been plotted as a function of depth. Habitability depends on physical and chemical constraints which are indicated on the right using color scales (green: highly favourable; red: hostile). Numbers refer to possible interior structures described in Panel (a) (Blanc *et al.* 2009).

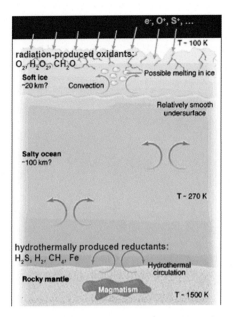

Figure 9.4 Scheme showing the possible existence of chemical energy sources on Europa (Stevenson 2000).

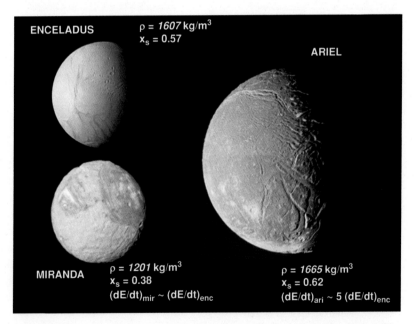

Figure 10.8 Pictures of Saturn's satellite Enceladus as well as of two Uranian satellites that benefit from about the same amount of tidal heating *(dE/dt)* as Enceladus (Figure 10.7). All three objects are characterized by surface properties suggesting ongoing or recent geological activity, in contrast to another Uranian satellite, Umbriel, presented in Figure 10.1. ρ = density; x_s = silicate mass fraction).

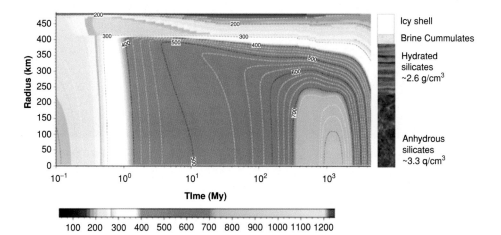

Figure 10.9 Thermal evolution model for Ceres (adapted from Castillo-Rogez and McCord 2010). The pressure reaches ~25 MPa at the base of the icy shell and ~140 MPa at the center of the core. This case assumes a time of formation of 3 Myr after the production of calcium–aluminum inclusions and fully hydrated silicate material following the separation of the rock phase from the water. Settling of the core is achieved by 10 Myr after the end of accretion. By 150 Myr, when the core temperature reaches ~750 K, the deep hydrated silicate starts dehydrating. In this example, we have assumed that 20% of potassium was removed from the core as a result of leaching in a hydrothermal context (see Castillo-Rogez and McCord 2010 and Castillo-Rogez and Lunine 2010 for discussion). Thus silicate dehydration is limited to the inner 200 km. The impacts of [40]K and prospective salt cumulates on the long-term evolution of the shell are not modeled. Under the assumption of conductive transfer, the ice shell would not be "frozen" today (i.e. in equilibrium with the upper boundary condition), and liquid may remain at depth provided that the eutectic temperature of the ocean is depressed by ~30 deg. However, convection onset should occur early (McCord and Sotin, 2005). Convective heat transfer would most probably yield much colder models. However, the long-term preservation of a deep ocean is primarily a function of the freezing temperature of the ocean, i.e. of the concentration and nature of the second-phase impurities part of the accreted planetesimals, or resulting from chemistry during differentiation.

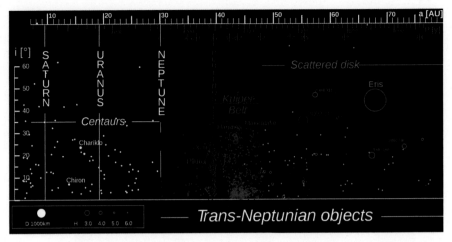

Figure 10.11 Distribution and relative sizes of trans-Neptunian objects (TNOs) discovered so far. Eris, which was the largest TNO as of 2010, is about the size of Europa. Source: http://en.wikipedia.org/wiki/Trans-Neptunian_object.

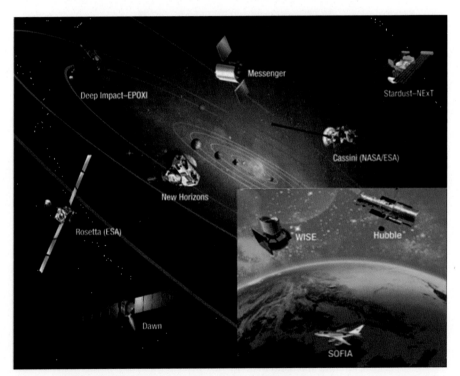

Figure 10.13 Ongoing deep space missions to small water-rich bodies, and space telescopes with the capability to survey and characterize small-body populations (collage of excerpts from NASA Science Mission Directorate 2010 Science Plan). See the appendix for information resources about these missions.

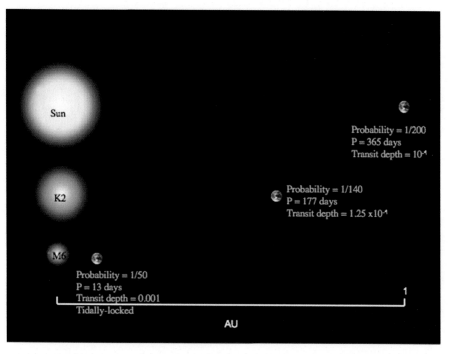

Figure 11.3 A schematic of transiting planets orbiting normal stars. The smaller the star, the closer the habitable zone is to the star, and the easier is the detection.

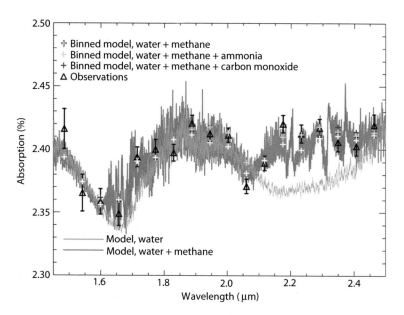

Figure 11.6 Transmission spectrum example: HD 189733. Hubble Space Telescope observations are shown by the black triangles. Two different models highlight the presence of methane in the planetary atmosphere. Reprinted by permission from Macmillan Publishers Ltd: *Nature*, Swain *et al.* (2008), copyright 2008. See also Gibson *et al.* (2011) for analysis of the same data with differing results.

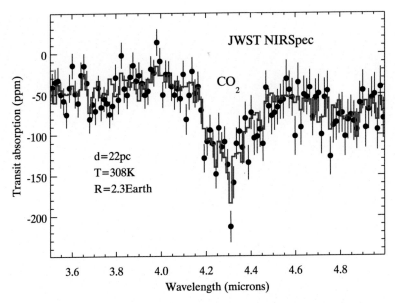

Figure 11.7 Simulated spectrum of a super-Earth as would be observed over many transits by the James Webb Space Telescope near-infrared spectrometer instrument. The points and error bars are the simulated spectrum of carbon dioxide absorption in a habitable super-Earth having $T = 308$ K and $R = 2.3$ R$_\oplus$, with the model spectrum overlaid (blue line). The aggregate SNR for this detection is SNR $= 28$ for 85 hours of in-transit observing, and the distance to this planetary system is $d = 22$ parsecs. Redrawn from Deming *et al.* (2009).

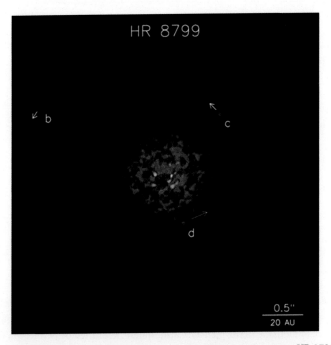

Figure 12.3 The planetary system surrounding the young star HR 8799 (from Marois *et al.* 2008). Three planets more massive than Jupiter are seen on distant orbits. Image courtesy of the National Research Council Canada, C. Marois & Keck Observatory.

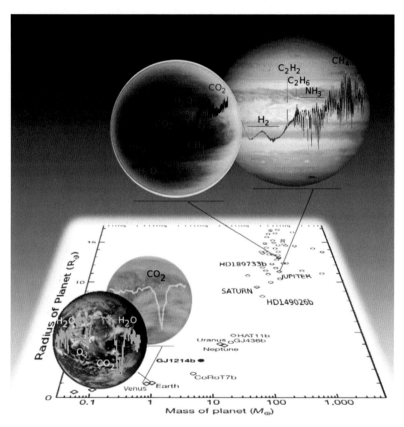

Figure 13.2 Planets can be very similar in mass and radius and yet be very different worlds, as demonstrated by these two pairs of examples. A spectroscopic analysis of the atmospheres is needed to reveal their physical and chemical identities.

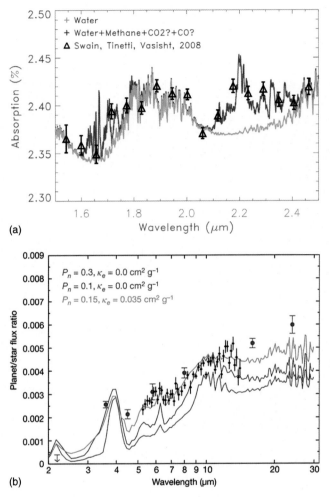

Figure 13.4 Observed and modeled spectra of the hot Jupiter HD 189733b.
(a) Transmission spectrum recorded with Hubble-NICMOS during the primary transit
of the planet. The spectrum can be well explained with water vapor and methane as
atmospheric components (Swain *et al.* 2008b). The low spectral resolution does not
allow the detection of CO or CO_2, detected on the "day-side" of the planet. (b) Emission
spectrum recorded with Spitzer-IRS during the secondary eclipse of the planet
indicating the presence of water vapor in the atmosphere (Grillmair *et al.* 2008).

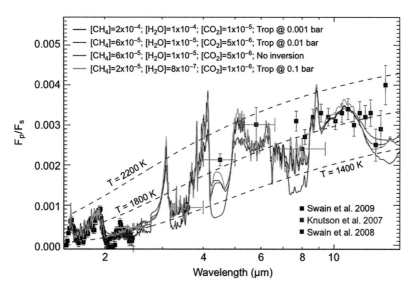

Figure 13.7 Emission photometry and spectroscopy data for HD 209458 b (Swain *et al.* 2009a). The near-infrared and mid-infrared eclipse observations are compared to synthetic spectra for four models that illustrate the range of temperature/composition possibilities consistent with the data. For each model case, the molecular abundance of CH_4, H_2O, and CO_2 and the location of the tropopause is given (see Figure 13.8). These serve to illustrate how the combination of molecular opacities and the temperature structure cause significant departures from a purely single-temperature thermal emission spectrum.

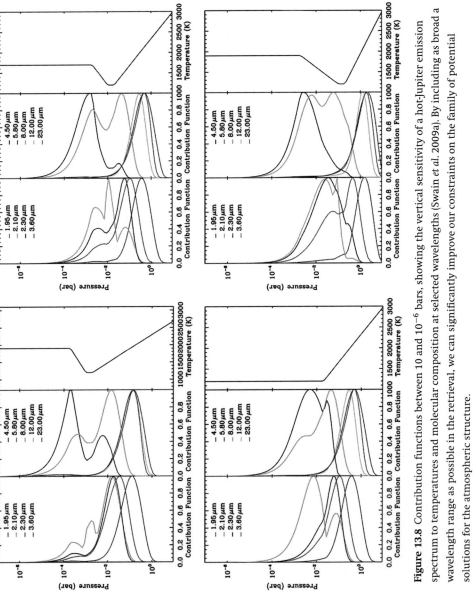

Figure 13.8 Contribution functions between 10 and 10^{-6} bars, showing the vertical sensitivity of a hot-Jupiter emission spectrum to temperatures and molecular composition at selected wavelengths (Swain *et al.* 2009a). By including as broad a wavelength range as possible in the retrieval, we can significantly improve our constraints on the family of potential solutions for the atmospheric structure.

Cloud, P. (1972). "A working model of the primitive Earth," *American Journal of Science*, Vol. 272, pp. 537–548.

Cohen. P. A., Kodner, R., and Knoll, A. H. (2009). "Large spinose acritarchs in Ediacaran rocks as animal resting cysts," *Proceedings of the National Academy of Sciences*, Vol. 106, pp. 6519–6524.

Dahl, T. W., Hammarlund, E., Gill, B. C. *et al.* (2010). "Devonian rise in atmospheric oxygen correlated to the radiations of terrestrial plants and large predatory fish," *Proceedings of the National Academy of Sciences*, Vol. 107, pp. 17853–18232.

Degnan, B. M., Verwoort, M., Larroux, C., and Richards, G. S. (2009). "Early evolution of Metazoan transcription factors," *Current Opinion in Genetics and Development*, Vol. 19, pp. 591–599.

del Giorgio, P. A. and Duarte, C. M. (2002). "Respiration in the open ocean," *Nature*, Vol. 420, pp. 379–384.

Diaz, R. J. and Rosenberg, R. (1995). "Marine benthic hypoxia: a review of its ecological effects and the behavioural responses of benthic macrofauna," *Oceanography and Marine Biology: An Annual Review*, Vol. 33, pp. 245–303.

Embley, T. M. and Martin, W. (2006). "Eukaryotic evolution, changes and challenges," *Nature*, Vol. 440, pp. 623–630.

Farquhar, J., Bao, H. M., and Thiemens, M. (2000). "Atmospheric influence of Earth's earliest sulfur cycle," *Science*, Vol. 289, pp. 756–758.

Fennel, K., Follows, M., and Falkowski, P. G. (2005). "The co-evolution of the nitrogen, carbon and oxygen cycles in the Proterozoic ocean," *American Journal of Science*, Vol. 305, pp. 526–545.

Fike, D. A., Grotzinger, J. P., Pratt, L. M., and Summons, R. E. (2006)."Oxidation of the Ediacaran ocean," *Nature*, Vol. 444, pp. 744–747.

Frei, R., Gaucher, C., Poulton, S. W., and Canfield, D. E. (2009). "Fluctuations in Precambrian atmospheric oxygenation recorded by chromium isotopes," *Nature*, Vol. 461, pp. 250–253.

Gaidos, E. (2010). "Lost in transition: the biogeochemical context of animal origins," in *Key Transitions in Animal Evoluion*, R. Desalle and B. Schierwater (eds.). Boca Raton, FL: CRC Press.

Gaidos, E., Dubuc, T., Dunford, M. *et al.* (2007). "The Precambrian emergence of animal life: a geobiological perspective," *Geobiology*, Vol. 5, pp. 351–373.

Guo, Q., Strauss, H., Kaufman, A. J. *et al.* (2009). "Reconstructing Earth's surface oxidation across the Archean–Proterozoic transition," *Geology*, Vol. 37, pp. 399–402.

Halverson, G. P., Hoffman, P. F., Schrag, D. P. *et al.* (2005). "Toward a Neoproterozoic composite carbon-isotope record," *Geological Society of America Bulletin*, Vol. 117, pp. 1181–1207.

Hanson, P. J., Edwards, N. T., Garten, C. T., and Andrews, J. A. (2000). "Separating root and soil microbial contributions to soil respiration: a review of methods and observations," *Biogeochemistry*, Vol. 48, pp. 115–146.

Heap, S. R. (2010). "Detecting biomarkers in exoplanetary atmospheres with terrestrial planet finder," *EAS Publications Series*, Vol. 41, pp. 517–520.

Hernandez-Leon, S. and Ikeda, T. (2005). "A global assessment of mesozooplankton respiration in the ocean," *Journal of Plankton Research*, Vol. 27, pp. 153–158.

Hjort, K. M., Goldberg, A. V., Tsaousis, A. D. *et al.* (2010). "Diversity and reductive evolution of mitochondria among microbial eukaryotes," *Philosophical Transactions of the Royal Society, Series B*, Vol. 365, pp. 713–727.

Hoffman, P. F. and Schrag, D. P. (2002). "The Snowball Earth hypothesis: testing the limits of global change," *Terra Nova*, Vol. 14, pp. 129–155.

Holland, H. D. (2006). "The oxygenation of the atmosphere and oceans," *Philosophical Transactions of the Royal Society, Series B*, Vol. 361, pp. 903–915.

Hurtgen, M. T., Arthur, M. A., and Halverson, G. P. (2005). "Neoproterozoic sulfur isotopes, the evolution of microbial sulfur species, and the burial efficiency of sulfide as sedimentary pyrite," *Geology*, Vol. 33, pp. 41–44.

Ilyin, A. (2009). "Neoproterozoic banded iron formations," *Lithology and Mineral Resources*, Vol. 44, pp. 78–86.

Javaux, E., Knoll, A. H., and Walter, M. R. (2001). "Ecological and morphological complexity in early eukaryotic ecosystems," *Nature*, Vol. 412, pp. 66–69.

Johnston, D. T., Wolfe-Simon, F., Pearson, A., and Knoll, A. H. (2009). "Anoxygenic photosynthesis modulated proterozoic oxygen and sustained Earth's middle age," *Proceedings of the National Academy of Sciences*, Vol. 106, pp. 16925–16929.

Johnston, D. T., Poulton, S. W., Dehler, C. *et al.* (2010). "An emerging picture of Neoproterozoic ocean chemistry: insights from the Chuar Group, Grand Canyon, USA," *Earth and Planetary Science Letters*, Vol. 290, pp. 64–73.

Kasting, J. F. (1987). "Theoretical constraints on oxygen and carbon dioxide concentrations in the Precambrian atmosphere," *Precambrian Research*, Vol. 34, pp. 305–329.

Kasting, J. F. (1992). "Proterozoic climate: the effect of changing atmospheric carbon dioxide concentration," in *The Proterozoic Biosphere*, J. W. Schopf and C. Klein (eds.). Cambridge: Cambridge University Press, pp. 165–168.

Kasting, J. F. (2005). "Methane and climate during the Precambrian era," *Precambrian Research*, Vol. 137, pp. 119–129.

Kasting, J. F., Whitmire, D. P., and Reynolds, R. T. (1993). "Habitable zones around main sequence stars," *Icarus*, Vol. 101, pp. 108–128.

Knoll, A. H. (1992). "Biological and biogeochemical preludes to the Ediacaran radiation," in *The Origin and Early Evolution of Metazoans*, J. Lipps and P. Signor (eds.). New York: Plenum.

Knoll, A. H. (2003). *Life on a Young Planet: The First Three Billion Years of Evolution on Earth*. Princeton, NJ: Princeton University Press.

Knoll, A. H. (2011). "The multiple origins of complex multicellularity," *Annual Review of Earth and Planetary Sciences*, Vol. 39, pp. 217–239.

Knoll, A. H., Hayes, J. M., Kaufman, J. *et al.* (1986). "Secular variation in carbon isotope ratios from Upper Proterozoic successions of Svalbard and East Greenland," *Nature*, Vol. 321, pp. 832–838.

Knoll, A. H., Kaufman, A. J., and Semikhatov, S. A. (1995). "The Proterozoic carbon isotope record: Mesoproterozoic carbonates from Siberia," *American Journal of Science*, Vol. 295, pp. 823–850.

Knoll, A. H., Javaux, E. H., Hewitt, D., and Cohen, P. (2006). "Eukaryotic organisms in Proterozoic oceans," *Philosophical Transactions of the Royal Society, Series B*, Vol. 361, pp. 1023–1038.

Knoll, A. H., Summons, R. E., Waldbauer, J., and Zumberge, J. (2007). "The geological succession of primary producers in the oceans," in *The Evolution of Primary Producers in the Sea*, P. Falkowski and A. H. Knoll (eds.). Burlington: VT: Elsevier, pp. 133–163.

Kump, L. R., Pavlov., A., and Arthur, M. A. (2005). "Massive release of hydrogen sulfide to the surface ocean and atmosphere during intervals of oceanic anoxia," *Geology*, Vol. 33, pp. 397–400.

Lasaga, A. C. and Ohmoto, H. (2002). "The oxygen geochemical cycle: dynamics and stability," *Geochimica et Cosmochimica Acta*, Vol. 66, pp. 361–381.

Lenton, T. M. and Watson, A. J. (2004). "Biotic enhancement of weathering, atmospheric oxygen and carbon dioxide in the Neoproterozoic," *Geophysical Research Letters*, Vol. 31, L05202.

Logan, G. A., Hayes, J. M., Hieshima, G. B., and Summons, R. E. (1995). "Terminal Proterozoic reorganization of biogeochemical cycles," *Nature*, Vol. 376, pp. 53–56.

Love, G. *et al.* (2009). "Fossil steroids record the appearance of Demospongiae during the Cryogenian period," *Nature*, Vol. 457, pp. 718–721.

Mangum, C. and Van Winkle, W. (1973). "Responses of aquatic invertebrates to declining oxygen conditions," *American Zoologist*, Vol. 13, pp. 529–541.

Martin, M. W., Grazhdankin, D. V., Bowring, S. A. *et al.* (2000). "Age of Neoproterozoic bilaterian body and trace fossils, White Sea, Russia; implications for metazoan evolution," *Science*, Vol. 288, pp. 841–845.

Martin, W., Rotte, C., Hoffmeister, M. *et al.* (2003). "Early cell evolution, eukaryotes, anoxia, sulfide, oxygen, fungi first, and a tree of genomes revisited," *IUBMB Life*, Vol. 55, pp. 193–204.

McDonald, A. E., Vanlerberghe, G. C., and Staples, J. F. (2009). "Alternative oxidase in animals: unique characteristics and taxonomic distribution," *Journal of Experimental Biology*, Vol. 212, pp. 2627–2634.

Millero, F. J. (2005). *Chemical Oceanography*. Boca Raton, FL: CRC.

Narbonne, G. M. (2004). "Modular construction of early Ediacaran complex life forms," *Science*, Vol. 305, pp. 1141–1144.

Pagani, M., Caldeira, K., Berner, R., and Beerling, D. J. (2009). "The role of terrestrial plants in limiting atmospheric CO_2 decline over the past 24 million years," *Nature*, Vol. 460, pp. 85–88.

Peterson, K. J., Cotton, J. A., Gehling, J. G., Pisani, D. (2008). "The Ediacaran emergence of bilaterians: congruence between the genetic and the geological fossil records," *Philosophical Transactions of the Royal Society, Series B*, Vol. 363, pp. 1435–1443.

Poulton, S. W., Fralick, P. W., and Canfield, D. E. (2004). "The transition to a sulphidic ocean similar to 1.84 billion years ago," *Nature*, Vol. 431, pp. 173–177.

Raich, J. W. and Schlesinger, W. H. (1992). "The global carbon dioxide flux in soil respiration and its relationship to vegetation and climate," *Tellus B*, Vol. 44, pp. 81–99.

Raich, J. W., Potter, C. S., and Bhagawati, D. (2002). "Interannual variability in global soil respiration, 1980–94," *Global Change Biology*, Vol. 8, pp. 800–812.

Rivkin, R. B. and Legendre, L. (2001). "Biogenic carbon cycling in the upper ocean: effects of microbial respiration," *Science*, Vol. 291, pp. 2398–2400.

Ruiz-Trillo, I., Burger, G., Holland, P. W. H. *et al.* (2007). "The origins of multicellularity: a multi-taxon genome initiative," *Trends in Genetics*, Vol. 23, pp. 113–118.

Runnegar, B. (1982). "Oxygen requirements, biology, and phylogenetic significance of the late Precambrian worm *Dickinsonia*, and the evolution of the burrowing habit," *Alcheringa*, Vol. 6, pp. 223–239.

Sakarya, O., Armstrong, K. A., Adamska, M. *et al.* (2007). "A post-synaptic scaffold at the origin of the animal kingdom," *PLoS ONE*, Vol. 2, e506.

Schlesinger, W. H. (1997). *Biogeochemistry: An Analysis of Global Change*. San Diego, CA: Academic Press.

Schuchert, P. (1993). "*Trichoplax adhaerens* (Phylum Placozoa) has cells that react with antibodies against the neuropeptide RF amide," *Acta Zoologica*, Vol. 74, pp. 115–117.

Scott, C., Lyons, T. W., Bekker, A. *et al.* (2008). "Tracing the stepwise oxygenation of the Proterozoic ocean," *Nature*, Vol. 452, pp. 456–459.

Shen, Y., Canfield, D. E., and Knoll, A. H. (2002). "The chemistry of Mid-Proterozoic oceans: evidence from the McArthur Basin, Northern Australia," *American Journal of Science*, Vol. 302, pp. 81–109.

Shen, Y., Knoll, A. H., and Walter, M. R. (2003). "Evidence for low sulphate and deep water anoxia in a Mid-Proterozoic marine basin," *Nature*, Vol. 423, pp. 632–635.

Shields, G. A. (2007). "A normalised seawater strontium isotope curve: possible implications for Neoproterozoic–Cambrian weathering rates and the further oxygenation of the Earth," *eEarth*, Vol. 2, pp. 35–42.

Slack, J. F., Grenne, T., Bekker, A. *et al.* (2007). "Suboxic deep seawater in the late Paleoproterozoic: evidence from hematitic chert and iron formation related to seafloor-hydrothermal sulfide deposits, Central Arizona, USA," *Earth and Planetary Science Letters*, Vol. 255, pp. 243–256.

Stolper, D. A., Revsbech, N. P., and Canfield, D. E. (2010). "Aerobic growth at nanomolar oxygen concentrations," *Proceedings of the National Academy of Sciences*, Vol. 107, pp. 18755–18760.

Tosca, N. J., Johnston, D. T., Mushegian, A. *et al.* (2010). "Clay mineralogy, organic carbon burial, and redox evolution in Proterozoic oceans," *Geochimica et Cosmochimica Acta*, Vol. 74, pp. 1579–1592.

Vaquer-Sunyer, R. and Duarte, C. M. (2008). "Thresholds of hypoxia for marine biodiversity," *Proceedings of the National Academy of Sciences*, Vol. 105, pp. 15452–15457.

Veizer, J. and Jansen, S. L. (1979). "Basement and sedimentary recycling and continental evolution," *Journal of Geology*, Vol. 87, pp. 341–370.

Wilson, J. P. *et al.* (2010). "Geobiology of the Paleoproterozoic Duck Creek Formation, Northwestern Australia," *Precambrian Research*, Vol. 179, pp. 135–149.

Wohlers, J., Engel, A., Zöllner, E. *et al.* (2009). "Changes in biogenic carbon flow in response to sea surface warming," *Proceedings of the National Academy of Sciences*, Vol. 106, pp. 7067–7072.

Xiao, S. and Knoll, A. H. (2000). "Phosphatized animal embryos from the Neoproterozoic Doushantuo Formation at Weng'an, Guizhou Province, South China," *Journal of Paleontology*, Vol. 74, pp. 767–788.

PART IV HABITABILITY OF THE SOLAR SYSTEM

8

Early Mars – Cradle or Cauldron?

ARMANDO AZUA-BUSTOS, RAFAEL VICUÑA, AND
RAYMOND PIERREHUMBERT

Mars constitutes one of the most interesting settings for astrobiological studies, not only due to its proximity to Earth, but also because it is conceivable that life may have originated in this seemingly barren planet. Mars has captured man's imagination since the time of Giovanni Schiaparelli, who in 1877 published a detailed map that became a standard reference in planetary cartography. Schiaparelli's original map showed a network of linear markings which went across the entire Martian surface joining different dark areas to one another. He referred to these lines as *canali* and named them after famous rivers. The Italian word *canale* (plural *canali*) was soon incorrectly translated to English as "canals," which denotes artificially made ducts. Being aware of this mistranslation, Schiaparelli stated that

> [T]hese names may be regarded as a mere artifice . . . After all, we speak in a similar way of the seas of the Moon, knowing very well that they do not consist of liquid masses.

Thus, the idea of artificially made water courses remained, implying that Mars was a planet harboring life. Later on, Percival Lowell fueled further speculations about possible Martian life forms in his book *Mars as the Adobe of Life* (1908), popularizing the view that these markings were manifestations of an intelligent civilization.

One can analyze the possibility of life on Mars taking into consideration the different geological ages of this planet (now under revision[1]). The Noachian period

---

1 Based on the latest findings, three new periods that largely, but not completely, overlap the conventional geological chronology, have been proposed: The Phyllosian, marked by the presence of phyllosilicates (clays with a large proportion of iron); the Theiikian (Greek for

Frontiers of Astrobiology, ed. Chris Impey, Jonathan Lunine and José Funes.
Published by Cambridge University Press. © Cambridge University Press 2012.

(named after Noachis Terra), which took place between 4.5 and about 3.7 billion years ago (Gya), is considered to be the warm and wet age of the planet. Extensive erosion by liquid water produced river valley networks whose marks have survived to the present time. There may even have been extensive lakes and oceans at this time, though evidence for such features is at present ambiguous. Thereafter, the Hesperian (named after Hesperia Planum) or volcanic period, from 3.7 to approximately 3.0 Gya, showed catastrophic releases of water that carved extensive outflow channels, with ephemeral lakes or seas. Finally, there is the Amazonian period (named after Amazonis Planitia), which extends from 3.0 Gya until today. It is considered the cold and dry period of Mars, with glacial/periglacial activity and minor releases of liquid water.

In an excellent and comprehensive recent review by Fairén *et al.* (2010), various Mars analogs on Earth for each of the Mars ages are discussed, thus providing a perspective for hypothesizing about the existence of life in each geological period. Much of the site descriptions over the following sections are based on this review and references therein, with emphasis on those in which reports about their microbiology are available. The discussion of climate physics and atmospheric evolution in this chapter is based for the most part on various results described by Pierrehumbert (2010).

Mars offers many lessons for the dawning era of exoplanet discovery, if only the secrets of her past climate evolution written in the geological record could be deciphered. Mars – and especially Early Mars under the Faint Young Sun – provides an archetype of planetary evolution at the very outer edge of the habitable zone about a star. Mars has only a tenth the mass of the Earth, and a third of its surface gravity; it is especially important to understand the role of small size in leading Mars to its present cold, dry nearly airless state. Would the Mars of today have had a more Earth-like climate if it were as massive as Earth? If it had several times the mass of Earth (as do many of the newly discovered extrasolar super-Earths)?

We will begin with a survey of the past ages of Mars, together with terrestrial analogs and their implications for the kind of life that could have emerged or survived at various points in Martian history. Then, we will discuss the evolution of the Martian atmosphere over time, and how this could have accounted for transitions in the Martian environment. There is no question that the Mars of the distant past was a different planet from the Mars of today, but radically different pictures have been put forward as to what that Early Mars was like. Some see Early Mars as a lush Eden-like planet bathed by gentle rains – an ideal cradle for nurturing life. Others see Early Mars as a mostly airless, dry, frigid place, punctuated

sulfate) characterized by the presence of occasional sulfates formed by a reaction with water, an indication of intense volcanic activity; and the Siderikian (Greek for ferric iron), marked by anhydrous ferric oxides and the absence of liquid water.

by great volcanic upheavals or giant impacts, which leave transient steam atmospheres or short-lived boiling acid outburst floods in their wake – more of a cauldron than a cradle. Further exploration of Mars will eventually allow us to tell which combination of the two extremes is closest to the truth. Meanwhile, we point out that it is not necessary for Early Mars to have been a watery Eden in order to have been hospitable to life; a quite broad range of hypothetical atmospheric conditions allow for a mostly cold desert world with transient warm, damp oases in which significant amounts of liquid water form near the surface.

Known analogs of the Noachian age

Known Earth analogs for the Noachian age are the North Pole Dome area in the Pilbara region of Western Australia, the evaporitic and acid sulfate lakes in Western Australia, the Río Tinto basin in Spain, and the cold acid drainage systems in the Canadian Arctic. The North Pole Dome is a structural dome that sits in the middle of the Pilbara Craton. It covers an area of about 500 square kilometers and is one of the two known untouched Archaean (3.6–2.7 Gya) crust remnants on Earth, as rocks of this age have largely been destroyed by plate tectonics. The North Pole Dome area contains some of the Earth's earliest stromatolites and microfossils, some of which are more than 3 billion years old. This area represents an analog for the period of phyllosilicate (from Greek φύλλον, *phyllon*, leaf, forming parallel sheets of silicate tetrahedra) formation on Mars. This group of minerals includes micas, chlorite, serpentine, talc, and clays. As clay minerals are known to form only in the presence of water, Mars phyllosilicates preserve a record of the interaction of water with rocks dating back to the Noachian epoch. One of the mechanisms proposed for the formation of phyllosilicates on Mars is hydrothermal activity. Since at least three hydrothermal events have been identified at the North Polar Dome between 3.5 and 3.2 Gya, the study of this site is of particular interest for the understanding of Mars' middle Noachian age. In addition, sulfur and nitrogen isotopic signatures measured in Dome rocks suggest the antiquity of sulfate-reducing bacteria and chemolithotrophs living in low-oxygen environments.

Río Tinto is a river that originates in the Sierra Morena mountains of Andalusia (Spain) and is part of the Iberian Pyritic Belt, a geological spot formed by hydrothermal activity at the seafloor during the Carboniferous. The Tinto River runs through ore deposits (sulfides of iron and copper) and its acidic waters (pH 2) contain a high concentration of ferric iron in solution. Although at first Río Tinto was thought to be the result of gold, silver, and copper mining operations, it has recently been discovered that, instead, it is a natural environment. Indeed, the acidity and high concentration of metals of this river are the result of iron and sulfur oxidizing microorganisms found in the sulfide deposits of the Iberian Pyritic Belt and along the river valley.

Río Tinto's chemistry is directly comparable to that of rocks found in the Meridiani Planum of Mars. Located south of Mars' equator, this site also contains hematite, large amounts of magnesium sulfate, and other sulfate-rich minerals such as jarosite. Evidence obtained by the Opportunity Rover indicates that Meridiani Planum was once saturated with liquid water of high salinity and acidity for a long period of time. On Earth, hematite is often formed in hot springs or in standing pools of water, indicative of ancient hot springs or liquid water containing environments.

There have been some studies dealing with the microbiology of Río Tinto. Photosynthetic algal biofilms contain communities of *Euglena*, *Pinnularia*, *Chlorella*, *Cyanidium*, and *Dunaliella*. Prokaryotes associated with these biofilms include acidophilic species related to the Alpha-, Beta-, and Gammaproteobacteria and to the phyla Nitrospira, Actinobacteria, Acidobacteria, and Firmicutes. Members of the Archaea domain have also been identified (Souza-Egipsy *et al.* 2008). These biofilms are characterized by the production of large amounts of exopolysaccharides (EPS) which can amount to up to 40% of the total biofilm dry weight. Heavy metals are the second main constituents of the EPS after carbohydrates, reaching up to 16% of total biofilm composition. EPS heavy metal composition closely resembles the metal composition of the water from which they are collected. Prokaryotic diversity of macroscopic filaments at the water surface has also been analysed. Dominant species are *Acidithiobacillus ferrooxidans*, *Leptospirillum ferrooxidans*, and *Acidiphilium spp.* Within the Gammaproteobacteria class, sequences related to the non-acidophilic genera *Aeromonas* and *Acinetobacter* have been identified. In addition, two other new phylotypes related to Gram-positive species from *Desulfosporosinus*, *Clostridium*, and *Mycobacterium* have been described (García-Moyano *et al.* 2007).

Analogs of the Hesperian age

The volcanic period of Mars is represented on Earth by several analogs. During this age a change in the forms of water (liquid, gaseous, solid) occurred, with Mars progressively turning drier and colder. The beginning of the Hesperian coincides with the end of the massive bombardment that prevailed during the Noachian, and there are also suggestions that the Martian magnetic field attenuated at this time. Geological features in the Hesperian suggest a substantially colder climate, with less ability to sustain liquid water at the surface, and a less Earth-like hydrological cycle. This climate transition was presumably caused by a loss of atmosphere, leading to reduced surface pressure and a reduced greenhouse effect. Loss of the magnetic field would facilitate loss of atmosphere to space by solar wind and ultraviolet-driven atmospheric erosion, but while such mechanisms certainly account for some part of the evolution towards the present thin atmosphere, it is exceedingly unlikely that they could bleed off enough atmosphere to account for the climate transition between the Noachian and Hesperian. Atmospheric evolution will be taken up in more detail a bit later in this chapter. Massive volcanism gave rise to outflow channels which resulted in episodic inundations of large parts of the lowlands. The conditions of Mars' surface at this age are assumed to be similar to those encountered in today's Earth Polar regions. Thus, analogous sites for the Hesperian age are the Axel Heiberg Island in the Canadian Arctic, Beacon Valley in Antarctica, and the North Greenland Eemian Ice Drilling Project site.

In the northern regions of the Canadian High Arctic, there are extensive areas of thick permafrost. The Axel Heiberg Island located in the Arctic Ocean of the Qikiqtaaluk region, Nunavut, Canada, is well known for its remarkable fossil forests dating to the Eocene. Polygons located along the Expedition Fiord in the island are formed due to contraction and expansion of the subsurface ice following changes in seasonal surface temperatures. These have been directly compared to similar processes in Martian permafrost characterized in detail by NASA's Phoenix. In these polygons, perennial springs that do not seem to be associated with volcanic heat sources flow through long stretches of permafrost. Two of these springs discharge a constant flow of brine throughout the year at temperatures ranging from -2 to 6.5 °C, despite air temperatures that fall well below -40 °C. These springs may arise from subsurface salt aquifers and show that liquid water is capable of reaching the surface in regions of thick, continuous permafrost without strong volcanic heating sources, a phenomenon that may have taken place on Mars too. Interestingly, images from NASA's Mars Global Surveyor suggest similar recent liquid water outflows in Mars' Southern Hemisphere. These flow features occur on surfaces with no craters and are assumed to have flowed less than a million years ago under the same extremely cold conditions seen on Mars now.

Microbial populations at Gypsum Hill and Colour Peak sites in Axel Heiberg Island are dominated by *Thiomicrospira*, a genus of the Gammaproteobacteria (Pollard *et al.* 2009). It is presumed that S-oxidizing microbial communities flourishing under the winter conditions survive due to H_2S trapped beneath the snow covering the runoff channels. The fact that these microbial structures can flourish under sub-zero temperatures via chemolithotrophic, phototrophic-independent biochemistry, is of particular interest for the research of subsurface waters that are hypothesized to exist on Mars. In another site in this island (Lost Hammer Spring), geochemical analysis has shown ambient temperatures of -4.8 °C and salt concentrations of 22–23%, with the emission of methane, CO_2, and H_2S, and the presence of viable microbial communities consisting of both bacteria and archaea. Archaeal signatures are related to the anaerobic methane group 1a with the closest relatives being detected in marine methane-seep sediments. Bacterial phylotypes are related to microbes previously detected in environments of Antarctica and the Arctic. Isolates cultured from this spring have proven to be active under extreme salinities and temperatures (>20% NaCl and $-10\,^{\circ}$C) under laboratory conditions, and low heterotrophic activities have also been detected in Lost Hammer sediments at $-10\,^{\circ}$C. A large number of novel bacteria have been isolated from these sites, with some of these isolates to be incubated in a Mars environmental simulator (SHOT Inc. Facility, Indiana, USA) to test the survivability and potential activity of these strains under Martian conditions.

In turn, the Beacon Valley is an ice-free region between Pyramid Mountain and Beacon Heights, in Victoria Land, Antarctica. Geological evidence suggests that this valley contains the oldest known ice on Earth (8–10 million years old). Similar to Axel Heiberg Island, it is partly covered by permafrost polygons, directly comparable to polygons that occur over buried ice in Utopia Planitia on Mars. A fair number of Protozoa, Nematodes Tardigrades, Flagellates, and Rotifers have been found in this valley (Bamforth *et al.* 2005). On the other hand, microorganisms colonize the pore spaces of exposed rocks. These cryptoendolithic communities are cyanobacterium and lichen dominated (Cowan 2009, Cary *et al.* 2010). Bacteria and eukaryotes are also identified, but no archaea. These include a member of the Proteobacteria that is potentially capable of aerobic anoxygenic photosynthesis and a distant relative of *Deinococcus* and fungi species with affinity to the *Dothideomycetidae* (Selbmann *et al.* 2005).

The aforementioned North Greenland Eemian Ice Drilling project aims at recovering ice cores from the Greenland Ice Sheet which date back to the Eemian period (140 000 years ago). Greenland is a good analog of the Martian North Polar Layered Deposits, since it shows comparable accumulation of materials. Studies at this site may help to understand how microorganisms could be preserved in the Martian North Pole sediments. In addition, this site could be used to develop drilling techniques to be used in similar sites on Mars. No information has been published yet pertaining to the microbiology of this site.

Earth analogs for the Amazonian age

The Amazonian period is characterized by extremely cold and dry conditions, impacted by the full unimpeded spectrum of solar radiation. With present-day mean surface pressures around 8 kPa and mean temperatures of $-60\,^{\circ}$C, liquid water is exceptional and transitory. Nonetheless, fluvial features have been identified in the middle to late Amazonian (Dickson *et al.* 2009). The Viking landers sent by NASA in the 1970s encountered soils at the landing sites in Chryse Planitia and

Figure 8.1 The hyperarid sector of the Atacama Desert in Chile. Image credit: Armando Azua-Bustos.

Utopia Planitia with the apparent absence of organic compounds and the possible presence of non-biological oxidants. The Phoenix lander failed to detect organic compounds too, but discovered that the soils contained a significant amount of perchlorate ions. The possible presence of perchlorates at the Viking landing sites might explain the failure in detecting organic compounds by this lander. There are no places on Earth comparable to the dry and cold actual conditions of Mars. However, there are extremely dry analogous environments. These are the Atacama Desert in northern Chile and the University Valley in Antarctica.

The Atacama Desert, located between 17° and 27° S latitude in northern Chile (Figure 8.1), is the driest and probably the oldest extant desert on Earth, having experienced extreme hyperaridity probably for the last 15 million years. The average annual rainfall in its central valley is about 1–2 mm. It is constrained on the east by the Andes Mountains and on the west by the Coastal Range. Navarro-Gonzalez *et al.* (2003) were the first to report the unique properties of soils in the extreme arid region core of the Atacama Desert. Samples analyzed from this region exhibited only trace levels of organic species and extremely low levels of culturable and non-culturable bacteria. Also, soils in the Atacama contain sulfates, chlorides, and perchlorates in concentrations imposing a high stress on its microorganisms. Soils of the Atacama contain the largest concentration of natural perchlorate found on Earth, which are similar to those detected by the Phoenix lander on Mars.

Perchlorates in the Atacama are probably generated by photochemical processes in the atmosphere.

To survive under these conditions, life adapted to very low air humidity, almost complete absence of rains, highly saline soils, and high solar radiation. Due to these physical and chemical environmental factors, extensive regions of the Atacama Desert seem to be virtually devoid of microbial life, with concentrations that are one or two orders of magnitude lower than in any other arid region on Earth.

There are several reports on the microbiology of the Atacama Desert, describing coastal, inland, and Andean located sites. Atacama surface and subsurface soils contain heterotrophic bacteria and microbial life that exhibits spatial heterogeneity (patchiness). A clear correlation between soil chemistry and microbial composition has not been found, a fact that is still not well understood. Bacterial numbers contained in surface soils from the most arid regions are extremely low and in most cases estimates are below the detection limit of the methods employed (Gómez-Silva 2009). Heterotrophic species isolated from soils at Yungay belong mainly to two distinct phylogenetic lineages, the Actinobacteria and Firmicutes, and there is also a small number of Proteobacteria. Bacteria belonging to the genera *Sphingomonas, Bacillus, Arthrobacter, Brevibacillus, Kocuria, Cellulomonas*, and *Hymenobacter* have also been identified. Bacterial diversity in surface and subsurface samples is similar, but not identical.

As for photosynthetic microorganisms at the driest zones of the Atacama, traces of cyanobacteria – *Halothece* phylotypes and associated heterotrophic bacteria – have been found living inside halite evaporate rocks (de los Ríos et al. 2010). Hypolithic biofilms under quartzes have also been described. Some are dominated by cyanobacteria (*Chroococcidiopsis* and *Nostocales* phylotypes), whereas in others a taxonomically diverse group of Chloroflexi prevails. Heterotrophic phylotypes common to these hypoliths are affiliated to desiccation-tolerant taxa within the Actinobacteria and Deinococci. These types of hypoliths have also recently been described at sites of the coastal range of the Atacama, where cyanobacteria, archaea, and heterotrophic bacteria can grow with fog as the main water source.

Gypsum (calcium sulphate), which has been detected abundantly in Mars, is widely present in the hyperarid area of the Atacama. Gypsum samples are colonized by photosynthetic microorganisms at crevices and pore spaces 1–3 mm below the crust surface in close association with heterotrophic bacteria. Cell material from enrichment cultures is dominated by *Chroococcidiopsis*.

Caves at the coastal range of the Atacama have also been examined. In one of these caves the presence of a member of the ancient eukaryote red algae *Cyanidium* group has been reported (Azúa-Bustos et al. 2009). This microorganism is a new member of a monophyletic lineage of mesophilic "cave" *Cyanidium*, distinct from other lineages which are thermo-acidophilic. It grows under extremely low photon flux levels. In a different cave, the first species of the well-known genus of green unicellular alga *Dunaliella* able to thrive in a subaerial habitat has been found. All previously reported members of this microalga are found in extremely saline aquatic environments. Interestingly, this microorganism grows on spiderwebs attached to the walls of the entrance-twilight transition zone of the cave, suggesting that it uses the air moisture condensing on the spiderweb silk threads as a source of water for photosynthesis (Azúa-Bustos *et al.* 2010).

University Valley in Antarctica

As mentioned above, another model of the Amazonian age of Mars is University Valley in Antarctica. Located above Beacon Valley, it shows a permanent snowfield and shallow ice-cemented ground on the valley floor. A comparison between the ice below its surface and the observed subsurface profile at the Phoenix landing site on Mars provides several similarities as there is a layer of dry permafrost above the ice-cemented ground. No information has been published pertaining to the microbiology of this particular valley.

Evolution of the Martian atmosphere

The Mars we know today has a very thin atmosphere consisting of almost pure CO_2. Its surface pressure fluctuates over the course of the year, as CO_2 condenses out on the winter pole each solstice and sublimates back into the atmosphere as each equinox is approached, but remains below 1% of the surface pressure of Earth's atmosphere. The typical atmospheric density at the surface is under 0.02 kg/m^3 – less than 2% of the value for Earth's atmosphere. Such a thin atmosphere has too little mass to carry much heat, and so the surface temperature variations of Mars have more in common with those of an airless body like the Moon than they do with the comparatively moderate fluctuations we experience on Earth. Because of the solid surface and thin atmosphere, neither the surface nor atmosphere provides much thermal inertia. In consequence, the surface warms almost instantaneously to come into equilibrium with the amount of solar energy absorbed at any given time and place, and at night-time cools rapidly until the temperature is so low that the surface can no longer radiate efficiently. In consequence, the noontime temperature shows considerable seasonal and geographic variation, while the night-time temperature depends comparatively little on location or time of year.

Even without taking into account any atmospheric greenhouse effect, the temperature at the subsolar point on Mars – the point where the Sun is directly overhead – is predicted to be 297 K (24 °C). The thin atmosphere actually helps such high temperatures to be realized, since the atmosphere carries little heat away

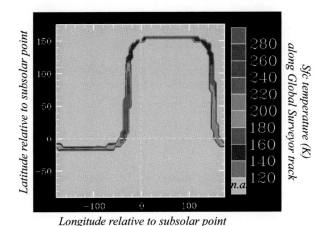

Longitude relative to subsolar point

Figure 8.2 Martian surface temperature along the track of the TES instrument on Mars Global Surveyor. Position is given as latitude/longitude relative to the subsolar point. A colour version of this illustration appears in the colour plate section between pp. 148–149.

from the subsolar point. The corresponding temperature on the Moon would be 397 K (110 °C), and that is indeed close to the maximum temperature observed on the Moon. That would also be the temperature of the subsolar point on Earth, were it not for the moderating effect of the atmosphere, which carries heat away from the subsolar point at the same time as it warms the global mean temperature through the greenhouse effect. The observed subsolar point temperature on Mars indeed moderately exceeds the theoretical no-atmosphere value stated above.

Thus, even on the Mars of today, the surface regularly heats up to temperatures that would melt ice. Given nearly ubiquitous permafrost, it might be thought that there are episodic near-surface damp patches. However, the low surface pressure on Mars keeps liquid water from accumulating at the surface. At the melting point of water ice, the saturation vapor pressure is 611 Pa, rising to 1000 Pa at 280 K. Since this mostly exceeds the surface pressure of the Martian atmosphere, any liquid that forms near the subsolar point would rapidly boil away. Boiling is more effective at removing mass than mere evaporation, because evaporation happens only at the surface of a body of water, whereas in boiling the high vapor pressure relative to ambient pressure allows bubbles of vapor to form in the interior. In consequence, the high temperature near the subsolar point serves as a surface dessication mechanism, rather than a means of accumulating water. This situation is different from that prevailing in terrestrial deserts such as the Atacama, where liquid is removed only by evaporation.

If the melt layer extended deep enough below the surface, then the high pressures there might allow liquid water to accumulate in pores. However, the low diffusivity of heat in rock limits the diurnal melt layer to a shallow range near the surface. With typical diffusivities, the melt layer would extend to depths of only 15–30 cm. At somewhat greater depths, the temperature approaches the diurnal average surface temperature, which is invariably below freezing on present day Mars.

If Mars had higher surface pressure in the past, then boiling at the subsolar point would have been suppressed. However, a denser atmosphere is a two-edged sword, since it also carries heat away from the subsolar point and makes it harder to reach the melting point there. For the atmosphere to be a net benefit in sustaining liquid water, it must have a sufficiently strong greenhouse effect to compensate for the cooling of the subsolar point. This is not much of an issue for Earth, where the bare subsolar point temperature is so high, but it is for Mars where the bare temperature only moderately exceeds freezing.

The extreme Martian obliquity cycle is an important factor bearing on the prospects for episodic melt layers both in the recent and distant past. Obliquity is the tilt of a planet's rotation axis relative to the normal to the plane of its orbit about the Sun. On Earth, obliquity only fluctuates within a few degrees about its current value of 23.4°, because of the stabilizing influence of our massive Moon. On Mars, obliquity fluctuates from under 15° to nearly 50°, and was in a high-obliquity phase as recently as six million years ago (Laskar *et al.* 2004). During high-obliquity stages, the long days and intense sunlight during high-latitude winter allow the seasonal warmth to penetrate to depths of several meters, which could in principle liberate large quantities of liquid water from frozen regolith. With the present Solar brightness and a thin atmosphere, the diurnal mean temperatures in a small polar region do get above freezing in the summer when the obliquity reaches 50°. To get more widespread liquid water without a thicker atmosphere, special circumstances must be invoked (e.g. deep craters with special illumination angles and high pressure at the bottom, or freezing point depression due to salt or other impurities). The obliquity cycle has nonetheless been implicated in at least some of the Amazonian aged fluvial features, though volcanism or geothermal heating may also play a role.

The essential puzzle of Martian climate is that Mars had liquid water flowing in considerable quantities across its surface during the Noachian and Hesperian, at a time when the Sun was considerably fainter than it is today. The extensive branching valley networks of the Noachian demand precipitation and a hydrological cycle involving the atmosphere. The outburst floods of the Hesperian do not in themselves require generally warm, wet conditions but they nonetheless place constraints on the environment because water would not flow far before freezing if the surface were too cold, and it would not flow far before boiling away if the atmospheric pressure were as low as it is today.

The prevailing hypothesis is that Mars had a more massive atmosphere in the past, either continuously or episodically, but lost it either through gradual or cataclysmic processes. Atmospheres are not a permanent, invariable endowment of a planet, but instead are dynamic entities subject to gain or loss by a variety of processes. These include outgassing of volatiles from the interior, delivery of volatiles by impacting bodies, loss of atmosphere to space through impact erosion or solar heating of the upper atmosphere, and loss due to chemical reactions at the surface which can bind up atmospheric constituents in minerals. All atmospheres are dynamic, but the Noachian and Hesperian crust suggests that the Early Martian atmosphere was subject to far more cataclysmic swings than Earth's. On the other hand, much more of the 4 billion year old Martian geological record survives as compared to Earth. Perhaps the preserved Noachian crust of Mars is a window into what the Early Earth was really like.

The volatiles available to form an atmosphere depend on the initial composition of the planet. If anything, Mars should have formed more volatile-rich than Earth, since its orbit is closer to the cold regions of the protoplanetary disk where volatiles can condense. Certainly, this picture is supported by the prevalence of sulfur compounds at the Martian surface. Some volatiles, notably nitrogen, reside predominantly in the atmosphere, whereas others like CO_2 or SO_2 can be bound up in minerals and regenerated through volcanic and tectonic activity. We can expect the very young Mars to have been endowed with plenty of water, an abundance of N_2, and the requisites for outgassing CO_2, SO_2, and perhaps methane.

Mars has very little N_2 left in its atmosphere – only 2.7% of its present thin atmosphere. The mass of N_2 per unit surface area is only four ten-thousandths of the amount in Earth's atmosphere, and Venus has about three times as much N_2 in its atmosphere as Earth. It is thus likely that Mars lost nearly all of its primordial nitrogen. Unlike CO_2 or other volatiles which can be sequestered in minerals and later regenerated, there is really no way to regenerate an N_2 atmosphere once it is lost. Therefore, if Mars went through several episodes of atmospheric loss in the early Noachian, almost all of Martian history is likely to have had an N_2 depleted atmosphere. This poses a very severe challenge for the emergence and survival of life as we know it, since known forms of life require nitrogen as a component of amino acids, DNA, and proteins.

The most readily available greenhouse gases for rocky planets like Earth, Venus, and Mars are CO_2 and water vapor, and the latter only comes into play when CO_2 or some other agent has warmed the planet enough to sustain appreciable water vapor content in the atmosphere. Almost unlimited amounts of CO_2 can outgas from the interior of a rocky planet (witness the 90 bar CO_2 atmosphere of Venus), but even allowing for a dense CO_2 atmosphere, it is not easy to account for a warm, wet equable Noachian climate. The reason is that for a planet at the distant orbit of

Mars, illuminated by the faint young Sun, CO_2 begins to condense once its surface pressure reaches a certain threshold value, and this limits further increases in the greenhouse effect. Moreover, once the pressure exceeds 2 bars the atmosphere itself scatters a great deal of solar energy back to space, and this further offsets the greenhose effect. Various exotic mechanisms have been proposed for keeping Early Mars warm and wet. These include a novel scattering greenhouse effect of CO_2 ice clouds, or accumulation of additional greenhouse gases such as SO_2. It is not clear at present whether any of this exotica actually works.

So what would Mars be like with a dense atmosphere rich in greenhouse gases? Let's consider a 2 bar pure CO_2 atmosphere (twice the surface pressure of Earth's atmosphere). Such an atmosphere would have enough thermal inertia to strongly damp the diurnal cycle, but it would not have enough thermal inertia to appreciably damp the seasonal cycle. Unless there is a very extensive ocean, most of the surface of Early Mars would have a continental climate, and nearly instantaneously come into equilibrium with the diurnal mean insolation at each point of the seasonal cycle. This contrasts with the mostly maritime climate of Earth, which responds primarily to the annual average insolation. At a solstice, the hottest point on Early Mars would be at the summer pole, and temperature would decay monotonically toward the winter pole. Even with just the conventional gaseous greenhouse effect of a 2 bar CO_2 atmosphere, an extensive region near the summer pole is well above freezing, at least when Mars is in a high-obliquity phase. This occurs even if one does not invoke additional greenhouse gases or exotic effects such as the scattering greenhouse effect of CO_2-ice clouds. Figure 8.3 shows a typical solstice pattern during a time of high obliquity. The summer extratropics are well above freezing; the long duration of the season would allow melting to a depth of several meters, and the high atmospheric pressure would allow liquid to accumulate on the surface without boiling away. The results in Figure 8.3 do not take atmospheric heat transport into account, which would even out the temperature pattern and transfer some heat into the cold winter hemisphere. Mild heat transport would make the liquid water region more extensive by spreading out excess heat from the pole, but if heat loss to the winter hemisphere becomes too strong, it can bring the summer hemisphere below freezing. In addition, the very cold equilibrium temperatures of the winter hemisphere imply that CO_2 condensation would occur there, and this could reduce the CO_2 content (and hence the greenhouse effect) of the summer hemisphere. Quantification of both points requires simulation with an atmospheric general circulation model, and will not be pursued here.

While we have emphasized CO_2 in the preceding, it should be noted that the sulfur cycle enters crucially in the Early Mars climate. We have earlier noted the sulfur-rich acidic tropical soils of Mars, which are indirectly created by volcanic SO_2 outgassing. SO_2 is itself a potent greenhouse gas, and in an anoxic atmosphere

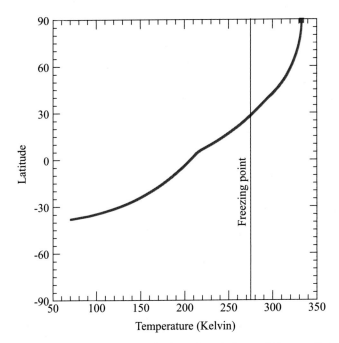

Figure 8.3 Temperature vs. latitude under solstice conditions on Early Mars, assuming the gaseous greenhouse effect of a 2 bar pure CO_2 atmosphere illuminated by a Sun which is 25% less luminous than at present. The calculation gives the local equilibrium temperature, which would be realized in the absence of atmospheric heat transport. In reality, some of the summer hemisphere warmth would be bled off into the cold winter hemisphere. The calculation was done for a time when the orbit is nearly circular and the obliquity is 50°. A planetary albedo (reflectivity) of 40% is assumed, to take account of the scattering from the dense atmosphere. The calculation was done using radiation results from Pierrehumbert (2010, Chapter 4) and insolation results from Pierrehumbert (2010, Chapter 7). The polar night region is masked out, because there would not be sufficient time for the temperature to cool to the equilibrium value in the course of the winter.

where it is not removed by conversion to sulfate, could potentially accumulate in the atmosphere to concentrations high enough to cause significant warming (Halevy *et al.* 2009). Further, the geochemistry associated with acidic solutions of SO_2 in water (sulfurous acid) inhibit the formation of carbonates, and facilitate the accumulation of CO_2 in the atmosphere (Halevy *et al.* 2007). Finally, and perhaps most importantly, when sulfate minerals can form, leading to sulfuric acid waters analogous to the Río Tinto, the freezing point of the liquid is severely depressed. For example, the freezing point of a 20 weight percent solution of sulfuric acid is −18 °C, falling all the way to −73 °C for a 40 weight percent solution. Hence, highly acidic waters could accumulate and flow on the surface of Mars even at temperatures where pure water would be frozen.

The post-Noachian evolution of the Martian climate is a race between the gradually brightening Sun, loss of atmosphere to space, and the freeze-out of Martian tectonic activity, which would limit the ability to recycle crustal carbonates into atmospheric CO_2 (if indeed any such recycling ever existed on Mars). The small size of Mars makes it susceptible to the long, sad slide into the senescent world we see today, because low gravity facilitates atmospheric loss to space and small planets cool off and become tectonically inactive more quickly. However, it is far from clear that such a decay of habitability is inevitable for all Mars-sized bodies; it could well have required a chance event – such as a giant impact – at an inopportune time.

When and how did Mars lose its dense atmosphere? Despite the low Martian gravity, it is actually not so easy to get rid of a 2-bar atmosphere. There are two main mechanisms in play. One is impact erosion, in which sufficiently large impactors accelerate significant portions of the atmosphere to escape velocity. However, impacts can also *deliver* volatiles, as well as cook existing volatiles out of the impacted crust. Moreover, the low gravity of Mars can lead to lower-energy impacts (depending on where the impactors come from), and this would tend to offset the role of low gravity in facilitating escape. The other mechanism in play consists of erosion of atmosphere from the top, driven by heating of the upper atmosphere through absorption of solar extreme ultraviolet, or by interaction with the solar wind. Unlike impact erosion, this form of atmosphere theft leaves its fingerprints behind at the scene of the crime, in the form of the residual atmospheric constituents being enriched in heavier isotopes. Isotopes only show the *proportion* of atmosphere lost to fractionating processes, and not the net mass lost. The isotopic evidence suggests that 90% of an initially unfractionated atmosphere was lost by top-of-atmosphere erosion over Martian history, but modeling of atmospheric escape suggests that it is hard to lose more than a few tenths of a bar of atmosphere by such processes, unless they occurred at the very beginning of the Noachian when extreme ultraviolet and solar wind fluxes may have been very high. The prevalent view is that Mars first lost 90% of its primordial (end Noachian?) atmosphere by non-fractionating impact erosion, and then lost 90% of the remaining 200 millibar atmosphere by fractionating top-of-atmosphere erosion processes, leaving only a bit to be taken care of by loss to carbonate formation at the planetary surface.

The conundrum of Martian atmosphere loss is that it is much easier to do very early in Martian history, when extreme ultraviolet flux, solar wind flux, and impact rates are all high. However, we can't afford to lose the atmosphere too early, because then it would be hard to account for Noachian fluvial features and

Hesperian outburst floods. While Mars is still tectonically active, an atmosphere could be regenerated after loss, but this becomes impossible after tectonic freeze-out. To make things even more problematic, the emplacement of Tharsis late in the Noachian is likely to have generated a fairly massive CO_2 atmosphere, and it is less easy to lose an atmosphere to space once one enters the Hesperian. Where did that atmosphere go? Could it have been lost to gradual carbonate formation in the episodic moist patches that survived into the Hesperian?

Segura *et al.* (2008) have put forth a picture of an inhospitable cauldron-like Early Mars, in which the planet spent most of its time locked in a nearly airless deep-freeze, punctuated by brief intervals during which bombardments evaporated enough water to create a steam atmosphere with a surface pressure of a bar or so. Such an atmosphere would condense out over the course of a decade, in response to radiative cooling. A serious, and perhaps fatal, flaw in this picture is that the rainout rate would only be on the order of a centimeter per day, and it is not at all clear that such gentle rains could account for the dramatic water-carved features seen on Mars in the short time they would persist. The cauldron picture can go beyond simple post-impact steam atmospheres, however. Mars may have lost and regenerated an atmosphere (including CO_2 and other volatiles as well as water vapor) many times during the Noachian, before it commenced its ultimate decline in the Hesperian and Amazonian.

Mars as a habitable desert world

It may be too much to ask of climate physics to give Early Mars a steadily warm, wet climate. But, given that life struggles bravely on in the cold, dry Atacama and the acid waters of the Río Tinto, demanding that Early Mars be a watery Eden may be setting the bar too high. With a moderately dense atmosphere, a moderate amount of additional greenhouse warming can permit seasonal surface liquid water, especially if one allows for the freezing point depression caused by the expected admixture of sulfuric acid in Martian waters. That might have been enough to get life going. The result would be a mostly dry desert world with episodic regional moist surface oases, rather like the world Arakis in Frank Herbert's novel *Dune*.

In accordance with the solstice temperature calculations for Early Mars given in the preceding section, we would have a planet that would mainly come alive during periods of high obliquity. During such periods, bitterly cold extratropical winters would alternate with hot summers liberating water in frozen soils and glaciers. This water would evaporate, and drizzle out elsewhere, causing additional moist patches. If the runoff is acidic, freezing point depression could lead to quite

extensive acid river and lake systems. Further, glacial melt water could sustain
ice-covered lakes, as in the Antarctic Dry Valleys.

Later in Martian history, the atmosphere becomes thinner and the greenhouse
effect becomes weaker, but the brighter Sun and lack of heat transport and reflectivity in the thin atmosphere allow more limited, shallow damp patches to persist
even to modern times. Whether or not life could adapt to the present conditions,
knowing whether it ever made a start at all during earlier ages of Mars would be
a major step forward in understanding the conditions for the emergence of life.
We won't know until we look, and the search is apt to require turning over more
than a few stones among the many that have so far been left unturned.

References

Azúa-Bustos, A., González-Silva, C., Mancilla, R. A. *et al.* (2009). "Ancient photosynthetic
    eukaryote biofilms in an Atacama Desert coastal cave," *Microbial Ecology*, Vol. 58,
    485–496.

Azúa-Bustos, A., González-Silva, C., Salas, L. *et al.* (2010). "A novel subaerial *Dunaliella*
    species growing on cave spiderwebs in the Atacama Desert," *Extremophiles*, Vol. 14,
    443–452.

Bamforth, S. S., Wall, D. H., and Virginia, R. A. (2005). "Distribution and diversity of soil
    protozoa in the McMurdo Dry Valleys of Antarctica," *Polar Biology*, Vol. 28,
    756–762.

Cary, S. C., McDonald, I. R., Barrett, J. E., and Cowan, D. A. (2010). "On the rocks: the
    microbiology of Antarctic Dry Valley soils," *National Review of Microbiology*, Vol. 8,
    129–138.

Cowan, D. (2009). "Cryptic microbial communities in Antarctic deserts," *Proceedings of
    the National Academy of Sciences*, Vol. 106, 19749–19750.

de Los Ríos, A., Valea, S., Ascaso, C. *et al.* (2010). "Comparative analysis of the microbial
    communities inhabiting halite evaporites of the Atacama Desert," *International
    Microbiology*, Vol. 13, 79–89.

Dickson, J. L., Fassett, C. I., and Head, J. W. (2009). "Amazonian-aged fluvial valley
    systems in a climatic microenvironment on Mars: melting of ice deposits on the
    interior of Lyot Crater," *Geophysical Research Letters*, Vol. 36, L08201,
    doi:10.1029/2009GL037472.

Fairén, A. G., Davila, A. F., Lim, D. *et al.* (2010). "Astrobiology through the ages of Mars:
    the study of terrestrial analogues to understand the habitability of Mars,"
    Astrobiology, Vol. 10, 821–843.

García-Moyano, A., González-Toril, E., Aguilera, A., and Amils, R. (2007). "Prokaryotic
    community composition and ecology of floating macroscopic filaments from an
    extreme acidic environment, Río Tinto (SW Spain)," *Systematic and Applied
    Microbiology*, Vol. 30, 601–614.

Gómez-Silva, B. (2009). "On the limits imposed to life by the hyperarid Atacama Desert in Northern Chile," in *Astrobiology: Emergence, Search and Detection of Life*, V. A. Basiuk (ed.). American Scientific Publishers, pp. 1–13.

Halevy, I., Zuber, M. T., and Schrag, D. P. (2007). "A sulfur dioxide climate feedback on Early Mars," *Science*, Vol. 18, 1903–1907.

Halevy, I., Pierrehumbert, R. T., and Schrag, D. P. (2009). "Radiative transfer in CO_2-rich paleoatmospheres," *Journal of Geophysical Research – Atmospheres*, Vol. 114, D18112. doi:10.1029/2009JD011915.

Laskar, J. *et al.* (2004). "Long term evolution and chaotic diffusion of the insolation quantities of Mars," *Icarus*, Vol. 170, 343–364.

Navarro-González, R., Rainey, F. A., Molina, P. *et al.* (2003). "Mars-like soils in the Atacama Desert, Chile, and the dry limit of microbial life," *Science*, Vol. 302, 1018–1021.

Pierrehumbert, R. T. (2010). *Principles of Planetary Climate*. Cambridge: Cambridge University Press.

Pollard, W., Haltigin, T., Whyte, L. *et al.* (2009). "Overview of analogue science activities at the McGill Arctic Research Station, Axel Heiberg Island, Canadian High Arctic," *Planetary and Space Science*, Vol. 57, 646–659.

Segura, T. L., Toon, O. B., and Colaprete, A. (2008). "Modeling the environmental effects of moderate-sized impacts on Mars," *Journal of Geophysical Research – Planets*, Vol. 113, E11007.

Selbmann, L., de Hoog, G. S., Mazzaglia, A. *et al.* (2005). "Fungi at the edge of life: cryptoendolithic black fungi from Antarctic desert," *Studies in Mycology*, Vol. 51, 1–32.

Souza-Egipsy, V., González-Toril, E., Zettler, E. *et al.* (2008). "Prokaryotic community structure in algal photosynthetic biofilms from extreme acidic streams in Río Tinto (Huelva, Spain)," *International Microbiology*, Vol. 11, 251–260.

9

Large Habitable Moons

Titan and Europa

ATHENA COUSTENIS AND MICHEL BLANC

Introduction

The "classical" criteria for habitability can be summarized as the presence
of liquid water, energy sources to sustain metabolism, and "nutrients" over a
period of time long enough to allow the development of life. The concept of a
"habitable zone" (HZ) around each star defines where water can be stable at the
surface as a result of the equilibrium temperature of the planet in the star's
radiation field. Liquid water may exist on the surface of planets orbiting a star at a
distance that does not induce tidal lock. But habitability conditions can be found
not only on the surfaces of Earth-like planets: a subsurface ocean within a planet
or the satellite of a gas giant may be habitable for some life form that may be very
different from Earth-like life. Indeed, icy surfaces may cover liquid oceans, move
and fracture like plate tectonics, and exsolute the internal material and energy
through an interconnected system. With the discovery of planets beyond the Solar
System and the search for life in exotic habitats such as Mars, Europa, Titan, and
Enceladus, habitability in general needs a better and broader definition.

Liquid water has been recognized as the best solvent for life to emerge and
evolve, although other possibilities have been suggested (e.g. Bains 2004). Water,
an abundant compound in our galaxy (e.g. Cernicharo and Crovisier 2005), is liquid
within a large range of temperatures and pressures and is also a strong polar–
nonpolar solvent. This dichotomy is essential for maintaining stable biomolecular
and cellular structures (Des Marais *et al.* 2002). A large number of organisms is
capable of living in water. However, in a body of pure water, life will probably
never spontaneously originate and evolve. This is because, while there are many

Frontiers of Astrobiology, ed. Chris Impey, Jonathan Lunine and José Funes.
Published by Cambridge University Press. © Cambridge University Press 2012.

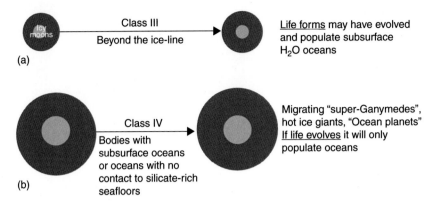

Figure 9.1 Illustration of icy satellites of gas giants. (a) Class III habitats: liquid water habitats or oceans beyond an ice layer which interacts with the silicate core or seafloor. Life may have originated within such subsurface habitats but surface conditions are hostile. (b) In Class IV habitats, the ocean layer is in contact with an ice floor (Lammer *et al.* 2009).

organisms living *in* water, none we know of is capable of living *on* water alone because life requires other essential elements such as nitrogen and phosphorus in addition to hydrogen and oxygen. Besides, no organism we know of is made entirely of water. So obviously "just water" is not an auspicious place for starting life and evolution in.

Planets with oceans may belong to three families, among "terrestrial" planets, hot giants, or giant planets' icy satellites. The question is if the liquid water existed for significant periods to have been biologically useful, because a frozen surface may not necessarily have inhibited life. If liquid layers exist below ice layers and these water-reservoirs are in contact with heat sources from the interior of the planet by radioactive decay, volcanic interactions, or hydrothermal activity, the planet may be considered a Class III habitat, as defined in Lammer *et al.* (2009) and shown in Figure 9.1(a). In the opposite case, we have a Class IV habitat (Figure 9.1b).

Habitability issues for outer planet satellites

Large satellites of gas giants at orbits beyond the ice-line, like Jupiter or Saturn, are known to include a large amount of water. Indeed, when the average density is around 1.8 g/cm^3, the largest moons are composed of almost 45% by weight of water. Thus, icy layers are very thick and one can assume that their thickness can be on the order of 600 km for Ganymede, Callisto, and Titan. Callisto, as Galileo data indicate, is probably not fully differentiated, which means that the icy layer might be thicker, but mixed with silicates. For Europa, a smaller icy layer

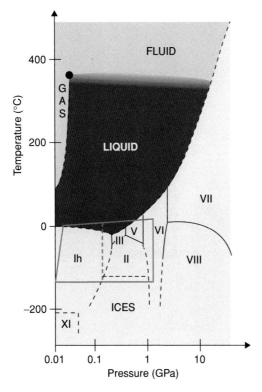

Figure 9.2 Phase diagram of the pure water system. Five ice polymorphs can exist in the pressure range relevant to icy moons (box). The dashed line symbolizes the highest pressure which can occur in the water layer of Europa (from Lammer *et al.* 2009).

(100 km) is hypothesized, because its amount of water is "only" 10% by weight. Additionally, Saturn's largest satellite, Titan, possesses a thick icy crust of ~120 km (Fortes *et al.* 2007). Understanding the internal structure of the water layer requires a good knowledge of the water phase diagram under pressure and temperature conditions shown in Figure 9.2.

This indicates that many ice polymorphs (e.g. Petrenko and Whitworth 1999) can exist in the pressure range 0–2 GPa, relevant to icy moons. As noted by Lammer *et al.* (2009), it is of fundamental importance for the icy moons of Jupiter and Saturn that the melting curve of the low-pressure ice Ih decreases with pressure, which is related thermodynamically to the fact that ice Ih is less dense than its liquid.

If the temperature is between 250 and 273 K, it is possible for a liquid layer to be located below an Ih ice layer (cases 3–4 in Figure 9.3). This case is compatible with large icy moons.

Class IV habitats and exoplanets where a water ocean is in contact with a thick ice layer ("ocean planets," see Léger *et al.* 2004) present a much better situation

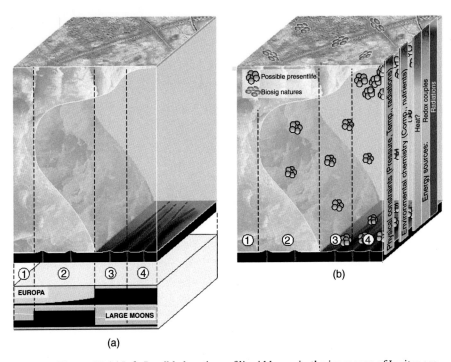

Figure 9.3 (a) Left: Possible locations of liquid layers in the icy moons of Jupiter are plotted here as a function of depth: (1) completely frozen; (2) three-layered structures impeding any contact between the liquid layer and the silicate floor; (3) thick upper icy layer (>10 km) and a thick ocean; (4) very thin upper icy layer (3–4 km). Structures 3–4 are most probable for Europa. The larger moons Ganymede and Callisto are located in the left region (1 or 2) where internal pressures are sufficient to allow for the formation of high-pressure ice-phases. Oceans in Ganymede and Callisto, if they exist, should be enclosed between thick ice layers (from Lammer *et al.* 2009). (b) Right: Present habitability of Europa. Possible locations of present life and biosignatures have been plotted as a function of depth. Habitability depends on physical and chemical constraints which are indicated on the right using color scales (green: highly favourable; red: hostile). Numbers refer to possible interior structures described in Panel (a) (Blanc *et al.* 2009). A colour version of this illustration appears in the colour plate section between pp. 148–149.

for the influx of organic material from outside the planet compared to bodies like Ganymede or Callisto. A problem encountered in Class IV habitats is, however, much more severe: that of the concentration of the necessary ingredients for life (Figure 9.3b). A planet whose surface is completely covered in water several kilometers deep with nothing to act as a concentrating "sponge" for organic chemistry is simply too vast for any two or more interesting molecules to meet. While a sea/ice system could in theory provide such a means of concentrating life's ingredients (Trinks *et al.* 2005), not even the starting concentrations needed for a system like that are likely to be reached in addition to the quite specific temperature conditions needed.

Of the large satellites of the gas giants, there are those that have a silicate-rich core and may house underground water deposits in direct contact with heat sources below their iced crust. Others are expected to have either liquid-water layers encapsulated between two ice layers, or liquids above ice (see Figures 9.3(a) and (b)). In the study of the emergence of life elements on such satellites, the time-scale is of the essence. If it is long enough, the liquid-water underground ocean may host life. Thus, the icy satellites of the outer planets of the Solar System, as well as the recently discovered exoplanets, host conditions which may inhibit the emergence of life precursors in isolating environments that can prevent the concentration of the ingredients necessary for life or the proper chemical inventory for the relevant biochemical reactions. Conversely, according to Trinks *et al.* (2005), a coupled sea/ice system could in theory provide the necessary conditions for the emergence of life on the primitive Earth. Additional laboratory experiments and *in situ* studies of deep subglacial isolated lakes in Antarctica (Kapitsa *et al.* 1996) would improve our understanding in this field, as the physical properties of deep subglacial lakes resemble those found on both Jupiter's moon Europa and Saturn's moon Enceladus (Bulat *et al.* 2009).

As a consequence, the satellites of the giant planets like Europa, Enceladus, Ganymede, or Titan are potential habitable environments and valid targets in the search for life with space missions and/or telescopes. However, life structures that do not influence the atmosphere of their host planet on a global scale will not be remotely detectable. In the Solar System's neighborhood, such potential habitats can then only be investigated with appropriately designed space missions. In the case of Europa, Enceladus, Ganymede, and Titan, the hypothesized internal liquid-water oceans add a significant advantage to the search for life in such environments, the ultimate goal of a new era of exploration of these four bodies (e.g. Blanc *et al.* 2009, Clark *et al.* 2009, Coustenis *et al.* 2009, Reh *et al.* 2008).

Europa

Looking at the pressure range encountered within icy moons (Figures 9.2 and 9.3), structures 1 and 2 are highly probable for the largest moons, while structures 1 and 3–4 are more probable for Europa. Europa is probably unique, because of its high rock/ice mass ratio, in that it is the only satellite on which a large ocean can be in contact with the silicate layer. On the other moons, the existence of an ocean necessarily implies the occurrence of a very thick high-pressure icy layer at the bottom, which impedes contact of the liquid with silicates.

Europa as a potential Class III habitat

Europa most likely represents the only Class III habitat in the Solar System, although recent findings that salty water exists below Enceladus's surface (Postberg *et al.* 2011) might elevate that target (see Chapter 10). Due to the difficult direct access to this potential habitat, the *in situ* exploration of subsurface oceans will be a long-term scientific endeavor. Class III habitats may have large astrobiological potential and could host life if some conditions exist simultaneously. Besides the correct pressure and temperature conditions for the presence of liquid-water, the planetary object should have adequate energy sources to support metabolic reactions, and chemical elements to provide nutrients for the synthesis of biomolecules (Figure 9.3).

The putative existence of a subsurface ocean depends of course on the evidence for liquid water. In the case of Europa, this possibility has been suggested based on measurements by the Galileo mission of the induced magnetic field (Kivelson *et al.* 1999), and on the interpretation of geological features. If such an ocean exists, we cannot today precisely determine its depth or its location, and the question as to whether there is ice convection within the icy crust leading to the possibility of an exchange of material between the surface and the ocean, remains open in the thin-crust model because of the large number of tectonic features such as cracks and faults visible on the surface. This is an important issue for habitability, because Europa's surface is subject to heavy bombardment by energetic particles from Jupiter's radiation belts, leading to the breaking up of water molecules and to the generation of hydrogen peroxide, H_2O_2, a strong oxidizing component which may be a source of chemical energy. Indeed, H_2O_2 tends to decompose exothermically into water and oxygen gas and it has been detected on Europa's surface in infrared and ultraviolet wavelength spectra (Carlson *et al.* 1999).

The determination of the topography and/or mass anomalies at the silicate core/liquid interface would provide hints on whether volcanism and hydrothermal activity may exist or have existed, in a way similar to mid-oceanic ridges on the Earth. Such activity releases into the ocean a variety of chemicals that could play a role in sustaining putative life forms at the ocean floor like those discovered 30 years ago in Earth's deep-sea hydrothermal vents.

Given the existence of an ocean, the question of habitability can be summarized by an inspection of the "triangle of habitability" (Hand *et al.* 2009). In addition to a billion-year-old ocean, it involves the presence of the key chemical elements for life (CHNOPS), and of energy sources. Hand *et al.* (2009) reviewed the possibilities of the presence and likely abundances of these key elements in Europa's ocean and seafloor, and reached a rather positive conclusion. Concerning the energy source, the presence of a H_2O_2 oxidizing chemistry at the

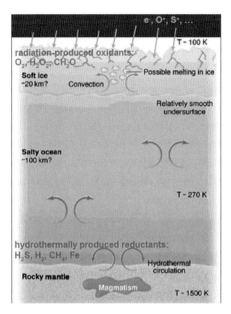

Figure 9.4 Scheme showing the possible existence of chemical energy sources on Europa (Stevenson 2000). A colour version of this illustration appears in the colour plate section between pp. 148–149.

surface due to the radiolysis of ice by radiation-belt particles, and of a reducing chemistry at the ocean floor if hydrothermal activity exists, open the possibility of the existence of a redox couple acting at the scale of the ocean. But this works only if oxidants from the surface can feed the ocean via a (hypothesized) partly permeable icy crust.

Characterizing Europa's ocean and its accessibility

The characterization of the ocean depth, thickness, and of the boundary of its interface with the silicate mantle will be possible only for a carefully designed Europa orbiter, such as has been the subject of numerous studies. Clearly, the *in situ* characterization of the oceanic chemical and possibly prebiotic properties will be a longer-term scientific and technological challenge. In this section we review the measurements of Europa that have high priority for a future mission, largely taken from Blanc *et al.* (2009).

Ocean. Although very probable, given the evidence from the Galileo geological measurements, the existence of a liquid internal ocean on Europa is still not proven, and its depth below the surface is unknown. Values range from hundreds of meters to kilometers. Outside direct exploration by an adequate payload aboard a lander, the most robust method to prove the existence of the ocean is to use an orbiter from which accurate gravimetric and altimetric measurements can be achieved. Indeed, the presence of an ocean strongly modifies both the amplitude of the crustal flexure and of nodding (libration) due to tidal forcing. Such measurements were not achieved with Galileo because of orbital and instrumental constraints (only flybys, no altimetry), but may become possible with a dedicated Europa orbiter. If the gravity potential is determined using a precise navigation system, it is possible to estimate the crustal flexure (which could be tens of meters) and prove the existence (or absence) of Europa's ocean independent of uncertainties in the material properties of the ice (Wahr *et al.* 2006). If in addition an altimeter were added to the system, it is possible to have direct and independent proof of the existence of the liquid layer by measuring the amplitude of surface motions due to tidal forcing. The *in situ* gravimetric and altimetric measurements at Europa will be supplemented by ultra-precise VLBI and Doppler tracking of the orbiter by the Earth-based network of radio telescopes.

Characterization of the liquid layer is more challenging. It requires the determination of its depth, thickness, and composition. The interface between the icy crust and the ocean is expected to be relatively flat, since it corresponds to a thermodynamic boundary between ice and liquid in which the liquid at the interface should be at constant temperature around the body. Furthermore, the flat topography of Europa is not in favor of large roots of ice below high mountains as we see on Earth. Two specific instruments can be considered for providing direct constraints on the icy crust thickness. First, a Ground Penetrating Radar (GPR) seems very attractive if the icy crust is relatively thin. This instrument could globally map the ice I/liquid interface, provided that the crust is not thicker than the penetration depth.

Another exciting option is a seismometer on a lander, which could provide, without ambiguity, the position of the ocean interface, although the measurement would be made at a single point on the surface and the cost of safely putting a lander on the surface is high. A third technique is based on the interpretation of the topography and gravity field at different degrees. It requires a laser or radar altimeter in addition to the determination of the gravity field by radio measurements, including VLBI, two-way Doppler, and delta-Doppler ranging techniques.

Mapping Europa's seafloor. If an internal ocean exists, it is probably in contact with a silicate core (Figure 9.4). These unique conditions on Europa allow water–rock interactions, especially in the presence of volcanism. The determination of the topography and/or of mass anomalies at the silicate core/liquid interface would provide hints as to whether volcanism exists or has existed. However, this topographic signal is extremely weak when viewed from orbit – mainly because the presence of the icy crust above Europa's liquid layer decouples the surface topography from the ocean floor. But some constraints could be obtained from accurate altimetry measurements and a precise determination of the gravity field at different altitudes. Moreover, magnetometer measurements would be useful, since the constraints provided on the induced magnetic field of Europa will help to define the depth of its magnetic source.

Small-scale features won't be detectable on the silicate seafloor, but medium-scale features (200 km) such as large volcanoes or long ridges will probably be detected if a microgradiometer is carried on the Europa orbiter. All of these data (gravimetry, altimetry, magnetometry) can be used in models of the internal structure that will eventually give constraints on the ocean floor topography. Determination of the ocean floor topography and the state of the iron core can only be achieved if the mission design provides a combination of very precise navigation, and altimetric, gravity, and magnetometric measurements of very high accuracy. Despite their difficulty, the results could have strong astrobiological implications.

Surface composition and chemistry. The Galileo mission has proven that the surface of Europa is not made of pure water ice. Whatever the origin of the non-water components, either endogenous or exogenous, their characterization is important. The NIMS (Near-Infrared Mapping Spectrometer) onboard the Galileo mission showed that the composition is different along some geological features. Because of the great controversy regarding the nature of non-ice components, compositional mapping of the surface must be redone, and over larger areas with higher spatial resolution, with a higher-spectral-resolution infrared mapping spectrometer.

In addition to a near-infrared mapping spectrometer, the main neutral species as well as many ion species of Europa's atmosphere and ionosphere can be easily measured with a mass spectrometer and dust analyzer on an orbiter at 200 km. Moreover, collecting dust expelled from the surface of Europa, and analyzing its composition, will constrain the origins of the building blocks of which Europa was formed, and the thermodynamic conditions prevailing during Europa's formation (through the possible abundance or absence of volatile species like CH_4, CO_2, NH_3).

Relating the composition of the surface of Europa to the interior is difficult because materials extruded from the interior are altered by depressurization, abrupt change in temperature, and particularly exposure to the high Jovian radiation environment at the orbit of Europa. These alteration processes can be studied in the laboratory, in particular the effect of radiation and energetic particles, but by their nature such processes make it extremely difficult if not impossible to piece together the original composition of the irradiated material. Finding places on the surface with an orbiter, for example fresh cracks, where ocean material is actively welling up, would be a discovery of tremendous importance to future exploration of Europa.

Global surface morphology and dynamics. The surface of Europa is very young: both crater-counting and estimation of erosion rates by sputtering (constrained by the measurements of H_2O escape rates by Galileo) suggest an age of Europa's surface between ten million and one billion years. This surface presents very intriguing tectonic and volcanic features. The size and geometry of linear features suggest surface motions enabled by soft ice or water encrusted with a thick ice layer close to the surface. Impact features suggest the existence of a low-viscosity layer at about 5–20 km in depth that may be either a liquid layer or a soft ice layer. For some of these intriguing features, Galileo also provided topographic estimates using the shadows of the elevated terrains. Interpretations of these features gave some of the strongest arguments in favor of a deep liquid ocean below the icy crust. One of the major goals of a Europa orbiter will be to study surface structures in more detail than was achieved by Galileo.

Both a high-resolution camera and a laser altimeter are key candidate instruments for the Europa orbiter. The full coverage of the surface from a low orbit will permit the acquisition of high-resolution images of geologic features. Laser altimetry measurements will bring new and valuable information on the topography, of considerable importance for understanding geologic processes and constraining the coupling between the ocean and the icy crust. Finally, such high-resolution analysis of the surface will help to define astrobiologically interesting landing sites for future space missions.

Surface–exosphere–magnetosphere interactions. The atmosphere of Europa is produced mainly from irradiation of its exposed outer surface by magnetospheric plasma (mainly sulfur and oxygen ions from Io) and UV photons. Its main component, molecular oxygen, has been indirectly observed by the Hubble Space Telescope. Oxidized constituents produced by irradiation may also provide key components for life support within the subsurface ocean. The atmospheric composition of Europa is determined by both the water and oxygen photochemistry in the near-surface region, escape of energetic atoms of oxygen and water into the Jovian magnetosphere, and exchange of water products with the porous regolith.

Water ice is evaporated by sputtering processes and sublimation. Generally the water sublimation rate should be extremely low. However, around the dayside equator of Europa the observed increase of the subliming away of water, 10 000 times faster than expected, means that sublimation could be a major process for water vapor production. "Boiling" water may also exist due to thermal and tidal flexing of cracks in the surface, resulting in an increased supply of water vapor to the atmosphere.

Europa's extremely tenuous atmosphere interacts with charged particles accelerated by Jupiter's powerful magnetic field. Energetic ions and electrons produced by this acceleration may be the principal chemical agents in layers close to the surface of moons that could alter species originating from the deep interior that have migrated to the surface. Such chemistry could create gradients in the oxidation state or acidity in the upper layers of the ocean that might serve as energy sources for life.

An orbiter could monitor the charged particles incident on Europa, the structure and composition of the exosphere and ionosphere, and the surface composition to establish the net production, loss, and exchange of key chemical species between the magnetosphere, the exosphere/ionosphere, the surface, and the subsurface.

Characterizing environments and searching for biosignatures. The possibility that Europa hosts a habitable ocean makes the search for life beneath its surface a high priority. Remote detection of biosignatures is very difficult, and *in situ* exploration is clearly preferable. Possible future mission concepts include as an option a surface module that could obtain environmental data (radiation, pH, eH, T, redox, chemistry) of the surface, and possibly of the near-surface properties. Geological evidence suggests that a link between the surface and the aqueous layer could have existed in the past or exists even now, and so the products of biology in the ocean might be present in freshly extruded material or may be trapped, protected, in the ice. More likely though is that the search for life within Europa will require an ambitious mission to probe beneath the ice crust, and would be a follow-on to a Europa orbiter.

Titan: Organic factory and habitat

Another planetary body that exhibits many similarities with the Earth (e.g. Coustenis and Taylor 2008), and can be considered as a possible Class IV habitat, is Saturn's largest satellite Titan. Titan is unique in the Solar System due to its extensive nitrogen atmosphere which is four times denser at the surface than our own atmosphere. It also hosts a rich organic chemistry and consists of a significant amount of CH_4. Titan's atmosphere is not in chemical equilibrium. Within it, a chemical factory initiates the formation of complex positive and negative ions in the high thermosphere as a consequence of magnetospheric–ionospheric–atmospheric interactions. The second most abundant constituent, methane, is dissociated irreversibly to produce hydrocarbons and nitriles.

Recent Cassini–Huygens discoveries reveal that Titan is rich in organics at high atmospheric levels, most probably contains a vast subsurface ocean, and has sufficient energy sources to drive chemical evolution (Coustenis and Taylor 2008).

Titan and the Earth

Although Titan is much colder than the Earth, it presents many similarities with our planet (see Figure 9.5). Made of the same main constituent, dinitrogen, Titan's atmosphere also has an equivalent structure from the troposphere to the ionosphere, and a surface pressure of 1.5 bars – the only such case of an extraterrestrial planetary atmospheric pressure close to that of the Earth.

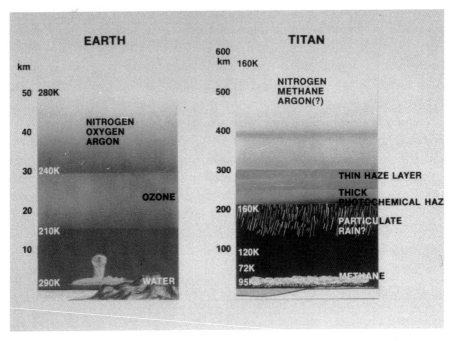

Figure 9.5 Atmospheric structure of Titan and the Earth (from Owen 2005).

Methane on Titan seems to play the role of water on the Earth, with a complex cycle that has yet to be fully understood. Methane can exist as a gas, liquid, and solid, since the mean surface temperature is almost 94 K (Fulchignoni *et al.* 2005), approaching the triple point of methane. Analogies can also be made between the current organic chemistry on Titan and the prebiotic chemistry, which was active on the primitive Earth. In spite of the absence of permanent bodies of liquid-water on Titan's surface, the chemistry is similar.

Moreover, Titan is the only planetary object, other than the Earth, with long-standing bodies of liquid on its surface (Stofan *et al.* 2007). The features range in size from less than 10 km^2 to at least 100 000 km^2. They are limited to the region poleward of 55°N. Currently, the instruments onboard the NASA/ESA Cassini–Huygens mission have identified and mapped almost 655 geological structures referred to as lakes and/or basins (Figure 9.6, Hayes *et al.* 2008).

Several of the organic processes which are occurring today on Titan form some of the organic compounds which are considered as key molecules in terrestrial prebiotic chemistry, such as hydrogen cyanide (HCN), cyanoacetylene (HC$_3$N), and cyanogen (C$_2$N$_2$). In fact, with several percent of methane in dinitrogen, the atmosphere of Titan is one of the most favorable atmospheres for prebiotic synthesis.

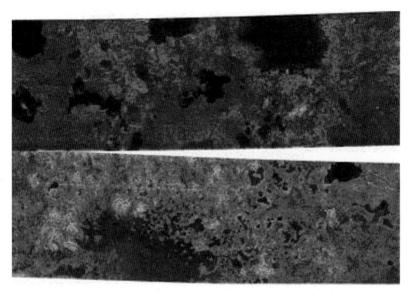

Figure 9.6 Lakes discovered on Titan by the Cassini–Huygens mission (courtesy NASA).

Organic chemistry on Titan

Titan has the largest abiotic organic factory in the Solar System. Indeed, the abundance of methane and its organic products in the atmosphere, seas, and dunes exceeds the carbon inventory in the Earth's ocean, biosphere, and fossil fuel reservoirs by more than an order of magnitude (Lorenz *et al.* 2008a). The Cassini Ion and Neutral Mass Spectrometer (INMS) showed that the process of aerosol formation appears to start more than 1000 km above the surface through a complex interplay of ion and neutral chemistry initiated by energetic photon and particle bombardment of the atmosphere (Waite *et al.* 2007), and includes polymers of high molecular weight – up to and certainly beyond C7 hydrocarbons. Measurements throughout the atmosphere have indicated the presence of numerous hydrocarbon and nitrile gases, as well as a complex layering of organic aerosols (tholins) that persists all the way down to the surface (Tomasko *et al.* 2005, Coustenis *et al.* 2007). Also, radar observations suggest that the ultimate fate of this aerosol "rain" is the generation of expansive organic dunes that produce an equatorial belt around the surface.

The energetic chemistry in the ionosphere produces large molecules like benzene, which begin to condense out at ~950 km, are detected as neutral hydrocarbons and nitriles in the stratosphere, and initiate the process of haze formation. As the haze particles fall through the atmosphere and grow, they become detectable with imaging systems such as the Cassini Imaging Science Subsystem (ISS) at ~500 km and are ubiquitous throughout the stratosphere. They are strong absorbers of solar UV and visible radiation and play a fundamental role in heating Titan's stratosphere and driving wind systems in the middle atmosphere, much as ozone does in the Earth's middle atmosphere. Eventually, these complex organic molecules are deposited on Titan's surface in large quantities, where data from Cassini's instruments hint at their existence. Hence the upper thermosphere is linked intimately with the surface and the intervening atmosphere.

Furthermore, the Gas Chromatograph and Mass Spectrometer (GCMS) on board the Huygens probe which successfully landed on Titan's surface in January 2005 did not detect a large variety of organic compounds in the low atmosphere. The mass spectra collected during the descent show that the medium-altitude and low stratosphere as well as the troposphere are poor in volatile organic species, with the exception of methane (Niemann *et al.* 2005). Condensation of these species on aerosol particles is a probable explanation for these atmospheric characteristics. Such particles, for which no direct data on their chemical composition were previously available, were analyzed by the Huygens Aerosol Collector and Pyrolyzer (ACP) instrument. ACP results show that the aerosol particles are made of refractory organics, which release HCN and NH_3 during pyrolysis. This supports the tholin hypothesis (Nguyen *et al.* 2007, and references therein). From these new *in situ* measurements it seems very likely that the aerosol particles are made of a refractory organic nucleus, covered with condensed volatile compounds (Israël *et al.* 2005). However, the nature and abundances of the condensates have not been measured. Even more importantly for astrobiology, neither the elemental composition nor the molecular structure of the refractory part of the aerosols has been determined.

Thus, analogies are obvious between the organic chemistry activity currently occurring on Titan and the prebiotic chemistry which was once active on the primitive Earth, prior to the emergence of life (e.g. McKay and Smith 2005). Schaefer and Fegley (2007) predict that Earth's early atmosphere contained CH_4, H_2, H_2O, N_2, and NH_3, similar to components used in the Miller–Urey synthesis of organic compounds, often related to the Titan and Enceladus atmospheric inventory. Furthermore, Trainer *et al.* (2006) looked at the processes that formed haze on Titan and on early Earth and found many similarities for what could have served as a primary source of organic material to the surface.

It is then possible that before the rise of atmospheric oxygen in the terrestrial atmosphere 2.5 Gyr ago, the abundance of methane gas was 10–20 times higher than today's value of 1.6×10^{-6} (Pavlov *et al.* 2003). Hence, if the atmospheric CO_2/CH_4 ratio had become equal to 10 at the mid-Archean era, an organic haze could have formed in this early environment (Pavlov *et al.* 2000, DeWitt *et al.* 2009). This hydrocarbon haze produced the anti-greenhouse effect which reduced the temperature of the atmosphere (Kasting and Howard 2006). Titan also hosts a thick methane-induced organic haze, similar to the one predicted for the early

Earth and, consequently, experiences the same anti-greenhouse effect (McKay *et al.* 1999). The absence of vast amounts of CO_2 on Titan is one of the major differences between the two atmospheric envelopes. On the other hand, hydrogen cyanide and other prebiotic molecules are among the starting materials for biosynthesis. The existence of hydrocarbons, and in particular acetylene and benzene, has really enlarged the borders of photochemical organic products.

McKay and Smith (2005) further note the astrobiological importance of the liquid hydrocarbon-filled lakes on Titan (Figure 9.6), since it is possible for a different form of life to exist in such environments. It has indeed been hypothesized that such a methanogenic life form consumes H_2 instead of O_2 and that this hydrogen decrease could be measured in the lower atmosphere. Two recent papers, Strobel (2010) and Clark *et al.* (2010), based on data from the Cassini Orbiter focus on the complex chemical activity on the surface of Titan. Strobel (2010) shows that hydrogen flows down through Titan's atmosphere and then somehow disappears on the surface. One of the most interesting phenomena occurring on Titan is that important quantities of atmospheric hydrogen precipitate and disappear when reaching the surface. This process resembles the oxygen consumption as occurring on the Earth although in Titan's case the element is hydrogen (Strobel 2010).

Even though this is not supportive of the theory of terrestrial-type life on Titan, it represents a hypothetical second form of life independent of the water-based life we know on the Earth. Strobel (2010) describes the densities of hydrogen in different parts of the atmosphere and the surface and finds a disparity in the hydrogen densities that lead to a flow down to the surface at a rate of about 10 000 trillion hydrogen molecules per second. This is about the same rate at which the molecules escape out of the upper atmosphere. Strobel (2010) states that it is not likely that hydrogen is stored in a cave or an underground space on Titan. Titan's surface is also so cold that a chemical process that involved a catalyst would be needed to convert hydrogen molecules and acetylene back to methane, even though overall there would be a net release of energy. The energy barrier could be overcome if there were an unknown mineral acting as the catalyst on Titan's surface.

Another possible indicator for some sort of life existing on Titan is the lack of acetylene on the surface since there is no clear evidence of this compound in the received data to date, while it is expected to have been deposited through the atmosphere. It has been suggested that this could be due to the fact that some form of life on the surface is using acetylene as an energy source (Clark *et al.* 2010). This theory is widely debated and controversial among the scientific community, especially due to suggestions of a non-biological origin of this phenomenon; however, it has the merit of proposing new interesting astrobiological theories.

In detail, Clark *et al.* (2010) mapped hydrocarbons on the surface from Cassini/VIMS data and found a lack of acetylene. McKay and Smith (2005) had identified acetylene as the best energy source for methane-based life on Titan.

While non-biological chemistry offers one possible explanation, these authors believe these chemical signatures bolster the argument for a primitive, exotic form of life or precursor to life on Titan's surface. According to one theory put forth by astrobiologists, the signatures fulfill two important conditions necessary for a hypothesized "methane-based life" which would consume not only methane but also hydrogen. However, another possibility is that sunlight or cosmic rays are transforming the acetylene in icy aerosols in the atmosphere into more complex molecules that would fall to the ground with no acetylene signature.

To date, methane-based life forms are only hypothetical. Scientists have not yet detected this form of life anywhere, though there are liquid-water-based microbes on the Earth that thrive on methane or produce it as a waste product. At Titan's low temperatures, a methane-based organism would have to use a substance that is liquid as its medium for living processes, but not water itself. Water is frozen solid on Titan's surface and much too cold to support life as we know it. The list of liquid candidates includes liquid methane and related molecules like ethane. While liquid-water is widely regarded as necessary for life, there has been extensive speculation published in the scientific literature that this is not a strict requirement. The new hydrogen findings are consistent with conditions that could produce an exotic, methane-based life form, but do not prove its existence.

Cloud systems within the size of terrestrial hurricanes (1000 km large) appear occasionally, while smaller systems exist on a daily basis. Titan's atmospheric methane may be supplemented by high-latitude lakes and seas of methane and ethane, which over time cycle methane back into the atmosphere where it rains out, creating fluvial erosion over a wide range of latitudes. With the current picture of Titan's organic chemistry, the chemical evolution of the main atmospheric constituents (N_2 and CH_4) produces mainly ethane, which accumulates on the surface or the near subsurface, eventually dissolved to form methane–ethane lakes and seas, and complex refractory organics that accumulate on the surface, together with condensed volatile organic compounds such as HCN and benzene.

In spite of the low temperature, Titan is not a congealed Earth: the chemical system is not frozen. Titan is an evolving planetary body and so is its chemistry. Once deposited on Titan's surface, the aerosols and their complex organic content may follow a chemical evolution of astrobiological interest. Laboratory experiments show that, once in contact with liquid water, Titan tholins can release many compounds of biological interest, such as amino acids (Khare *et al.* 1986). These processes could be particularly favorable in zones of Titan's surface where cryovolcanism is occurring. The N_2-CH_4 byproducts in Titan's atmosphere eventually end up as sediments on the surface, where they presently accumulate at a rate of roughly 0.5 km in 4.5 Gyr.

Thus, one can envision the possible presence of such compounds on Titan's surface or near subsurface. Long-term chemical evolution is impossible to study in the laboratory: *in situ* measurement of Titan's surface thus offers a unique opportunity to study some of the many processes which could have been involved in prebiotic chemistry, including isotopic and enantiomeric fractionation (Nguyen *et al.* 2007).

Even with the detection of the large lakes in the north, no viable source was detected by Cassini at or near the surface to resupply the full methane atmospheric abundance. Cryovolcanic outgassing has been hypothesized, yet over what time-scales and through which internal processes are unknown even though several areas are currently believed to have been formed under the influence of cryo-volcanism. Cassini–Huygens also found that the balance of geologic processes – impacts, tectonics, cryovolcanism, fluvial, aeolian – is somewhat similar to the Earth's, more so than for Venus or Mars. Titan may well be the best analog of an active terrestrial planet in the sense of our home planet, albeit with different working materials and conditions.

In particular, the presence of benzene seems extremely interesting, as it is the only polycyclic aromatic hydrocarbon (PAHs) discovered on Titan today. The presence of PAHs in Titan's atmosphere is very important as they may contribute to the synthesis of biological building blocks. Moreover, the combination of the liquid deposits on the surface of Titan and the low temperature could host the proper environment for this biosynthesis. Recent laboratory experiments showed that aromatic compounds could be plausibly produced on icy surfaces (Menor-Salván *et al.* 2008). Benzene was first detected at 674 cm^{-1} based on Infrared Space Observatory data (Coustenis *et al.* 2003) with a mixing ratio of 4×10^{-10}, was then also detected in the thermosphere (950–1150 km) from the analysis of Cassini/INMS data (Waite *et al.* 2007), and firmly in the stratosphere (100–200 km) at all latitudes by Cassini Composite Infrared Spectrometer measurements (Flasar *et al.* 2005, Coustenis *et al.* 2007, 2010a). Moreover, benzene has been tentatively identified on Titan's surface by Huygens Gas Chromatograph Mass Spectrometer measurements (Niemann *et al.* 2005).

Although Titan lacks oxygen and sufficiently high temperatures, as primitive Earth did, different evolutionary pathways from the Earth must have been fol-lowed on Titan and polyphenyls may possibly have been created (Delitsky and McKay 2010). The abundances of hydrocarbons are higher on Titan than those on the Earth by a factor of about 10^2–10^4. Moreover, the temporal variations of the hydrocarbon traces on Titan experience a full cycle during the Titan year (Coustenis *et al.* 2010b) and are probably influenced by local or regional sources and sinks. The degree of complexity that can be reached from organic chemistry in the absence of permanent liquid-water bodies on Titan's surface is still unknown, but it could be quite high.

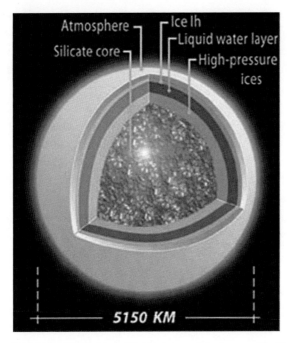

Figure 9.7 Illustration of Titan's internal structure with a liquid ocean between two subsurface-ice layers (Tobie *et al.* 2005).

Interior models for Titan and a possible subsurface ocean

The presence of an internal ocean is supported by internal models, radar and gravity Cassini measurements (Lorenz *et al.* 2008b), and the Huygens Atmospheric Structure Instrument experiment (HASI) experiment. Beghin *et al.* (2009) recently interpreted the extremely low-frequency electric signal recorded by HASI as a Schumann resonance between the ionosphere and a modestly conducting ocean roughly 50 km below the surface.

Thermal evolution models suggest that Titan may have an ice crust between 50 and 150 km thick, lying atop a liquid-water ocean a couple of hundred kilometers deep, with some amount (a few to 30%, most likely ~10%) of ammonia dissolved in it, acting as an antifreeze (Lorenz *et al.* 2008b). Beneath lies a layer of high-pressure ice (Figure 9.7). The presence of ammonia, from which Titan's nitrogen atmosphere was presumably derived, distinguishes Titan's thermal evolution from that of Ganymede and Callisto. Cassini's measurement of a small but significant asynchronicity in Titan's rotation is most straightforwardly interpreted as a result of decoupling the crust from the deeper interior by a liquid layer (Lorenz *et al.* 2008b).

Definite detection of this ocean of water and ammonia under an icy layer can be provided by the Radio Science Subsystem aboard Cassini, by measuring the principal components of Titan's and Enceladus' gravitational potential of Titan and Enceladus (Rappaport *et al.* 1997, 2007). This will provide important constraints on the satellites' internal differentiation.

Although the chemical reactions that lead to life on the Earth take place in liquid-water, the reactions themselves are almost entirely between organics. The study of organic chemistry is therefore an important, and arguably richer, adjunct to the pursuit of liquid-water in the Solar System. Titan's organic inventory is impressive, and carbon-bearing compounds are widespread across the surface in the form of lakes, seas, dunes, and probably sedimentary deltas at the mouths of channels. Thus, all the ingredients which are supposed to be necessary for life to appear and even develop – liquid-water, organic matter, and energy – seem present on Titan.

Once deposited on Titan's surface, the aerosols and their complex organic content may follow a chemical evolution of astrobiological interest. Laboratory experiments show that Titan tholins can release many compounds of biological interest, such as amino acids, once in contact with liquid-water. This seems even possible with water ice. Those processes could be particularly favorable in zones of Titan's surface where cryovolcanism is occurring and hydrothermal processes are present. The Huygens Descent Imager-Spectral Radiometer (DISR) showed that the reflectance spectra of Titan's surface were different from any other surface in the Solar System (Tomasko *et al.* 2005). Some of the characteristics of these spectra can be explained by the presence of tholin-like material on the surface. The VIMS instrument has also identified tholin color properties as compatible with the global surface properties of Titan. One can envision the possible presence of such compounds on Titan's surface or in the subsurface through exchange atmosphere–interior processes that could bring the organics into contact with the liquid-water. At the beginning of Titan's history, this hypothetical ocean may have been in direct contact with the atmosphere and with the internal bedrock, offering interesting analogies with the primitive Earth, and the potential implication of hydrothermal vents in terrestrial prebiotic chemistry.

Consequently, it cannot be excluded that life may have emerged on or in Titan. In spite of the extreme conditions in this environment, life may have been able to persist and even to adapt, knowing that the current conditions (pH, temperature, pressure, salt concentrations) may be compatible with life in extreme conditions as we know it on the Earth. However, the detection of potential biological activity in the current internal Titan water ocean seems very challenging. Titan's overall density requires it to have roughly equal proportions of rock and ice. The extent of its differentiation (ice from rock, rock from metal) constrains temperatures in the early Saturnian nebula: Titan was almost certainly warm enough to allow some differentiation into a rocky core with a water/ice envelope, but whether an iron or iron–sulfur core formed is not known.

However, besides the well-known Earth-like life, other forms of living organisms have been speculated to exist on Titan (as discussed previously).

Conclusions

Europa is currently our major target for finding an internal liquid-water ocean among the giant-planet satellites. More dedicated exploration is required before this can be established and characterized, but the astrobiological potential has been recognized for quite some time now and begs for further investigation.

A significant geophysical difference that is evident when one compares Titan and Europa is that on Titan the liquid-water layer, if real, would not be presently in contact with a silicate core. The surface of Titan appears (like Mars or Europa) as an unlikely location for extant or present-time life, at least for terrestrial-type life. However, it has been noted (Fortes 2000) that Titan's internal water ocean might support terrestrial-type life, which had been introduced there previously or formed when liquid-water was in contact with silicates early in Titan's history. McKay and Smith (2005) have noted that there are photochemically derived sources of free energy on Titan's surface, which could support life that would have to be an exotic type of life using liquid hydrocarbons as solvents. In a similar vein, Stoker *et al.* (1990) observed that terrestrial bacteria could in fact satisfy their energy and carbon needs by "eating" tholin. In this sense, a methane-rich atmosphere may act as a "poor-planet's photosynthesis," providing a means to capture the free energy from ultraviolet light and make it available for metabolic reactions.

It is clear that Titan's organic chemistry and the presence of a subsurface ocean remain to be investigated. In particular, joint measurements of large-scale and mesoscale topography and gravitational field anomalies on Titan from both an orbiter and an aerial platform would impose important constraints on the thickness of the lithosphere, the presence of mass anomalies, at depth, and any lateral variation of the ice mantle thickness. It is astrobiologically essential to confirm the presence of such an internal ocean, although the water layer may not be in contact with the silicate core like Europa. However, the detection of potential biological activity in the putative liquid mantle seems very challenging.

Future exploration

To explore Europa, Ganymede, and the Jovian system (all high-priority targets recognized by the Decadal Survey), EJSM had been proposed and studied by ESA and NASA for an L-class (Flagship) mission study in 2008–2010, following previous proposals (Blanc *et al.* 2009). The EJSM mission would study, among many other science tasks, with two orbiters, the habitability of Europa and Ganymede in the context of the Jupiter system and its links with the formation scenario and key coupling processes (see http://opfm.jpl.nasa.gov/europajupitersystemmissionejsm/).

Following budgetary restrictions in 2010, NASA abandoned the Jupiter Europa Orbiter (JEO) project within EJSM, and ESA reformulated the mission, which became JUICE, one orbiter alone to explore Ganymede and the other two icy Galilean moons Europa and Callisto (http://sci.esa.int/science-e/www/object/index.cfm?fobjectid=42291) and to investigate the emergence of habitable worlds around giant planets. The mission has now been selected after an open competition with two other missions to become the first Large Class mission to be implemented within ESA's Comic Vision plan. For technical and budgetary reasons, having on board the orbiter a surface element (e.g. a penetrator) that could perform some *in situ* characterization of the surface ice is difficult. However, there are projects for a future lander, currently under study by the Russian Space Agency (Roskosmos), focusing on Europa and/or Ganymede as a planetary object and a potential habitat. The JUICE mission will nevertheless provide important information by establishing the existence of Ganymede's subsurface ocean, defining its main surface characteristics, and even attempting to map the topography of its icy seafloor. The proposed JUICE mission will thus perform detailed spacecraft investigations of Jupiter and its system in all its complexity with particular emphasis on Ganymede as a planetary body and potential habitat. The investigations of the neighboring moons, Europa and Callisto, and of the gas giant will complete a comparative picture of the Galilean moons.

The Cassini–Huygens mission is a remarkable success, answering many outstanding questions about the Saturnian system and Titan in particular. As for many successful missions, the key contributions of Cassini may be the questions raised rather than those answered. An important limitation of Cassini, with respect to Titan science, is the insufficient spatial coverage allowed by its orbit around Saturn. While measurements have highlighted the complexity of Titan's atmosphere and magnetic environment, the coverage has been insufficient to achieve a full understanding. The minimum possible flyby altitude of 950 km and the uneven latitudinal coverage have limited our ability to explore the full set of atmospheric chemical processes. Opportunities for occultation have been rare, thus gaps remain in the magnetospheric downstream region. The single vertical profile of the atmosphere taken by Huygens limits our understanding of horizontal transport and latitudinal variations.

The surface of Titan, as revealed by both the Huygens probe and Cassini orbiter, offers us an opportunity to stretch our current models in an effort to explain the presence of dunes, rivers, lakes, cryovolcanos, ridges, and mountains in a world where the rocks are composed of water ice rather than silicates and the liquid is methane or ethane rather than liquid-water, but the limited high-resolution spatial coverage limits our view of the range of detailed geological processes ongoing on this body. The exciting results from the Huygens post-landing measurements are limited to a fixed site, short time-scales, and do not allow for direct subsurface access and sampling.

The two major themes in Titan exploration – the methane cycle as an analog to the terrestrial hydrological cycle and the complex chemical transformations of organic molecules in the atmosphere – make Titan a very high priority if we are to understand how volatile-rich worlds evolve and how organic chemistry and planetary evolution interact on large spatial and temporal scales. Both are of keen interest to planetology and astrobiology.

The intriguing discoveries of geological activity, excess warmth, and outgassing on Enceladus (due perhaps to the ejection of water and organics from subsurface pockets bathed in heat, or by some other mechanism), mandate a follow-up investigation to that tiny Saturnian world that can only be achieved with high-resolution remote observations, and detailed *in situ* investigations of the near-surface south polar environment.

These questions will remain unanswered until a new mission to Titan and the Saturnian system is launched. The concept of such a future mission has been studied extensively by ESA and NASA. The so-called Titan Saturn System Mission (TSSM) (http://www.lesia.obspm.fr/cosmicvision/tssm/tssm-public/) would have focused on enhancing our understanding of Titan's and Enceladus' atmospheres, surfaces, and interiors, determining the pre- and proto-biotic chemistry that may be occurring on both objects, and deriving constraints on the satellites' origin and evolution, both individually and in the context of the complex Saturnian system as a whole (Coustenis *et al.* 2009).

Many internal processes play crucial roles in the evolution of Titan and Enceladus. The formation and replenishing of Titan's atmosphere and the jet activity at Enceladus' South Pole are intimately linked to the satellites' interior structures and dynamics. Open issues are listed below:

- to determine their present-day structure and levels of activity;
- to determine whether the satellites underwent significant tidal deformation, and whether they possess intrinsic or induced magnetic fields and significant seismicity;
- to identify heat sources, internal reservoirs of volatiles (in particular methane and ammonia), and eruptive processes.

Beyond the icy satellites of Solar System gas giants, Class III/IV habitats may also be discovered by measuring the mean densities of discovered super-Earth-type exoplanets. Selsis *et al.* (2007) studied this possibility for present-day transit missions in space like CoRoT (CNES) and Kepler (NASA) in combination with ground-based Doppler velocimetry measurements HARPS (ESO) and possible future instruments. These authors studied the sources of uncertainty on the planetary density related to those on the mass determination by radial velocity measurements, the determination of the stellar radius, and the photometric measurement during the transits. They found that with the presently available instruments, the accuracy of radial velocity measurements is the main uncertainty and limiting factor for expected detections by CoRoT and Kepler.

As a consequence the determination of the nature of such planets seems only possible if they are in an adequate domain related to the host star magnitude (Selsis *et al.* 2007). Their study showed that if each star in the CoRoT field had a 6–10 Earth-mass planet at ~0.10 AU, the number of detections with CoRoT would be several hundred. On the other hand, the absence of detection would indicate that such planets are not present at the level of ~1%.

However, a definite answer to these open questions is related to the photometric precision of CoRoT and Kepler. New generations of radial velocity instruments are necessary, which can make accurate measurements of faint stars. In that case, the identification of ocean planets could be done on a significantly larger stellar sample.

References

Bains, W. (2004). "Many chemistries could be used to build living systems," *Astrobiology*, Vol. 4, pp. 137–167.

Beghin, C., Canu, P., Karkoschka, E. *et al.* (2009). "New insights on Titan's plasma-driven Schumann resonance inferred from Huygens and Cassini Data," *Planetary and Space Science*, Vol. 57, pp. 1872–1888.

Blanc, M. *et al.* (2009). "LAPLACE: a mission to Europa and the Jupiter system for ESA's cosmic vision programme," *Experimental Astronomy*, Vol. 23, pp. 849–892.

Brown, R. H., Soderblom, L. A., Soderblom, J. M. *et al.* (2008). "The identification of liquid ethane in Titan's Ontario Lacus," *Nature*, Vol. 454, pp. 607–610.

Bulat, S., Alekhina, I., and Petit, J. R. (2009). "Life detection strategy for subglacial Lake Vostok, Antarctica: lessons for Jovian moon Europa." *Geochimica et Cosmochimica Acta Supplement*, Vol. 73, p. A173.

Carlson, R. W., Anderson, M. S., Johnson, R. E. *et al.* (1999). "Hydrogen peroxide on the surface of Europa," *Science*, Vol. 283, pp. 2062–2064.

Cernicharo, J. and Crovisier, J. (2005). "Water in space: the water world of ISO," *Space Science Reviews*, Vol. 119, pp. 29–69.

Clark, K., Stankov, A., Pappalardo, R. T. *et al.* (2009). "Europa Jupiter System Mission, joint summary report," NASA/ESA. JPL D-48440 and ESA-SRE (2008) 1.

Clark, R. N., Curchin, J. M., Barnes, J. W. *et al.* (2010). "Detection and mapping of hydrocarbon deposits on Titan," *Journal of Geophysical Research*, Vol. 115, 10005.

Coustenis, A. and Taylor, F. W. (2008). *Titan: Exploring an Earthlike World*, Series on Atmospheric, Oceanic and Planetary Physics, Volume 4. Singapore: World Scientific.

Coustenis, A., Salama, A., Schulz, B. *et al.* (2003). "Titan's atmosphere from ISO mid-infrared spectroscopy," *Icarus*, Vol. 161, pp. 383–403.

Coustenis, A., Achterberg, R. K., Conrath, B. J. *et al.* (2007). "The composition of Titan's stratosphere from Cassini/CIRS mid-infrared spectra," *Icarus*, Vol. 189, pp. 35–62.

Coustenis, A. *et al.* (2009). "TandEM: Titan and Enceladus mission," *Experimental Astronomy*, Vol. 23, pp. 893–946.

Coustenis, A., Bampasidis, G., Vinatier, S. *et al.* (2010a). "Titan trace gaseous composition from CIRS at the end of the Cassini-Huygens prime mission," *Icarus*, Vol. 207, pp. 461–476.

Coustenis, A., Bampasidis, G., Nixon, C. *et al.* (2010b). "Titan's atmospheric chemistry and its variations," in *Titan Through Time: A Workshop On Titan's Past, Present and Future*, p. 68.

Delitsky, M. L. and McKay, C. P. (2010). "The photochemical products of benzene in Titan's upper atmosphere," *Icarus*, Vol. 207, pp. 477–484.

Des Marais, D. J., Harwit, M. O., Jucks, K. W. *et al.* (2002). "Remote sensing of planetary properties and biosignatures on extrasolar terrestrial planets," *Astrobiology*, Vol. 2, pp. 153–181.

DeWitt, H. L., Trainer, M. G., Pavlov, A. A. *et al.* (2009). "Reduction in haze formation rate on prebiotic Earth in the presence of hydrogen," *Astrobiology*, Vol. 9, pp. 447–453.

Flasar, F. M. (1983). "Oceans on Titan?," *Science*, Vol. 221, pp. 55–57.

Flasar, F. M. *et al.* (2005). "Titan's atmospheric temperatures, winds, and composition," *Science*, Vol. 308, pp. 975–978.

Fortes, A. D. (2000). "Exobiological implications of a possible ammonia–water ocean inside Titan," *Icarus*, Vol. 146, pp. 444–452.

Fortes, A. D., Grindrod, P. M., Trickett, S. K., and Vočadlo, L. (2007). "Ammonium sulfate on Titan: possible origin and role in cryovolcanism," *Icarus*, Vol. 188, pp. 139–153.

Fulchignoni, M. *et al.* (2005). "In situ measurements of the physical characteristics of Titan's environment," *Nature*, Vol. 438, pp. 785–791.

Hand, K. P., Chyba, C. F., Priscu, J. C. *et al.* (2009). "Astrobiology and the potential for life on Europa," in *Europa*, R. T. Pappalardo, W. B. McKinnon, and K. K. Khurana (eds.), with the assistance of René Dotson with 85 collaborating authors. Tucson: University of Arizona Press, p. 589.

Hayes, A., Aharonson, O., Callahan, P. *et al.* (2008). "Hydrocarbon lakes on Titan: distribution and interaction with a porous regolith," *Geophysical Research Letters*, Vol. 35, pp. 09204–09204.

Israel, G., Szopa, C., Raulin, F. *et al.* (2005). "Complex organic matter in Titan's atmospheric aerosols from in situ pyrolysis and analysis," *Nature*, Vol. 438, pp. 796–799.

Kapitsa, A. P., Ridley, J. K., de Q. Robin, G. *et al.* (1996). "A large deep freshwater lake beneath the ice of Central East Antarctica," *Nature*, Vol. 381, pp. 684–686.

Kasting, J. F. and Howard, M. T. (2006). "Atmospheric composition and climate on the early Earth," *Philosophical Transactions of the Royal Society, Series B: Biological Sciences*, Vol. 361, pp. 1733–1742.

Khare, B. N., Sagan, C., Ogino, H. *et al.* (1986). "Amino acids derived from Titan tholins," *Icarus*, Vol. 68, pp. 176–184.

Kivelson, M. G., Kurana, K. K., Stevenson, D. J. *et al.* (1999). "Europa and Callisto: induced or intrinsic fields in a periodically varying plasma environment," *Journal of Geophysical Research*, Vol. 104, pp. 4609–4625.

Lammer, H., Bredehoft, J., Coustenis, A. *et al.* (2009). "What makes a planet habitable?," *Astronomy and Astrophysics Review*, Vol. 17, pp. 181–249.

Leger, A., Selsis, F., Sotin, C. *et al.* (2004). "A new family of planets? Ocean-planets," *Icarus*, Vol. 169, pp. 499–504.

Lorenz, R. D., Mitchell, K. L., Kirk, R. L. *et al.* (2008a). "Titan's inventory of organic surface materials," *Geophysical Research Letters*, Vol. 35, p. 02206.

Lorenz, R. D., Stiles, B. W., Kirk, R. L. *et al.* (2008b). "Titan's rotation reveals an internal ocean and changing zonal winds," *Science*, Vol. 319, pp. 1649–1651.

McKay, C. P. and Smith, H. D. (2005). "Possibilities for methanogenic life in liquid methane on the surface of Titan," *Icarus*, Vol. 178, pp. 274–276.

McKay, C. P., Lorenz, R. D., and Lunine, J. I. (1999). "Analytic solutions for the antigreenhouse effect: Titan and the early Earth," *Icarus*, Vol. 137, pp. 56–61.

Menor-Salván, C., Ruiz-Bermejo, M., Osuna-Esteban, S. *et al.* (2008). "Synthesis of polycyclic aromatic hydrocarbons and acetylene polymers in ice: a prebiotic scenario," *Chemistry and Biodiversity*, Vol. 5, pp. 2729–2739.

Nguyen, M. J., Raulin, F., Coll, P. *et al.* (2007). "Carbon isotopic enrichment in Titan's tholins? Implications for Titan's aerosols," *Planetary and Space Science*, Vol. 55, pp. 2010–2014.

Niemann, H. B., Atreya, S. K., Bauer, S. J. *et al.* (2005). "The abundances of constituents of Titan's atmosphere from the GCMS instrument on the Huygens probe," *Nature*, Vol. 438, pp. 779–784.

Owen, T. (2005). "Planetary science: Huygens rediscovers Titan," *Nature*, Vol. 438, pp. 756–757.

Pavlov, A. A., Kasting, J. F., Brown, L. L. *et al.* (2000). "Greenhouse warming by CH_4 in the atmosphere of early Earth," *Journal of Geophysical Research*, Vol. 105, pp. 11981–11990.

Pavlov, A. A., Hurtgen, M. T., Kasting, J. F., and Arthur, M. A. (2003). "Methane-rich Proterozoic atmosphere?," *Geology*, Vol. 31, pp. 87–90.

Petrenko, V. F. and Whitworth, R. W. (1999). *Physics of Ice*. Oxford: Oxford University Press.

Postberg, F., Schmidt, J. Hillier, J. *et al.* (2011). "A salt-water reservoir as the source of a compositionally stratified plume on Enceladus," *Nature*, Vol. 474, pp. 620–622.

Rappaport, N., Bertotti, B., Giampieri, G., and Anderson, J. D. (1997). "Doppler measurements of the quadrupole moments of Titan," *Icarus*, Vol. 126, pp. 313–323.

Rappaport, N. J., Iess, L., Tortora, P. *et al.* (2007). "Mass and interior of Enceladus from Cassini data analysis," *Icarus*, Vol. 190, pp. 175–178.

Reh, K., Lunine, J., Matson, D. *et al.* (2008). "TSSM final report on the NASA contribution to a joint mission with ESA," JPL D-48148. NASA Task Order NMO710851.

Schaefer, L. and Fegley, B. (2007). "Outgassing of ordinary chondritic material and some of its implications for the chemistry of asteroids, planets, and satellites," *Icarus*, Vol. 186, pp. 462–483.

Segura, A. and Kaltenegger, L. (2010). "Search for habitable planets," in *Astrobiology: Emergence, Search and Detection of Life*, V. Basiuk and C. A. Valencia (eds.). American Scientific Publishers.

Selsis, F., Chazelas, B., Bordé, P. *et al.* (2007). "Could we identify hot ocean-planets with CoRoT, Kepler and Doppler velocimetry?," *Icarus*, Vol. 191, pp. 453–468.

Stevenson, D. (2000). "Planetary interiors," in *Encyclopedia of Astronomy and Astrophysics*, P. Murdin (ed.). Bristol: Institute of Physics Publishing, pp. 1823–1823.

Stofan, E. R., Elachi, C., Lunine, J. I. *et al.* (2007). "The lakes of Titan," *Nature*, Vol. 445, pp. 61–64.

Stoker, C. R., Boston, P. J., Mancinelli, R. L. *et al.* (1990). "Microbial metabolism of tholin," *Icarus*, Vol. 85, pp. 241–256.

Strobel, D. F. (2010). "Molecular hydrogen in Titan's atmosphere: implications of the measured tropospheric and thermospheric mole fractions," *Icarus*, Vol. 208, pp. 878–886.

Tobie, G., Grasset, O., Lunine, J. I. *et al.* (2005). "Titan's internal structure inferred from a coupled thermal-orbital model," *Icarus*, Vol. 175, pp. 496–502.

Tomasko, M. G. *et al.* (2005). "Rain, winds and haze during the Huygens probe's descent to Titan's surface," *Nature*, Vol. 438, pp. 765–778.

Trainer, M. G., Pavlov, A. A., DeWitt, H. L. *et al.* (2006). "Organic haze on Titan and the early Earth," *Proceedings of the National Academy of Sciences*, Vol. 103, pp. 18035–18042.

Trinks, H., Schröder, W., and Biebricher, C. (2005). "Ice and the origin of life," *Origins of Life and Evolution of Biospheres*, Vol. 35, pp. 429–445.

Wahr, J. M., Zuber, M. T., Smith, D. E., and Lunine, J. I. (2006). "Tides on Europa, and the thickness of Europa's icy shell," *Journal of Geophysical Research*, Vol. 111, E12005.

Waite, J. H., Young, D. T., Cravens, T. E. *et al.* (2007). "The process of tholin formation in Titan's upper atmosphere," *Science*, Vol. 316, pp. 870–875.

10

Small Habitable Worlds

JULIE CASTILLO-ROGEZ AND JONATHAN LUNINE

Introduction

The astrobiological relevance of small bodies has been acknowledged for several decades with regard to their role in delivering volatiles to Earth and the inner Solar system (see Lunine 2006 for a review). However, until recently these objects were considered too small to sustain a deep liquid layer and hydrothermal activity over the long term. The last decade has been marked by a dramatic evolution of our understanding of small bodies, from observational constraints and theoretical arguments. The discoveries of geological activity on Saturn's satellite Enceladus and Pluto's satellite Charon have prompted theoreticians to develop new approaches for modeling the interiors of these objects, some of which are larger and/or warmer than Jupiter's satellite Europa, considered an archetype of a potentially habitable icy world. The purpose of this chapter is to evaluate the habitability potential of certain small bodies, i.e. their potential for sheltering life, whether life could develop in these environments, or was brought in from a different source.

This chapter focuses on large wet asteroids, small icy satellites, and trans-Neptunian objects (e.g. see the representatives of each class in Figure 10.1). We evaluate the occurrence in each class of objects of certain parameters that determine their capacity to sustain habitable conditions: the energy necessary to support chemical activity and chemical conditions amenable to the thriving of living organisms. The latter aspect is difficult to fully fathom as life has been found in the most surprising environments and based on unexpected nutrient systems. The question of the origin of life in favorable environments is considered in Chapter 2.

Frontiers of Astrobiology, ed. Chris Impey, Jonathan Lunine and José Funes.
Published by Cambridge University Press. © Cambridge University Press 2012.

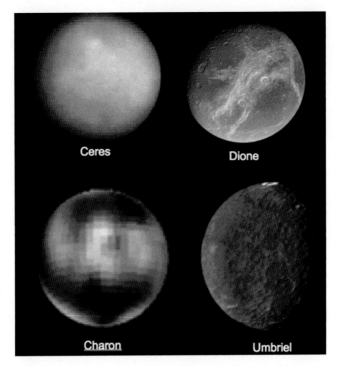

Figure 10.1 Representatives of the objects considered in this chapter. Ceres and Charon are representatives of the planetesimal belts located in the inner and outer Solar System, respectively. Dione and Umbriel are two satellites from the Saturnian and Uranian systems, respectively. While all four objects are about the same size and mass, their surface temperature ranges from ~200 K (Ceres) to ~50 K (Charon), with important implications as to the respective habitability potential of these objects.

Decades of ground-based observations and visits by spacecraft have demonstrated the astrobiological significance of large icy satellites, e.g. Europa and Titan, which are addressed in this book by Coustenis and Blanc in Chapter 9. However, observational constraints on geophysical and geochemical properties of small bodies are limited to ground-based and space telescope observations. Pluto, discovered in 1930, turned out to be the first of a class of bodies that orbit near and beyond the orbit of Neptune – the trans-Neptunian objects (TNOs). While the number of known TNOs has been increasing ever since, only a few of these objects are known well enough to permit the discussion of their geophysical evolution and habitability potential.

While asteroids are closer to us, their chemistry and physical structures are poorly constrained because these objects are covered by a layer of externally processed material (caused by exposure to ultraviolet light, to particle radiation, and to micrometeoroid bombardment). As a result, even the state of water (free vs. water bound in hydrated silicates) in wet asteroids is still a matter of debate. We

Table 10.1 *Properties of the main representatives of the classes of objects considered in this chapter. The maximum temperature quoted for asteroids is subsolar (Lim et al. 2005). 1 AU = 1.5 × 10⁶ km.*

Class	Name	Mean radius (km)	Density (g/cm³)	Distance to the Sun (AU)	Mean surface temperature (K)	Max. surface temperature (K)	Surface composition
Ast.	1 Ceres	476.2	2.078	2.77	~167	235	C-class, magnesite and brucite
Ast.	2 Pallas	544	2.4–2.8	2.77	~164	265	B-class
Ast.	10 Hygeia	431	2.0–2.2	3.14	~164	247	C-class
Ast.	65 Cybele	273	2.3–2.8	3.43	~150	?	C-class, water ice, organics
Sat.	Enceladus	252.1	1.607	9.5	75	145	Water ice
Sat.	Dione	561.7	1.476	9.5	87	?	Water ice, CO_2
Sat.	Miranda	236	~1.2	19	~59	~86	Water, CO_2
Sat.	Ariel	579	~1.66	19	~60	~70	Water, CO_2
TNO	Triton	1353	2.1	30	38	?	Water, N_2, CH_4
TNO	Orcus	~500	1.2–1.8	39.2	~44	?	Water, CH_4, tholins
TNO	Pluto	1153	~2.03	39.4	~44	~55	Water, N_2, CH_4
TNO	Charon	~603	~1.65	39.4	~53	?	Water, ammonia
TNO	Makemake	~750	?	45.3	30–35	?	Methane, ethane, tholins, N_2?
TNO	Eris	~1300	2.3–2.5	67.67	~43	~55	CH_4
TNO	Sedna	~800	2 ?	~510	12	?	Water, N_2, CH_4, tholins

shall know more in the next decade about the characteristics of these classes of objects, as several missions are currently en route to some of their largest representatives. Their study will also contribute to better understanding the distribution and nature of volatiles in the Solar System and thus test the dynamical models leading to the current Solar System architecture.

In this study, we review the observational record about the physical and chemical properties of these classes of objects. We also compare and contrast the properties of some representatives of these families against those of large icy satellites viewed as astrobiological targets. Details about the technical aspects of the modeling of specific objects can be found in Hussmann *et al.* (2006) and Desch *et al.* (2009) for TNOs, Castillo-Rogez and McCord (2010) for the large asteroid Ceres, Matson *et al.* (2009) for icy satellites, and Spencer *et al.* (2009) for Enceladus, as well as other resources listed in the appendix to this chapter.

We wrap up this chapter with an overview of the exploration landscape for the decade ahead, with a brief description of prospective large projects and space

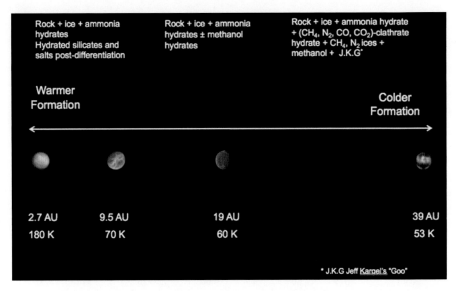

Figure 10.2 Relative position and surface temperature of the objects presented in Figure 10.1 with respect to the Sun. The temperature is one of the key parameters determining the composition of volatiles accreted in these objects (updated from Kargel and Lunine 1998).

missions dedicated to the study of small bodies. We will address how these projects bear the capability to test some of the scenarios discussed in this chapter.

Ingredients for habitability

We describe the ingredients that determine the habitability potential of icy objects and assess their occurrence in various families of small bodies. We note, like Frank and Mojzsis (2010), the "broad spectrum of environmental tolerance" of living organisms that spans a temperature range from −15 to 120 °C, pH from 0 to 13.5, and salinities greater than 35%.

Liquid water

The primary ingredient common to all habitable objects is the presence of water, and more specifically the existence, at least at some point in the history of these objects, of liquid water. Second-phase volatiles are also believed to play an important role in the geophysical evolution of icy objects, with ammonia hydrates and methanol as the most probable (e.g. Kargel and Lunine 1998; Figure 10.2). The potential role of clathrate hydrates[1] in trapping and storing volatiles in planetesimals condensed in the outer Solar System has been noted

---

1 Clathrates are made up of a cage of water molecules surrounding a gas molecule, for example simple organic molecules like ethane and methane (see Choukroun et al. 2011 for a review.)

since the 1970s, but the existence of these compounds in outer solar system plan-etesimals remains to be confirmed.

In the asteroid belt, water is present primarily in the form of water of hydra-tion, although recent discoveries of free water ice at the surfaces of 24 Themis (Rivkin and Emery 2010) and 65 Cybele (Licandro *et al.* 2011), and its putative role in driving activity on main-belt comets, strongly suggest that ice is buried below a thick layer of rocky debris (regolith) in many other asteroids. The most recent developments of cosmochemical and astrophysical models suggest that water condensed with ammonia became chemically trapped in hydrated silicates up till the edge of the outer main belt (Dodson-Robinson *et al.* 2009). The potential role of ammonia hydrates as antifreeze compounds, by depressing the melting temperature of an ocean down to 176 K, had been pointed out in the 1980s and served as a strong argument for the preservation of a deep ocean in the large satellite Titan. The most recent developments of the so-called Nice model (Walsh *et al.* 2011), which is the current leading scenario for the dynamical evolution of the Solar System toward its current architecture, suggest that most/all volatiles in the inner Solar System, and necessarily a large fraction of the organics, were sup-plied from the outer Solar System during an early stage of penetration of Jupiter deep into the inner Solar System.

A warm interior

The second key ingredient for the preservation of conditions amenable to living organisms is the availability of energy. It is necessary first to promote differentiation of a rocky core from the ice phase, and then to generate a deep liquid layer and sustained endogenic activity. The long-term potential preservation of warm conditions in ice is thus a function of the initial heat budget, long-term heat sources, as well as the ability of these objects to retain heat.

Small bodies did not benefit much from accretional heating, i.e. coming from the conversion of kinetic energy from accreting planetesimals (Matson *et al.* 2009). The most significant original heat source in these objects comes from the decay of short-lived radioisotopes (Figure 10.3), provided that they formed within 5 Myr after the production of calcium–aluminum inclusions (CAIs). These inclu-sions are taken by many workers as a reference point in time for the appearance of the first Solar System solids. This in turn provides a time-scale for comput-ing the concentration of the main short-lived radioisotopes of aluminum and iron, ^{26}Al and ^{60}Fe (Castillo-Rogez *et al.* 2009). Here the superscript gives the mass number (number of protons and neutrons) in the isotope of the named element.

Short-lived radioisotopes, i.e. whose decay time-scale is a few Myr or less, are believed to have played a major role in the early evolution of meteorite parent bodies, driving hydrothermal activity in those enriched in water (e.g. Grimm and

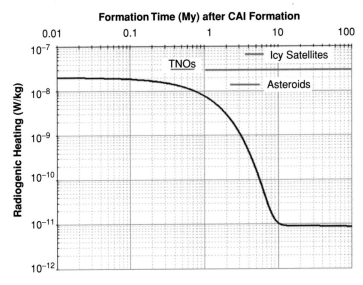

Figure 10.3 Heat available from radioisotope decay, integrated over 4.5 Gyr, as a function of the time of formation with respect to the formation of calcium–aluminum inclusions. Cosmochemical studies have shown that when these inclusions were formed, their fraction of live ^{26}Al with respect to ^{27}Al (the most abundant isotope) was about 5.5×10^{-5} (Wasserburg and Papanastassiou 1982). Details about the calculation of that figure can be found in Castillo-Rogez et al. (2009). Formation time-frames suggested for the objects considered in this study are also indicated (after Cuzzi and Weiden Schilling (2006) for asteroids, Kenyon et al. (2008) for TNOs, and Castillo-Rogez et al. (2009) for Saturn's satellite Iapetus).

McSween 1989, Castillo-Rogez and McCord 2010). Their impact on the early history of TNOs is more limited because the time-scale of accretion of the large TNOs of interest to this study is longer than the period during which ^{26}Al was abundant (Figure 10.3). Whether short-lived radioisotopes could have played a role in the early geophysical history of icy satellites is a current matter of debate, directly tied to the time-scale of satellite system formation. While their occurrence may help explain some features observed in the Saturnian satellite system (Castillo-Rogez et al. 2007a, b), further observational evidence is required in order to confirm that the giant planet systems formed and got organized in a few Myr. The heat pulse associated with the decay of ^{26}Al would drastically affect the internal structure and chemistry, with long-term consequences (Castillo-Rogez et al. 2007b).

Long-term heating comes from three main heat sources: light from the Sun, decay of long-lived radioisotopes, which are common to all objects, and tidal heating, specific to icy satellites and binary asteroids and TNOs. Long-lived radioisotope decay is associated with a heating peak at about one billion years ("1 Gyr") after accretion, due to ^{40}K decay, which may promote ice melting in bodies that did

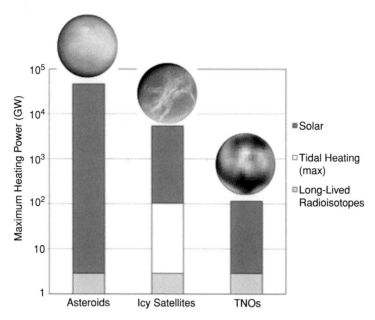

Figure 10.4 Maximum heat power per square meter available to the various categories of objects considered in this study from three sources: solar, tidal, and long-lived radioisotope decay. Only icy satellites involved in orbital resonances, whose eccentricities are continuously excited, are likely to benefit from tidal heating as a long-term heat source. The reference size and density used for the calculation of radioactive decay heat are those of Ceres/Dione/Charon.

not benefit from ^{26}Al decay heat. Tidal heating is a function of the amount of stress exerted by tides on a body (the "tidal stress"), determined by the orbital properties (eccentricity and semi-major axis), as well as the internal composition and structure of the body.

A comparison of the heat budget available to each family of object as a function of their distance to the Sun and the existence of a primary source of tidal stress is featured in Figure 10.4. One should note that solar energy is by far the most significant heat source and acts in keeping the surface temperature of asteroids at a point where liquid water may exist, provided that it is saturated in low-eutectic impurities.

Heat retention is a function of the size of the objects (the bigger, the warmer), and the mechanism driving heat transfer from the interior to the surface. Heat may be transferred by conduction or by solid-state convection of material from the interior to the surface (Figure 10.5). Heat transferred by conduction is a strong function of the thermal conductivity of the material. Volatile and soluble impurities act in decreasing the thermal conductivity, by up to a factor of 4 in the case of

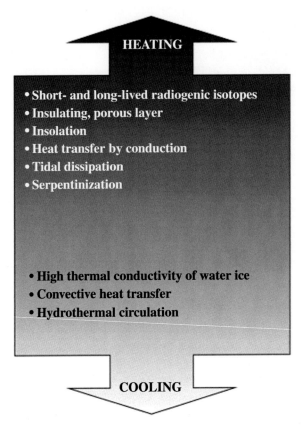

Figure 10.5 Summary of the main parameters and processes promoting warming or driving the cooling of water-rich bodies.

ammonia hydrates and one order of magnitude in the case of hydrated salts (Ross and Kargel 1998).

Convective onset is determined by the buoyancy of warm material at depth and its ability to deform, a function of its viscosity. If it can be initiatied, then convection is a very efficient heat transfer mode accompanied by a heat flow at least one order of magnitude greater than that possible by conduction. While McCord and Sotin (2005) demonstrated that convection is expected in large, wet asteroids, Ruiz (2005) on the other hand suggested that convective onset may be impeded in outer Solar System bodies, owing to their very low surface temperatures. This may contribute to keeping interiors warm over time-scales longer than 1 Gyr.

Promoting internal activity

Following accretion, the chemistry of these objects can evolve as a consequence of several processes, sketched in Figure 10.5. The most significant is hydrothermal activity whose main chemical reaction is the serpentinization of silicate material (reactions of water with rock under heat and pressure), accompanied by the release of major elements in the form of soluble salts, and the racemization of amino acids (Cohen and Coker 2000). Serpentinization also results in the release of hydrogen that influences the redox conditions of the environment, depending on the pressure achieved in the oceans of these objects, thus as a function of their size (e.g. Sleep *et al.* 2004). Hydrogen is expected to escape from objects smaller than ~200 km in radius, while it may remain under pressure in larger bodies, hence promoting reducing conditions. Aqueous alteration is believed to have been widespread in the parent bodies of carbonaceous chondrites (Keil 2000, Brearley 2006). Its occurrence in objects beyond the orbit of Neptune, the trans-Neptunian objects (TNOs), is a matter of debate because the largest of these objects probably formed by slow accretion over hundreds of Myr due to the low surface density of material in that region of the Solar System (Kenyon *et al.* 2008). According to Kenyon *et al.* (2008), only the smallest of these objects (100 km radius) formed in a few Myr. As a matter of fact, Saturn's irregular satellite Phoebe, believed to be a captured TNO (Johnson and Lunine 2005), shows hydrated silicates (Clark *et al.* 2008).

The second major type of internal activity is associated with the evolution of the icy shell. Ice freezing is accompanied by dimensional changes (e.g. Nimmo 2004) resulting in the development of stress that can drive tectonic activity. This may also lead to the release of liquid water from the ocean pressurized by the growing shell (Manga and Wang 2007). Convection is associated with the thinning of the outer icy crust (that overlays a soft icy mantle), which may result in the relaxation of craters and rejuvenating of the surface (resurfacing). Convection and tectonic stresses may also be associated with cryovolcanic activity, as observed at Enceladus' South Polar Terrains (SPT). Evidence for resurfacing and cryovolcanism is a strong indication of a warm interior and the possible presence of a liquid layer (Figure 10.7).

Habitable small bodies

We focus on representatives from the various families of objects whose main characteristics were described in the previous section.

Icy satellites

The main heat sources in icy satellites come from long-lived radioisotope decay and, for some of them, from tidal heating. The heat budget for each source is sketched in Figure 10.7 in the form of the abundance of rock (mass fraction) and the potential for tidal heating, expressed as a function of the ratio k^2/Q. The tidal Love number is small, of the order of 10^{-3} for frozen satellites. However, the presence of a deep liquid layer results in increased deformation of the outer shell decoupled

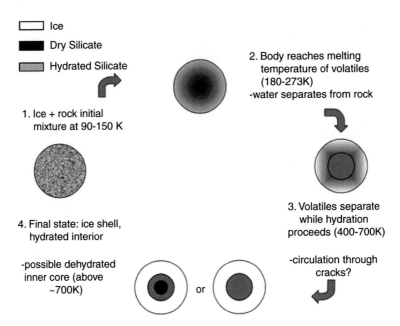

Figure 10.6 Sketch of the main stages in the geophysical evolution of small bodies (adapted from Schmidt and Castillo-Rogez 2011). The sequence is typical of that expected for wet asteroids. Internal melting is expected in colder objects if they accreted a significant amount of [26]Al and/or are subject to tidal heating.

from the rocky core, and thus in increased dissipation. Hence, satellites that could undergo melting of their volatiles are likely to be more dissipative than objects that remained frozen. However, only a few satellites, such as Enceladus, could benefit from outstanding resonance orbital conditions that allow them to remain stressed over long periods of time. In the absence of resonance, tidal dissipation leads to orbital circularization (eccentricity dampens) and thus the fading of that heat source, followed by rapid freezing of the interior.

Endogenic activity may be expressed at the surface by thinning of the rigid outer lid, or lithosphere (convection), crater relaxation, or even resurfacing if lavas (liquid water, or cryolava, in the case of icy bodies) could be extruded to the surface as a result of volcanic or tectonic activity. This is the reason why heavily cratered objects are expected to have been subject to little internal activity. The link between surface geology and available heat sources is illustrated in Figure 10.7. This figure shows that the objects that present the youngest surface (sign of warm interior and endogenic activity) systematically combine a high tidal heating potential and high silicate mass fraction. This may indicate that a minimum number of long-lived radioisotopes is necessary in order to bring the ice temperature to a threshold above which the ice becomes dissipative enough with positive feedback. As a notable example, Mimas is subject to the largest tidal stress among all regular

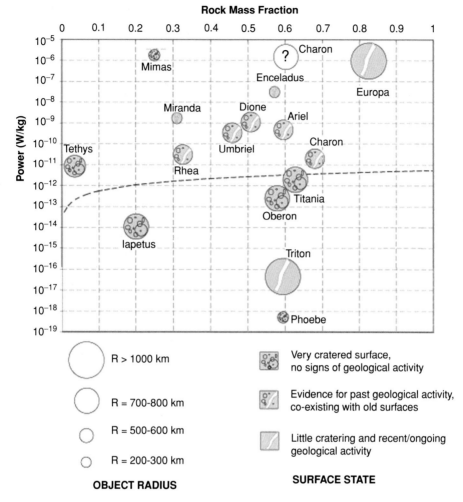

Figure 10.7 Energy budget available to icy satellites expressed as a function of their rock mass fraction (i.e. content in radioisotopes) and tidal heating (power produced per kg, multiplied by k_2/Q). The global geological state of the various satellites is sketched as a marker of endogenic activity and the possible presence of a deep liquid layer at some point in the evolution of these objects.

satellites, while it displays a heavily cratered surface and no signs of internal activity whatsoever. This suggests that its content in silicates was too small to warm up the interior to a point where it became dissipative.

Most small icy satellites can be discarded as possible hosts for a deep ocean (Titania, Oberon) at present. Hussmann *et al.* (2006) suggested that all medium-sized satellites could be stratified in a rocky core and icy shell and thus hold a deep ocean for a while, but it is not clear that these objects could differentiate in the first place, due to the lack of major heat sources. Tidal heating is likely to be a small contributor to these objects, due to their large semi-major axes. This is also illustrated by their large eccentricities, evidence for limited tidal evolution. Still, the lack of detailed observational constraints makes it difficult to fully characterize that class of medium-sized satellites.

Despite its small size, Enceladus exhibits an impressive geological activity at its south pole, as well as a relatively young surface. A power anomaly of up to 17 gigawatts (GW) has been measured by the Cassini Infra-Red Spectrometer (Abramov and Spencer 2009), associated with four ~180-km long rifts, the Tiger Stripes (Spitale and Porco 2007). The Cassini Ion and Neutral Mass Spectrometer (INMS) has detected and identified species and interestingly many of them may be stored in the form of clathrate hydrates (Waite *et al.* 2010), which prompted Kieffer *et al.* (2006) to suggest that Enceladus is frozen and its icy component is dominated by clathrate hydrates, compounds of water ice in which other species are trapped. However, frozen salts in the E-ring spewed by south polar jets (Postberg *et al.* 2009) suggest the presence of liquid (Glein and Shock 2010), possibly just below the surface (Abramov and Spencer 2009).

Future targeted flybys by the Cassini Orbiter may provide constraints on the state of liquid water, especially thanks to two passes in which the distribution of mass inside the body will be measured. Still, the presence of clathrate hydrates in the outer icy shell is possible especially if methane or other small molecules are available to stabilize it. As to the source of the intense heating at the south pole, it remains to be identified. Its origin bears important implications for the astrobiological potential of the satellite. A surficial source associated with friction along the Tiger Stripes and water sublimation (Nimmo *et al.* 2007) is rapidly radiated to space. On the other hand, dissipation in a deep ocean resulting from tidal forcing due to the obliquity of the icy moon (Tyler 2008) may be a major and deep heat source contributing to the preservation of liquid water. The occurrence of mobile-lid convection has been suggested (Barr 2008), which involves the exchange of material between the deep interior and the surface and could supply the interior in organics contained in dust infalls (Parkinson *et al.* 2008).

A couple of objects share similar properties with Enceladus in terms of physical properties, available tidal heating, and relatively recent surface (Figure 10.8). These are Uranus' satellites Ariel and Miranda whose surfaces are lightly cratered and show unusual signatures of tectonic activity. However, we lack further observational constraints to support the discussion about the possible occurrence of liquid water in these objects.

Large wet asteroids

C-class asteroids dominate the mid- and outer main belt. Most of these objects are small and porous (e.g. 253 Mathilde), and may be remnants of

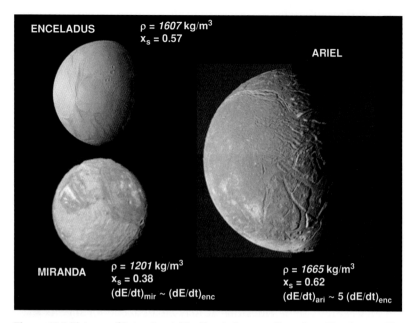

Figure 10.8 Pictures of Saturn's satellite Enceladus as well as of two Uranian satellites that benefit from about the same amount of tidal heating (dE/dt) as Enceladus (Figure 10.7). All three objects are characterized by surface properties suggesting ongoing or recent geological activity, in contrast to another Uranian satellite, Umbriel, presented in Figure 10.1. ρ = density; x_s = silicate mass fraction). A colour version of this illustration appears in the colour plate section between pp. 148–149.

planetesimals that formed the inner Solar System objects, planetesimals coming from the outer Solar System (Walsh *et al.* 2011), or the break-up products of larger objects. The main belt also counts about 20 C-class bodies that are hundreds of kilometers in diameter. The largest of these objects is the dwarf planet Ceres, which is about the size of Saturn's satellite Dione and Pluto's satellite Charon (Figure 10.1).

Although the signature of water of hydration has been detected at the surface of these objects since the 1970s (e.g. Lebofsky 1978), the nature of water at depth is still a matter of debate. This is illustrated by the ongoing debate around the nature of Ceres' interior: the rocky model by Zolotov (2009) and the "warm icy satellite" model by McCord and Sotin (2005) and Castillo-Rogez and McCord (2010). Several recent observations have shed new light on this question. First the discovery of main-belt comets by Hsieh and Jewitt (2006) whose activity has been explained by the sublimation of water exposed as a consequence of small impacts (see Lisse *et al.* 2012 for a review). Then, the discovery of water ice, as well as organics, at the surface of 24 Themis (Rivkin and Emery 2010, Campins *et al.* 2010) and 65 Cybele (Licandro 2011) suggests that free water is stable, at least in that region of the belt (>3 AU), at least for some period of time consistent with theoretical models (Schorghofer 2008).

The potential presence of free water in the subsurface of asteroids has important consequences. Salts (identified in meteorites) can decrease the water ice melting temperature and may result in the formation of films of liquid water a few nanometers thick, as has been suggested for Mars (Möhlmann et al. 2010, Boxe et al. 2012). Several recent studies suggest that volatiles accreted in the inner Solar System included a fraction of ammonia (5–7% by weight of the volatile phase), either as a natural result of the pressure and temperature expected in the inner Solar System nebula (Dodson-Robinson et al. 2009), or supplied from the outer Solar System (Mousis and Alibert 2005, Walsh et al. 2011). The latter model also implies the accretion of other exotic species (methanol, clathrate hydrates). This idea is supported by the abundance of nitrogen, up to 5 wt.% in certain chondrites (Pearson et al. 2006). As discussed earlier, ammonia and some other materials can depress the freezing temperature of ices at temperatures below the surface temperature of these objects.

Ceres exhibits some clues that it may be an abode for life. Its shape appears hydrostatically relaxed and indicates that the asteroid is differentiated in a rocky core and icy shell (Thomas et al. 2005, Castillo-Rogez and McCord 2010). Ceres' surface displays a mixture of brucite $(Mg,Fe(OH)^2)$ and magnesite $(MgCO_3)$ (Milliken and Rivkin 2009), and as such is the third object at the surface of which carbonates have been detected, after Earth and Mars. Also, A'Hearn and Feldman (1992) discovered a transitory event of OH-outgassing, which may be the signature of cryovolcanic activity, or, alternatively of sublimation resulting from a cratering event. That observation has not been reproduced despite several observational campaigns (Olivier Mousis, personal communication). Theoretical models indicate that Ceres could have been the object of two phases of hydrothermal activity (Figure 10.9). First, during accretion and differentiation, as a consequence of short-lived radioisotope decay, provided that it accreted within a few Myr after the formation of CAIs. The second event may be occurring nowadays, thanks to the development of cracks in the core subsequent to its cooling (Vance et al. 2007). The thermal model presented in Figure 10.9, typical of a model consistent with available observational constraints (shape and density), suggests that Ceres is not yet frozen and could have preserved a deep ocean until the present if it is enriched in compounds significantly depressing its eutectic temperature, such as chlorides (Wynn-Williams et al. 2001). In Figure 10.10, we compare the habitability potential of Ceres against that of Europa based on some of the criteria defined in the previous section. The main belt counts a few other large asteroids that appear mostly relaxed to a hydrostatic state and exhibit outstanding features.

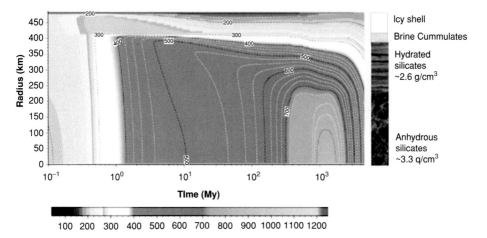

Figure 10.9 Thermal evolution model for Ceres (adapted from Castillo-Rogez and McCord 2010). The pressure reaches ~25 MPa at the base of the icy shell and ~140 MPa at the center of the core. This case assumes a time of formation of 3 Myr after the production of calcium–aluminum inclusions and fully hydrated silicate material following the separation of the rock phase from the water. Settling of the core is achieved by 10 Myr after the end of accretion. By 150 Myr, when the core temperature reaches ~750 K, the deep hydrated silicate starts dehydrating. In this example, we have assumed that 20% of potassium was removed from the core as a result of leaching in a hydrothermal context (see Castillo-Rogez and McCord 2010 and Castillo-Rogez and Lunine 2010 for discussion). Thus silicate dehydration is limited to the inner 200 km. The impacts of [40]K and prospective salt cumulates on the long-term evolution of the shell are not modeled. Under the assumption of conductive transfer, the ice shell would not be "frozen" today (i.e. in equilibrium with the upper boundary condition), and liquid may remain at depth provided that the eutectic temperature of the ocean is depressed by ~30 deg. However, convection onset should occur early. (McCord and Sotin, 2005). Convective heat transfer would most probably yield much colder models. However, the long-term preservation of a deep ocean is primarily a function of the freezing temperature of the ocean, i.e. of the concentration and nature of the second-phase impurities part of the accreted planetesimals, or resulting from chemistry during differentiation. A colour version of this illustration appears in the colour plate section between pp. 148–149.

Pallas, for example, has a surface temperature (subsolar) of 252 K at perihelion (Lim *et al.* 2005). As a reference, the addition of hydrated salts to water can decrease the freezing temperature down to 238 K (e.g. Marion *et al.* 2005). The spectral properties of that asteroid are typical of those generally found on comets, and a genetic link between the comet-turned-asteroid Phaeton and Pallas has recently been suggested (de Leon *et al.* 2010). A contrast in spectral properties between a large basin and the rest of Pallas' surface (Schmidt *et al.* 2009) is also evidence

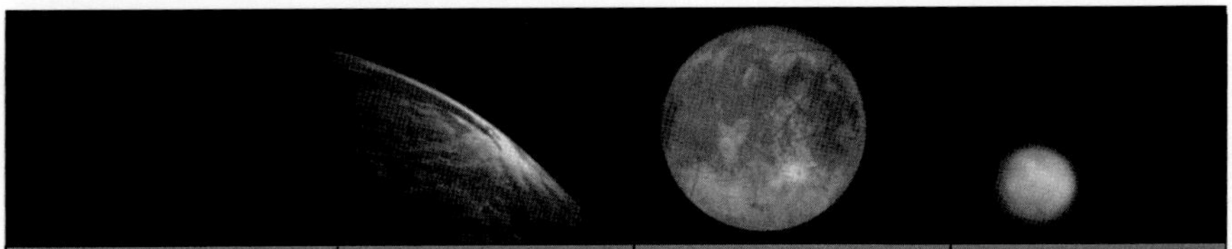

Property	Earth	Europa	Ceres
Water	✔	✔	✔
Liquid Water	✔	✔	Maybe
Surface Temperature	Average: 300 K	50-110 K	130-200 K (below thermal skin)
Geothermal Energy	Average: 0.057 W/m²	10^{11} W (~0.004 W/m²)	3×10^9 W (~0.001 W/m²)
Other Key Heat Sources	Solar	Tidal	Solar
Hydrothermal Vents	✔	Possible	In the past
Water Shell Thickness	~4 km	130 km	70 km
Crust Thickness	N/A	< 5 km?	< 5 km?
Activity in Shell?	✔	Maybe (Stagnant Lid)	✔
Source of Nutrients	✔	From hydrothermal?	From hydrothermal?
Surface Conditions	Nice	Radiations	UV photolysis, sputtering

Figure 10.10 Comparison of key properties of Ceres, Europa, and Earth that can help to assess the habitability potential of these objects.

for chemical stratification inside the asteroid. Unfortunately, Pallas is difficult to reach by spacecraft, owing to its very high inclination (~34 deg.). An indirect constraint on the internal structure of large wet asteroids, and thus on their thermal evolution, may come from the study of the Themis family. That family results from the break-up of a parent that was about 200 km in radius. Castillo-Rogez and Schmidt (2010), pointed out that, prior disruption, that object had differentiated a rocky core, a water-dominated shell, and a layer enriched in salts and likely organics. The approximately 1000 fragments of that family sample various layers of the original body, and their characterization offers the prospect to evaluate the products of chemistry in that class of objects.

Trans-Neptunian objects

These objects are remnants from the planetesimal belt originally located between the orbits of Uranus and Neptune. This category includes Kuiper Belt dwarf planets, as well as large Centaurs. In contrast to their inner Solar System counterparts, the surface temperature of these objects is of the order of 40–50 K, which warrants the condensation of eutectic-forming ammonia hydrates, as well as clathrates or, as an alternative, of amorphous ice, as well as an abundance of

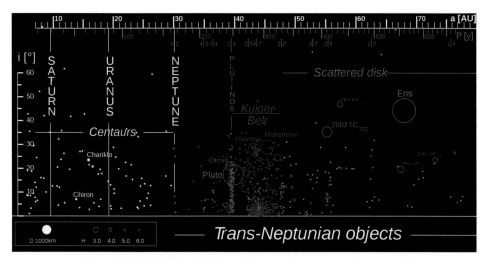

Figure 10.11 Distribution and relative sizes of trans-Neptunian objects (TNOs) discovered so far. Eris, which was the largest TNO as of 2010, is about the size of Europa. Source: http://en.wikipedia.org/wiki/Trans-Neptunian_object. A colour version of this illustration appears in the colour plate section between pp. 148–149.

organics. Fresh ammonia-rich deposits have been found at the surface of Charon (Cook *et al.* 2007, Merlin *et al.* 2010) and suspected at the surface of some of the small TNOs (Barucci *et al.* 2008). If future observations confirmed the extent of ongoing cryovolcanic activity on these objects, then they would represent the biggest reservoir of potentially habitable objects in the Solar System (Figure 10.11).

In the absence of major heat sources, chemistry is then the key parameter determining the habitability potential of these objects. Several models suggest that the presence of ammonia would promote the differentiation and long-term preservation of a deep ocean (e.g. Ruiz and Fairen 2005, Desch *et al.* 2009). The model by Desch *et al.* (2009) relies on the assumption that the cores of these objects are dominated by hydrated silicates with low thermal conductivity, as a way to limit the loss of radiogenic heating. This requires these objects to form on a relatively short time-scale for [26]Al to be abundant enough and promote aqueous alteration of the rocky material. Hydrated silicates detected at the surface of Saturn's satellite Phoebe, believed to be a captured TNO, have been interpreted as evidence for the existence of hydrothermal conditions (temperature of 300 °C or more) at some point in Phoebe's evolution (Cruikshank *et al.* 2008). Another indirect observational constraint comes from the detection of hydrated silicates in exoplanetary systems, which prompted Morris and Desch (2009) to suggest that silicate hydration is a pervasive process in planetesimals and proto-planets. The long formation time-scale of 400 km objects or more in radius is much longer than the time-frame during which [26]Al was abundant (Kenyon *et al.* 2008). However, medium-sized

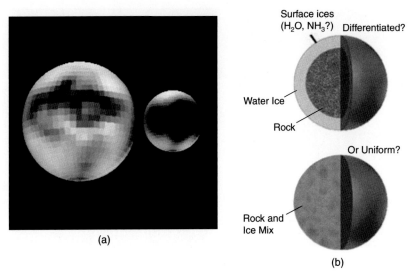

Figure 10.12 (a) Surfaces of Pluto and Charon reconstituted from photometric observations. This image shows the hemisphere of Pluto that always faces Charon, and the hemisphere of the satellite that faces away from the planet (image courtesy of Marc W. Buie/Southwest Research Institute). More images and descriptions can be found at http://www.boulder.swri.edu/~buie/pluto/plutomap1.html. (b) Possible interior models suggested for Charon. The current internal state is in part determined by the abundance of low-eutectic volatiles accreted in the object and of its thermophysical properties (source: http://en.wikipedia.org/wiki/Charon_(moon)).

TNOs could have formed from the accretion of large planetesimals, like Phoebe, that were large enough to undergo ^{26}Al-driven melting and hydrothermal activity.

Although many large dwarf planets have been identified in the Kuiper Belt, we focus the rest of this section on three objects that have attracted much attention: Neptune's satellite Triton, which was observed by Voyager 2, and Pluto and Charon, targets of the New Horizons Mission (Figure 10.12.) Triton and Pluto are slightly smaller than Europa, while Charon is about the size of Ceres. All three objects contain a silicate mass fraction of 50–60%. Geophysical modeling by Shock and McKinnon (1993) suggests that following its capture, Triton underwent a phase of intense tidal heating triggering internal melting accompanied by hydrothermal activity. Although tidal heating is currently not a major heat source in that satellite, as its eccentricity is now fully damped, it may still contain a liquid layer thanks to the presence of low-eutectic compounds (e.g. methanol) and be currently active (Stern and McKinnon 2000). Ruiz and Fairén (2005) noted the potential role of a low-thermal-conductivity layer dominated by nitrogen ice that would insulate heat loss from the interior and enable a deep ocean to exist a few tens of kilometers below the surface.

Pluto and Charon are also believed to have sheltered a deep ocean at some point in their history, and possibly at present (Ruiz and Fairén 2005; Robuchon and Nimmo 2011). Pluto is about the same size as Europa and Triton, both of which have been pointed out as astrobiological targets. Desch *et al.* (2009) suggested that cryovolcanic activity on Charon is evidence for the preservation until the present of a deep ocean, thanks to the presence of antifreezing compounds (e.g. ammonia hydrates). Besides, the dynamical properties of Charon are striking. Pluto's axis, and thus Charon's orbit, is tilted at an angle of 119.6 degrees. Both objects are locked in synchronous rotation and face each other. Gilmour and Sephton (2004) noted that in that peculiar configuration, the torques exerted on Charon by Pluto and the Sun are large, and they suggested that the potential for tidal heating in that object could be significant. Detailed modeling remains to be performed in order to evaluate if tidal heating could explain why Charon is so peculiar in comparison to Pluto, whose surface is dominated by nitrogen and methane ice.

Thus, by the standards commonly used to assess the astrobiological potential of icy bodies, Pluto and Charon may qualify as astrobiological targets. However, interior models for these objects have not so far accounted for the complex chemistry predicted by cosmochemical models that may involve methanol, clathrate hydrates, nitrogen and methane ices, and organics. The products of aqueous chemistry accompanying melting and differentiation may thus be extremely rich, but is poorly constrained at present. Groundwork is needed, from an experimental and theoretical point of view, to enable the development of more realistic evolution models.

Summary and future exploration

In our tour of the Solar System small-body populations, we have identified different categories of objects that may be abodes for life. Obviously, proximity to the Sun is a key parameter. Increased evidence for the presence of water ice in the main belt, possibly in Ceres, implies that water could be close to its melting temperature, depressed by the presence of impurities (salts), close to the surface, a situation that is particularly propitious for the preservation of live forms. The potential for hydrogeochemistry promoted by short-lived radioisotope decay is another important feature of these objects, which is illustrated by the abundance of organics in carbonaceous chondrites.

The habitability potential of objects in the outer Solar System is primarily determined by their content in low-eutectic impurities. Liquid water may have been, and may still be present in these objects, but the outcome of hydrogeochemistry at the low temperatures characteristic of ammonia- and methanol-rich environments remains to be quantified through theoretical and experimental investigations.

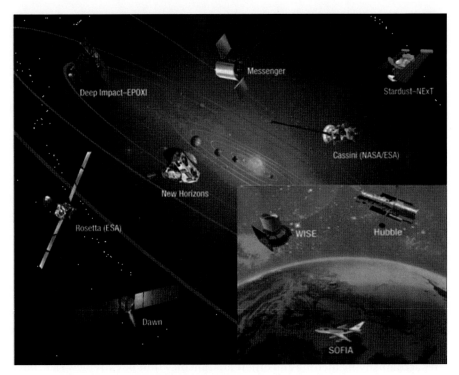

Figure 10.13 Ongoing deep space missions to small water-rich bodies, and space telescopes with the capability to survey and characterize small-body populations (collage of excerpts from NASA Science Mission Directorate 2010 Science Plan). See the appendix for information resources about these missions. A colour version of this illustration appears in the colour plate section between pp. 148–149.

Enceladus is a notable exception because of its intense tidal heating output that warrants conditions amenable to a deep ocean, whether low-eutectic species are present or not.

Several missions are currently on their way to some of these objects (Figure 10.13). The Cassini Orbiter is going to dedicate several targeted flybys of Enceladus during the Equinox mission, including two gravity passes that will bring constraints on Enceladus' interior structure, and possibly shed some light on the heat sources responsible for its current geological state. Further measurements with the INMS instrument are necessary in order to confirm the composition measured during the first flybys, which includes light organics.

The Dawn mission that was launched in 2007 will reach Ceres in 2015, after a first stop at Vesta in July 2011. The primary phase of the mission will dedicate six months in orbit around Ceres and will provide the first complete, high-resolution mapping of the surface composition and geology of two asteroids. Measurements of Ceres' gravity field up to degree 10 will also yield constraints on the degree of chemical differentiation of the interior, and possibly help identify the form under which water is stable in the asteroid. The Gamma Ray and Neutron Diffraction instrument bears also the capability to detect the loss of volatiles from the interior. More details on the Dawn Mission, its payload, and the constraints it could bring about Ceres can be found in a dedicated issue of *Space Science Reviews*, e.g. McCord *et al.* (2011).

The New Horizons mission is on its way to the Pluto–Charon system, which it will also reach in 2015. That mission bears the capability to map the geological properties and spectral properties over a wide range of wavelengths (infrared and ultraviolet), and to characterize the nature of particles ejected from these objects. The spacecraft will then search for Kuiper Belt objects in the relative vicinity of the Pluto–Charon system. Thus in 2015 two spacecraft will reach two bodies of similar size and density at two extremes of the Solar System (5.5 billion kilometers apart), representing the coldest and the warmest icy body populations. Observations returned by these missions will help to test some of the hypotheses presented in this chapter, for example the respective impacts of surface temperature and composition on habitability.

Research on small bodies is also supported by active observational campaigns dedicated to the characterization of the spectral properties of these objects, as a clue to their composition and thermal state. Observations with the Hubble Space Telescope have yielded the first resolved images of Ceres' and Pallas' surfaces (Thomas *et al.* 2005, Schmidt *et al.* 2009), while mapping by the Wide Infrared Survey has brought to light the intriguing distribution of C-class asteroids across the main belt (Masiero *et al.* 2011).

Besides space-borne and astronomical exploration, there is an important need for fundamental research to support the study of small bodies. Since the recognition of their astrobiological significance is recent, fundamental research is lagging behind the observation-based hopes that some of the closest of these objects could shelter habitable conditions. Of primary importance, the kinetics and products of chemical reactions occurring at temperatures below the water freezing point needs to be experimentally measured. The impact of short-lived radioisotopes on the chemistry of outer Solar System planetesimals also remains to be quantified in order to assess the amount of hydrogeochemical processing, especially of organics, which could take place prior to accretion into larger objects. The question of the state of water across the main belt is of primary importance and may be addressed by studies combining Solar System dynamics, accretional models, and geophysical evolution. However, the current databases lack information on the thermophysical and thermodynamical properties of ice mixtures that contain a significant fraction of clathrate hydrates and volatile species such as methane, nitrogen, and organics. Although the acquisition of such data is challenging, it will be necessary for the development of models aiming to properly simulate heat transfer and geological activity on Pluto, Charon, and Kuiper Belt objects.

Acknowledgement. JCR is supported by the Jet Propulsion Laboratory, California Institute of Technology, under contract to NASA. JIL is supported by the program "Incentivazione alla mobilita' di studiosi straineri e italiani residenti all'estero" of Italy. Government sponsorship is acknowledged. This chapter is dedicated to the memory of Angioletta Coradini (1945–2011), a pioneer of planetary exploration and a leading scientist in the area of the origin and evolution of small bodies.

Appendix: Web resources and books

Websites for ongoing space missions

http://saturn.jpl.nasa.gov/
http://dawn.jpl.nasa.gov/
http://pluto.jhuapl.edu/
http://wise.ssl.berkeley.edu/mission.html
http://science.nasa.gov/planetary-science/focus-areas/small-bodies-
of-the-solar-system/

NASA Science Mission Directorate Science Plan

http://science.nasa.gov/media/medialibrary/2010/08/30/
2010SciencePlan_TAGGED.pdf

Minor Planet Center

http://www.minorplanetcenter.org/iau/mpc.html

Small Bodies Assessment Group

http://www.lpi.usra.edu/sbag/

NASA resources for educators

http://www.nasa.gov/audience/foreducators/topnav/schedule/
extrathemes/F_Asteroids_Comets_Meteorites_Extra.html
http://www.nasa.gov/audience/forkids/kidsclub/flash/index.html

NASA Astrobiology Institutes

http://astrobiology.nasa.gov/nai/

Books and reviews

Planetary Tectonics, editors T. Watters and R. Schultz, Cambridge
University Press, 530 pp., 2009, ISBN: 9780521765732.

Science of Solar System Ices, editors M. Gutipati and J. Castillo-Rogez,
Springer, in press.

Space Science Reviews – Special issue on the *Dawn* Mission, Russell *et al.*
(2011).

The Solar System Beyond Neptune, editors Barucci *et al.*, University of Arizona
Press, Tucson.

Space Science Reviews – Special issue on processes in icy bodies, Grass
et al. (2010).

Space Science Reviews – Special issue on the New Horizons Mission,
Russell *et al.* (2009).

Saturn from Cassini-Huygens, editors M. Dougherty, L. Esposito, and
S. Krimigis, Springer, New York (2009), ISBN:
978–1–4020–9216–9

Pluto and Charon: Ice Worlds on the Ragged Edge of the Solar System, 2nd
edition. Wiley-VCH, 2005, by A. S. Stern and J. Mitton.

References

Abramov, O. and Spencer, J. R. (2009). "Endogenic heat from Enceladus' south polar
fractures: new observations, and models of conductive surface heating," *Icarus*,
Vol. 199, pp. 189–196.

A'Hearn, M. F. and Feldman, P. D. (1992). "Water vaporization on Ceres," *Icarus*, Vol. 98,
pp. 54–60.

Allen, D. E. and Seyfried, W. E. (2004). "Serpentinization and heat generation:
constraints from Lost City and rainbow hydrothermal systems," *Geochimica et
Cosmochimica Acta*, Vol. 68, pp. 1347–1354.

Barr, A. C. (2008). "Mobile lid convection beneath Enceladus' south polar terrain,"
Journal of Geophysical Research, Vol. 113, E07009. doi:10.1029/2008JE003114.

Barucci, M. A., Merlin, F., Guilbert, A. *et al.* (2008). "Surface composition and
temperature of the TNO Orcus," *Astronomy and Astrophysics*, Vol. 479, pp. L13–L16.

Boxe, C. S., Hand, K. P., Nealson, K. H. *et al.* (2012). "Adsorbed water and thin liquid
films on Mars: implications for life," *International Journal of Astrobiology*, Vol. 11,
pp. 169–175.

Brearley, A. J. (2006). The action of water, in *Meteorites and the Early Solar System II*,
D. S. Lauretta and H. Y. McSween Jr. (eds.). Tucson: University of Arizona Press,
pp. 584–624.

Busarev, V. V., Dorofeeva, V. A., and Makalkin, A. B. (2003). "Hydrated silicates on
Edgeworth–Kuiper objects – probable ways of formation," *Earth, Moon, and Planets*,
Vol. 92, pp. 345–375, doi:10.1023/B:MOON.0000031951.59946.97.

Campins, H., Hargrove, K., Pinilla-Alonso, N. et al. (2010). "Water ice and organics on the surface of the asteroid 24 Themis," Nature, Vol. 464, pp. 1320–1321.

Canup, R. M. (2005). "A giant impact origin of Pluto–Charon," Science, Vol. 28, pp. 546–550.

Castillo-Rogez, J. and Lunine, J. I. (2010). "Evolution of Titan's rocky core constrained by Cassini observations," Geophysical Research Letters, Vol. 37, L20205.

Castillo-Rogez, J. C. and McCord, T. B. (2010). "Ceres' evolution and present state constrained by shape data," Icarus, Vol. 205, pp. 443–459.

Castillo-Rogez, J. C. and Schmidt, B. E. (2010). "Geophysical evolution of the Themis family parent body," Geophysical Research Letters, Vol. 37, L10202, doi:10.1029/2009GL042353.

Castillo-Rogez, J., Matson, D. L., Sotin, C. et al. (2007). "Iapetus' geophysics: rotation rate, shape, and equatorial ridge," Icarus, Vol. 190, pp. 179–202. doi:10.1016/j.icarus.2007.02.018.

Castillo-Rogez, J. C., Johnson, T. V., Lee, M. H. et al. (2009). "^{26}Al decay: heat production and a revised age for Iapetus," Icarus, Vol. 204, pp. 658–662.

Choukroun, M., Chevrier, V., Kieffer, S. et al. (2011). "Clathrate hydrates in the Solar System," in The Science of Solar System Ices, M. S. Gudipati and J. C. Castillo-Rogez (eds.). Springer.

Clark, R. N., Brown, R. H., and Jaumann, R. (2005). "Compositional maps of Saturn's moon Phoebe from imaging spectroscopy," Nature, Vol. 435, pp. 66–69.

Cohen, B. A. and Coker, R. F. (2000). "Modeling of liquid water on CM meteorite parent bodies and implications for amino acid racemization," Icarus, Vol. 145, pp. 369–381.

Collins, G. C. and Barr, A. C. (2008). "Tectonics and interior structure of Pluto: predictions from the orbital evolution of the Pluto–Charon system," American Geophysical Union, Fall Meeting 2008, abstract #P51C–1425.

Cook, J. C., Desch, S. J., Roush, T. L. et al. (2007). "Near-infrared spectroscopy of Charon: possible evidence for cryovolcanism on Kuiper Belt objects," Astrophysical Journal, Vol. 663, pp. 1406–1419.

Cruikshank, D. P., Wegryn, E., Dalle Ore, C. M. et al. (2008). "Hydrocarbons on Saturn's satellites Iapetus and Phoebe," Icarus, Vol. 193, pp. 334–343.

Cuzzi, J. N. and Weidenschilling, S. J. (2006). "Particle-gas dynamics and primary accretion," in Meteorites and the Early Solar System, D. Lauretta and H. McSween (eds.). Tucson: University of Arizona Press, pp. 353–381.

de Leon, J., Campins, H., Tsiganis, K. et al. (2010). "Origin of the near-Earth asteroid Phaethon and the Geminids meteor shower," Bulletin of the American Astronomical Society, Vol. 42, p. 1058.

Desch, S. J., Cook, J. C., Doggett, T. C., and Porter, S. B. (2009). "Thermal evolution of Kuiper Belt objects, with implications for cryovolcanism," Icarus, Vol. 202, pp. 694–714.

Dodson-Robinson, S. E., Willacy, K., Bodenheimer, P. et al. (2009). "Ice lines, planetesimal composition and solid surface density in the solar nebula," Icarus, Vol. 200, pp. 672–693.

Frank, E. A. and Mojzsis, S. J. (2010). "Cumulative ocean volume estimates of the Solar System," *American Geophysical Union, Fall Meeting 2010*, abstract #P33B-1572.

Gilmour, I. and Sephton, M. A. (2004). *An Introduction to Astrobiology*. Cambridge: Cambridge University Press.

Glein, C. R. and Shock, E. L. (2010). "Sodium chloride as a geophysical probe of a subsurface ocean on Enceladus," *Geophysical Research Letters*, Vol. 37, L09204.

Grimm, R. E. and McSween, H. Y. (1989). "Water and the thermal evolution of carbonaceous chondrite parent bodies," *Icarus*, Vol. 83, pp. 244–280.

Hsieh, H. and Jewitt, D. (2006). "A population of comets in the main asteroid belt," *Science*, Vol. 312, p. 561.

Hussmann, H. *et al.* (2006). "Subsurface oceans and deep interiors of medium-sized outer planet satellites and large trans-Neptunian objects," *Icarus*, Vol. 185, pp. 258–273.

Johnson, T. V. and Lunine, R. A. (2005). "Saturn's moon Phoebe as a captured body from the outer Solar System," *Nature*, Vol. 435, pp. 69–71.

Kargel, J. S. and Lunine, J. I. (1998). "Clathrate hydrates on Earth and in the Solar System," in *Solar System Ices*, B. Schmitt *et al.* (eds.). Norwell, MA: Kluwer Academic Press, pp. 97–117.

Keil, K. (2000). "Thermal alteration of asteroids: evidence from meteorites," *Planetary and Space Science*, Vol. 48, pp. 887–903.

Kenyon, S. J., Bromley, B. C., O'Brien, D. P., and Davis, D. R. (2008). "Formation and collisional evolution of Kuiper Belt objects," in *The Solar System Beyond Neptune*, M. A. Barucci, H. Boehnhardt, D. P. Cruikshank, and A. Morbidelli (eds.). Tucson: University of Arizona Press, pp. 293–313.

Kieffer, S. W., Lu, X., Bethke, C. M. *et al.* (2006). "A clathrate reservoir hypothesis for Enceladus' south polar plume," *Science*, Vol. 314, pp. 1764–1766.

Lebofsky, L. A. (1978). "Asteroid 1 Ceres – evidence for water of hydration," *Monthly Notices of the Royal Astronomical Society*, Vol. 182, pp. 17P–21P.

Licandro, J., Campins, H., Kelley, M. *et al.* (2011). "65 Cybele: detection of small silicate grains, water-ice, and organics," *Astronomy and Astrophysics*, Vol. 525, doi:10.1051/0004-6361/201015339.

Lim, L. F., McConnochie, T., Bell, J. III, and Hayward, T. (2005). "Thermal Infrared (8–13 μm) spectra of 29 asteroids: the Cornell Mid-Infrared Asteroid Spectroscopy (MIDAS) survey," *Icarus*, Vol. 173, pp. 385–408. doi:10.1016/j.icarus.2004.08.005.

Lisse, C., Bar-Nun, A., Laufer, D. *et al.* (2012). "Cometary ices," in *The Science of Solar System Ices*, M. Gudipati and J. Castillo-Rogez (eds.). Berlin: Springer.

Lunine, J. I. (2006). "Origin of water ice in the Solar System," in *Meteorites and the Early Solar System*, Vol. II, D. S. Lauretta and H. Y. McSween Jr. (eds.). Tucson: University of Arizona Press, pp. 309–319.

Manga, M. and Wang, C.-Y. (2007). "Pressurized oceans and the eruption of liquid water on Europa and Enceladus," *Geophysical Research Letters*, Vol. 34, L07202.

Marion, G. M., Kargel, K. S., Catling, D. C., and Jakubowski, S. D. (2005). "Effects of pressure on aqueous chemical equilibria at subzero temperatures with applications to Europa," *Geochimica et Cosmochimica Acta*, Vol. 69, pp. 259–274.

Masiero, J., Mainzer, A., Grav, T. *et al.* (2011). "Main Belt asteroids with WISE/NEOWISE I: preliminary albedos and diameters," *Astrophysical Journal*, accepted.

Matson, D. L., Castillo-Rogez, J. C., McKinnon, W. B. *et al.* (2009). "The thermal evolution and internal structure of Saturn's midsize icy satellites," in *Saturn after Cassini–Huygens*, R. Brown, M. Dougherty, L. Esposito, T. Krimigis, and H. Waite (eds.). Chapter 18. doi:10.1007/978-1-4020-9217-6_18.

McCord, T. B. and Sotin, C. (2005). "Ceres: evolution and current state," *Journal of Geophysical Research*, Vol. 110, E05009.

McCord, T. B., Castillo-Rogez, J. C., and Rivkin, A. S. (2011). "Ceres: its origin, evolution and structure and Dawn's potential contribution," *Space Science Reviews*, Vol. 163, pp. 63–76.

Merlin, F., Barucci, M. A., de Bergh, C. *et al.* (2010). "Chemical and physical properties of the variegated Pluto and Charon surfaces," *Icarus*, Vol. 210, pp. 930–943.

Milliken, R. E. and Rivkin, A. S. (2009). "Brucite and carbonate assemblages from altered olivine-rich materials on Ceres," *Nature Geoscience*, Vol. 2, pp. 258–261.

Möhlmann, D. (2010). "The three types of liquid water in the surface of present Mars," *International Journal of Astrobiology*, Vol. 9, pp. 45–49.

Morbidelli, A., Bottke, W. F., Nesvorny, D., and Levison, H. F. (2009). "Asteroids were born big," *Icarus*, Vol. 204, pp. 558–573.

Morris, M. A. and Desch, S. J. (2009). "Phyllosilicate emission from protoplanetary disks: is the indirect detection of extrasolar water possible?," *Astrobiology*, Vol. 9, pp. 965–978.

Mousis, O. and Alibert, Y. (2005). "On the composition of ices incorporated in Ceres," *Monthly Notices of the Royal Astronomical Society*, Vol. 358, pp. 188–192.

Nimmo, F. (2004). "Stresses generated in cooling viscoelastic ice shells: application to Europa," *Journal of Geophysical Research*, Vol. 109, E12001.

Nimmo, F., Spencer, J. R., Pappalardo, R. T., and Mullen, M. E. (2007). "Shear heating as the origin of the plumes and heat flux on Enceladus," *Nature*, Vol. 447, pp. 289–291.

Parkinson, C. D., Liang, M. C., Yung, Y. L., and Kirschvink, J. L. (2008). "Habitability of Enceladus: planetary conditions for life," *Origin of Life and Evolution in the Biosphere*, Vol. 38, pp. 355–369.

Pearson, V. K., Sephton, M. A., Franchi, I. A. *et al.* (2006). "Carbon and nitrogen in carbonaceous chondrites: elemental abundances and stable isotopic compositions," *Meteoritics and Planetary Science*, Vol. 41, pp. 1899–1918.

Postberg, F., Kempf, S., Schmidt, J. *et al.* (2009). "Sodium salts in E-ring ice grains from an ocean below the surface of Enceladus," *Nature*, Vol. 459, pp. 1098–1101.

Rivkin, A. S. and Emery, J. P. (2010). "Detection of ice and organics on an asteroidal surface," *Nature*, Vol. 464, pp. 1322–1323.

Robuchon, G. and Nimmo, F. (2011). "Thermal evolution of Pluto and implications for surface tectonics and a subsurface ocean," *Icarus*, Vol. 216, pp. 426–439.

Ross, R. G. and Kargel, J. S. *et al.* (1998). "Thermal conductivity of solar system ices, with special reference to Martian polar caps," in *Solar System Ices*, B. Schmitt *et al.* (eds.). Norwell, Massachusetts: Kluwer, pp. 32–66.

Ruiz, J. and Fairén, A. G. (2005). "Seas under ice: stability of liquid-water oceans within icy worlds," *Earth, Moon, and Planets*, Vol. 97, pp. 79–90.

Schenk, P. and Zahnle, K. (2007). "On the negligible surface age of Triton," *Icarus*, Vol. 192, pp. 135–149.

Schmidt, B. E. and Castillo-Rogez, J. C. (2011). "Constraints on the internal structure and thermal evolution of 2 Pallas," *Icarus*, submitted.

Schmidt, B. E., Thomas, P. C., Bauer, J. M. *et al.* (2009). "The shape and surface variation of 2 Pallas from the Hubble Space Telescope," *Science*, Vol. 326, pp. 275–279.

Schorghofer, N. (2008). "The lifetime of ice on main belt asteroids," *Astrophysical Journal*, Vol. 682, pp. 697–705.

Shock, E. L. and McKinnon, W. B. (1993). "Hydrothermal processing of cometary volatiles – applications to Triton," *Icarus*, Vol. 106, pp. 464–477.

Sleep, N., Meibom, A., Fridriksson, T. *et al.* (2004). "H_2-rich fluids from serpentinization: geochemical and biotic implications," *Proceedings of the National Academy of Sciences*, Vol. 101, pp. 12818–12823.

Spencer, J. R., Barr, A. C., Esposito, L. W. *et al.* (2009). "Enceladus: an active cryovolcanic satellite," in *Saturn after Cassini–Huygens*, R. Brown *et al.* (eds.). New York: Springer, pp. 683–722.

Spitale, J. N. and Porco, C. C. (2007). "Association of the jets of Enceladus with the warmest regions on its south-polar fractures," *Nature*, Vol. 449, pp. 695–697.

Stern, S. A. and McKinnon, W. B. (2000). "Triton's surface and impactor population revisited in light of Kuiper Belt fluxes: evidence for small Kuiper Belt objects and recent geological activity," *Astronomical Journal*, Vol. 119, pp. 945–952.

Thomas, C., Parker, J. W., McFadden, L. A. *et al.* (2005). "Differentiation of the asteroid Ceres as revealed by its shape," *Nature*, Vol. 437, pp. 224–226.

Travis, B. J. and Schubert, G. (2005). "Hydrothermal convection in carbonaceous chondrite parent bodies," *Earth and Planetary Science Letters*, Vol. 240, pp. 234–250.

Tyler, R. H. (2008). "Strong ocean tidal flow and heating on moons of the outer planets," *Nature*, Vol. 456, pp. 770–772.

Vance, S., Harnmeijer, J., Kimura, J. *et al.* (2007). "Hydrothermal systems in small ocean planets," *Astrobiology*, Vol. 7, pp. 987–1005.

Waite Jr., J. H., Lewis, W. S., Magee, B. *et al.* (2009). "Ammonia, radiogenic Ar, organics, and deuterium measured in the plume of Saturn's icy moon Enceladus," *Nature*, Vol. 460, pp. 487–490.

Walsh, K. J., Morbidelli, A., Raymond, S. N. *et al.* (2011). "Origin of the Asteroid Belt and Mars' small mass," *Bulletin of the American Astronomical Society*, Vol. 42, p. 947.

Wasserburg, G. J. and Papanastassiou, D. A. (1982). "Some short-lived nuclides in the early Solar System," in *Essays in Nuclear Astrophysics*, C. A. Barnes, D. D. Clayton and D. N. Schramm (eds.). New York: Cambridge University Press, p. 77.

Weidenschilling, S. J. and Cuzzi, J. (2006). "Accretion dynamics and timescales: relation to chondrites," in *Meteorites and the Early Solar Systems II*, D. Lauretta and H. Y. McSween (eds.). Tucson: University of Arizona Press, pp. 473–485.

Wong, M. H. *et al.* (2008). "Oxygen and other volatiles in the giant planets and their satellites," in *Oxygen in the Solar System*, G. J. MacPherson, and V. A. Chantilly (eds.).

Mineralogical Society of America. *Reviews in Mineralogy and Geochemistry*, Vol. 68, pp. 241–246.

Wynn-Williams, D. D., Cabrol, N. A., Grin, E. A. *et al.* (2001). "Brines in seepage channels as eluants for subsurface relict biomolecules on Mars," *Astrobiology*, Vol. 1, pp. 165–184.

Zolotov, M. Yu. (2009). "On the composition and differentiation of Ceres," *Icarus*, Vol. 204, pp. 183–193.

PART V EXOPLANETS AND LIFE IN THE GALAXY

11

Searches for Habitable Exoplanets

SARA SEAGER

Introduction

For thousands of years people have wondered about the existence of habitable worlds. We frame the discussion in terms of a hierarchical series of ancient questions: "Do other Earths exist?" and "Are they common?" and "Do any have signs of life?" With hundreds of known exoplanets of increasingly smaller mass and size, we are on the verge of answering these questions. Thousands of years from now, people will look back and see as one of the most significant, positive accomplishments of our early twenty-first century society the first discoveries of exoplanets, and the human foray into finding and characterizing habitable worlds.

How to discover Earths

Do other Earths exist? Our Galaxy, the Milky Way, has about 100 billion stars. The universe has upwards of 100 billion galaxies. The chance therefore that another Earth exists is extremely high, even if Earths are rare.

Finding Earths to confirm their logical existence, however, is another matter. We will not believe that other Earths exist until we have robust evidence. The biggest challenge in detecting another Earth is that Earth-like planets are miniscule compared to their adjacent parent star. Our own Earth is so much smaller than the Sun (100 times), so much less massive (1000 times), and so much less bright (10^7–10^{10}, depending on wavelength). Any planet-detection technique (see Figure 11.1) is challenged to find another Earth.

Frontiers of Astrobiology, ed. Chris Impey, Jonathan Lunine and José Funes.
Published by Cambridge University Press. © Cambridge University Press 2012.

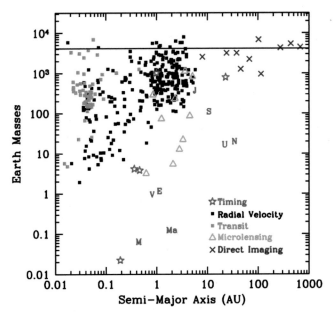

Figure 11.1 Known planets as of January 2010. The solid line is the conventional upper mass limit for the definition of a planet. Data taken from http://exoplanet.eu/.

Definitions

A pedagogical impediment to answering the question of the existence of other Earths is the definition of "other Earth." The reason is simply that, in the short term, we are able to identify Earth-size or Earth-mass planets in orbits close to the star, but not *Earth-like* planets.

To illuminate the nomenclature issue further, consider the two planets Venus and Earth. Venus and Earth are both about the same size and mass – and would appear equivalent to most exoplanet indirect detection techniques (astrometry, radial velocity, transit observation, or microlensing). Yet Venus is completely hostile to life due to the strong greenhouse effect and resulting high surface temperatures (over 700 K). In contrast to Venus, Earth has the right surface temperature for liquid-water oceans and is teeming with life. This is why, in describing the search for habitable planets, we must take care discriminating among Earth-size, Earth-mass, and Earth-like exoplanets.

A potentially habitable planet is one that is neither too hot or too cold nor too big or too small, but one that is just right. Too hot, and the liquid-water oceans will evaporate away (like Venus). Too cold, and the water will freeze into ice (like Mars). Too big, and the planet will have an immense atmosphere of hydrogen and helium, creating an interior too hot to support life (like Jupiter). Too small, and the planet will not be able to hold on to its protective atmosphere (like Mercury).

Habitable planet – A habitable planet is a terrestrial planet on whose surface liquid water can exist in steady state. This definition presumes that extraterrestrial life, like Earth life, requires liquid water for its existence. Both the liquid water, and any life that depends on it, must be at the planet's surface in order to be detected remotely. This, in turn, requires the existence of an atmosphere with a surface pressure and temperature suitable for liquid water. (Moons of exoplanets might also be habitable.)

Potentially habitable planet – A potentially habitable planet is one whose orbit lies within the habitable zone, broadly construed, and has a solid or liquid surface. This includes planets that have high eccentricities, but whose semi-major axis is within the habitable zone. A definition for a potentially habitable planet implies that some planets within the habitable zone may not actually be habitable.

Earth-Like planet – An Earth-like planet or Earth analog is a habitable planet of approximately 1 $M_\oplus$ and 1 $R_\oplus$ in an Earth-like orbit about a Sun-like star. Earth-like planets are not necessarily habitable planets, nor vice versa; it depends on the context of usage. While an Earth-like planet is used to describe a planet similar in mass, radius, and temperature to Earth, the term Earth twin is usually reserved for an Earth-like planet with liquid-water oceans and continental landmasses.

Habitable zone – The habitable zone is the region around a star in which a planet may maintain liquid water on its surface, i.e. it is the zone in which Earth-like planets may exist. Its boundaries may be defined empirically (based on the observation that Venus appears to have lost its water some time ago and that Mars appears to have had surface water early in its history) or with models.

For other exoplanet subcategories[1] see Seager and Lissauer (2010).

Techniques for detecting Earths

Four hundred years after the invention of the telescope, six different techniques for finding exoplanets exist. Each exoplanet technique favors a different semi-major axis range. In addition, each technique measures one of the planetary size (with respect to the star size), mass (with respect to the star mass), or brightness. Figure 11.1 shows the known exoplanets as of March 2011 with symbols indicating the discovery technique.[2]

The majority of the known and confirmed exoplanets have been discovered by the Doppler technique, which measures the star's line-of-sight motion as the star orbits the planet–star common center of mass (see e.g. Butler *et al.* 2006,

---

1 NASA's Kepler team has the following definitions: Earth-size ($R_p < 1.25\ R_\oplus$), super-Earth size ($1.25\ R_\oplus < R_p < 2\ R_\oplus$), Neptune-size ($2\ R_\oplus < R_p < 6\ R_\oplus$), and Jupiter-size ($6\ R_\oplus < R_p < 15\ R_\oplus$). Borucki *et al.* (2011).

2 http://exoplanet.eu/

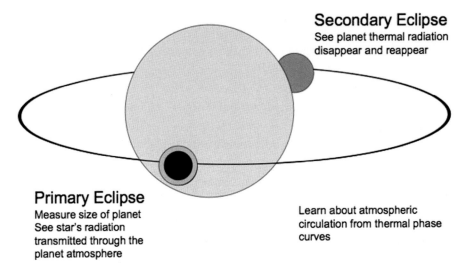

Secondary Eclipse
See planet thermal radiation
disappear and reappear

Primary Eclipse
Measure size of planet
See star's radiation
transmitted through the
planet atmosphere

Learn about atmospheric
circulation from thermal phase
curves

Figure 11.2 Schematic of a transiting exoplanet showing primary and secondary eclipses and what is learned from observations at those phases.

Udry *et al.* 2007). While most planets discovered with the Doppler technique are massive planets akin to Jupiter or Neptune, the new frontier is discovery of super-Earths (loosely defined as planets with masses between 1 and 10 $M_\oplus$). About a dozen planets with $M < 10$ $M_\oplus$ and another dozen with 10 $M_\oplus$ $< M < 30$ $M_\oplus$ have been reported from the radial-velocity technique. As a potentially big step forward, three independent teams are monitoring Alpha Centauri B for a super-Earth in the habitable zone (D. Fischer, private communication). New calibration techniques, namely the frequency comb (e.g. Li *et al.* 2008), are being developed for higher radial velocity precision that may pave the way to Earth-mass planets with the radial-velocity technique.

The transit technique finds planets by monitoring thousands of stars, looking for the small drop in brightness of the parent star that is caused by a planet crossing in front of the star. At the time this chapter went into production, over 200 transiting planets are known. Due to selection effects, transiting planets found from ground-based searches are limited to small semi-major axes (see Figure 11.1). Nonetheless, transiting planets offer unique possibilities for follow-up observations (Figure 11.2). NASA's Kepler Space Telescope (see below, and Borucki *et al.* 2011) is revolutionizing the field of exoplanets with over 1200 transiting planet candidates. With Kepler, the transit technique will enable discovery of Earth-size transiting planets in Earth-like orbits about Sun-like stars. Although the Kepler Earth-size, Earth-like orbit transiting planets are too faint for follow-up, Kepler will establish their frequency.

Gravitational microlensing has recently emerged as a powerful planet-finding technique, discovering nine planets, two belonging to a scaled-down version of our own Solar System (Gaudi *et al.* 2008). Microlensing could find an Earth-mass planet in an Earth-like orbit about a Sun-like star with the next generation. Microlensing planet discoveries will collectively provide a statistical census. But, because the events are not repeatable, and the lens star and planet are so distant, follow-up observations of the planet are not possible.

The astrometry planet search technique has not yet discovered any exoplanets (despite a checkered history of retracted discoveries; see Jayawardhana 2011, Bean *et al.* 2010). Ground-based search programs are nevertheless in operation (e.g. Boss *et al.* 2009).

The timing discovery method includes both pulsar timing (Wolszczan and Frail 1992) and time perturbations of stars with stable oscillation periods (Silvotti *et al.* 2007). The only exoplanet detection technique that has demonstrated detection of Earth-mass planets in Earth-like orbits is pulsar timing (Figure 11.1). Pulsar planets, however, orbit dead neutron stars and the powerful, poisonous radiation emitted by the pulsars would be death to life as we know it. While five pulsar planets are known in two different systems, the current crop of millisecond pulsars has been exhaustively searched and a larger pool from putative next-generation radio searches is needed to expand the pulsar planet population (Wolszczan and Kuchner 2010).

Direct imaging is currently able to find young or massive planets with very large semi-major axes. The mass of directly imaged planets (Kalas *et al.* 2008, Marois *et al.* 2008) is inferred from the measured flux based on evolution models, and is hence uncertain. Although extremely challenging, direct imaging is the best path to finding and characterizing Earths by spectral observations, as described later in this chapter.

Expanding the definition of "Earth"

The discovery of a true Earth analog is immensely challenging, making it at present many years off in the future. Yet, a different kind of habitable planet is within reach – if only we are willing to extend our definition of Earth. The extension is to a big Earth orbiting close to a small star. Furthermore, to find a planet that can be further characterized, we seek a transiting planet until the direct-imaging technique can reach down to planets near Earth size or Earth mass.

All life on Earth requires liquid-water and so a natural requirement in the search for habitable exoplanets is a planet with the right surface temperature to have liquid-water. Terrestrial-like planets are heated externally by the host star, so that a star's "habitable zone" is based on distance from the host star. Small stars

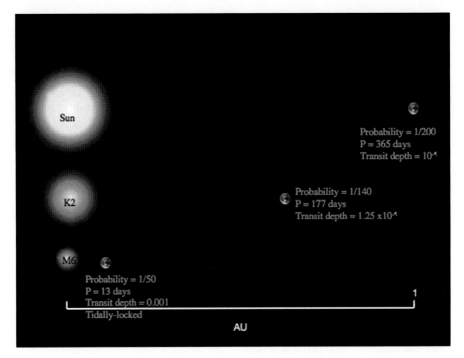

Figure 11.3 A schematic of transiting planets orbiting normal stars. The smaller the star, the closer the habitable zone is to the star, and the easier is the detection. A colour version of this illustration appears in the colour plate section between pp. 148–149.

have a habitable zone much closer to the star as compared to Sun-like stars, owing to the small stars' much lower energy output than the Sun (see Figure 11.3).

The chance of discovering a super-Earth transiting a low-mass star in the immediate future is huge. Observational selection effects favor the discovery of super-Earths orbiting in the habitable zones of M stars compared to Earth analogs in almost every way. The magnitude of the planet transit signature is related to the planet-to-star area ratio. Low-mass stars can be 2–10 times smaller than a Sun-like star, improving the total transit depth signal from about 1/10 000 for an Earth transiting a Sun, to 1/2500 or 1/100 for the same-sized planet. A planet's equilibrium temperature scales as $T_{eq} \sim T_*(R_*/a)^{1/2}$, where a is the planet's semi-major axis. The temperature of low-mass stars is about 3500–2800 K (for stars smaller than the Sun by 2 and 10 times, respectively). The habitable zone of a low-mass star would therefore be 4.5–42 times closer to the star compared to the Sun's habitable zone. To measure a planet mass, the radial-velocity semi-amplitude, K, scales as $K \sim (aM_*)^{-1/2}$, and the low-mass star masses are 0.4–0.06 times that of the Sun. Obtaining a planet mass is therefore about 3–30 times more favorable for an Earth-mass planet orbiting a low-mass star compared to an Earth–Sun analog. The transit probability scales as R_*/a. The probability for a planet to transit in a low-mass star's habitable zone is about 2.3–5%, much higher than the low 0.5% probability of an Earth–Sun analog transit. Finally, from Kepler's third law, a planet's orbital period, P, scales as $P \sim a^{3/2}/M_*^{1/2}$, meaning that the period of a planet in the habitable zone of a low-mass star is 7–90 times shorter than the Earth's one-year period, and the planet transit can be observed often enough to build up a signal (as compared to an Earth analog once-a-year transit). A super-Earth larger than Earth (and up to about 10 $M_\oplus$ and 2 $R_\oplus$) is even easier to detect, because of its larger transit signal and mass signature, than Earth. (See Nutzman and Charbonneau (2008) for more discussion on the benefit of targeted-star searches for transiting planets orbiting in habitable zones of M stars, and see Deming et al. (2009) for a description of the proposed TESS space-based survey transiting planet yield.)

Debate on how habitable a planet orbiting close to an M star can actually be is an active topic of research (see the reviews by Scalo et al. 2007 and Tarter et al. 2007). Some previously accepted "show stoppers" are no longer considered serious impediments. Atmospheric collapse due to cold night-side temperatures on a tidally locked planet will not happen as long as the atmosphere is thick enough (0.1 bar) for atmospheric circulation to redistribute absorbed stellar energy to heat the night side (Joshi et al. 1997). Short-term stellar variability due to large-amplitude star spots could change the planet's surface temperature by up to 30 K in the most severe cases (Joshi et al. 1997, Scalo et al. 2007), but even some terrestrial life can adapt to freeze–thaw scenarios. Bursts of UV radiation due to stellar flares could be detrimental for life, but the planet's surface could be protected by a thick abiotic ozone layer (Segura et al. 2005), or alternatively life could survive by inhabiting only the subsurface.

Other concerns about the habitability of planets orbiting close to M stars have not yet been resolved. Flares and UV radiation could erode planet atmospheres (Lammer et al. 2007), especially because the active phase of M stars can last for billions of years (West et al. 2008). Tidally locked planets in M-star habitable zones will be slow rotators; a weak magnetic field due to an expected small dynamo effect will not protect the atmosphere from erosion. Planets accreting in regions that become habitable zones of M-dwarf stars form rapidly (within several million years); the planet may not have time to accrete volatiles (e.g. water) that are present in the proto-planetary disk much farther away from the star (Lissauer 2007).

Observers never limit their plans based on theoretical reasoning, and so the search for life on planets in the habitable zones of M stars will proceed. Nevertheless, the transiting super-Earths in the habitable zones are highly valuable as the nearest-term potentially habitable planets. Eventually the pendulum will swing back from M stars to true Earth analogs. For the rest of the chapter, we focus on the true Earths orbiting Sun-like stars.

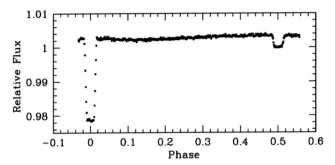

Figure 11.4 Transiting planet example: Infrared light curve of HD 189733a and b. The flux in this light curve is from the star and planet combined. The first dip (from left to right) is the transit and the second dip is the secondary eclipse. The error bars have been suppressed for clarity. Data from Knutson *et al.* (2007).

Are Earths common?

The answer to how common other Earths are is on its way. NASA's Kepler Space Telescope (Borucki *et al.* 2010, 2011) aims to find the frequency of Earth-size planets in Earth-like orbits about Sun-like stars. Kepler will do this by staring at a field of Sun-like stars for 3.5 years.

NASA's Kepler Space Telescope is a Discovery-class mission designed to determine the frequency of Earth-size planets in and near the habitable zone of solar-type stars. The Kepler Mission, launched in March 2009, uses transit photometry. The instrument is a wide field-of-view (115 deg^2) photometer comprising a 0.95 m effective aperture Schmidt telescope feeding an array of 42 CCDs that continuously and simultaneously monitor the brightness of up to 170 000 stars. A comprehensive discussion of the characteristics and on-orbit performance of the instrument and spacecraft is presented in Koch *et al.* (2010). The Kepler mission recently reached a milestone toward meeting its primary goal: the discovery of its first rocky planet, Kepler-10b, a $1.146^{+0.033}_{-0.036}$ R$_\oplus$ and $4.56^{+1.17}_{-1.29}$ M$_\oplus$ planet in an 0.84-day period orbit about a Sun-like star (Batalha *et al.* 2011; see, for example, Figure 11.4). Kepler planet discoveries, planet candidates, and statistical analysis for the first four months of data is presented by Borucki *et al.* (2011).

Kepler has been designed to determine the frequency of other Earths. Kepler has not been designed to tell us that an individual star has a planet the size of Earth. In other words, Kepler is taking a statistical approach to tell us the frequency of other Earths. Kepler has over 1200 planet candidates. To turn a planet candidate into a conventionally accepted planet means verification of the Kepler planet signature by a mass measurement. Traditionally this is done by radial-velocity follow-up measurements, but most Kepler stars with small planets will be too faint for this approach. A large number of multiple-planet systems have been discovered and when the transit times are perturbed, masses of the planets can be determined or constrained. The problem with planet candidates is false positives. Other astrophysical phenomena, namely categories of binary stars, can mimic a transiting planet signature. Kepler will estimate its false positive rate in order for meaningful statistics to be derived from the large and growing pool of Kepler planet candidates.

Recent, careful attempts have estimated the frequency of Earth-mass planets (0.5–2 $M_\oplus$) planets at about 1 in 4 for periods of less than 50 days (Howard *et al.* 2010). Related constraints for planets larger than Earth measure the occurrence of planets with periods less than 50 days and with radii 2–4 Earth radii at 0.130 ± 0.008 (Howard *et al.* 2012). For planets with orbital periods closer to Earth's 365.25 days, we must be patient and await Kepler's findings after accruing more data.

How to find biosignatures on other Earths

Biosignatures

Exoplanets are too distant for us to travel to or send probes to. We therefore plan to observe the planets remotely, using sophisticated space telescopes to observe their atmospheres for signs of life. An atmospheric biosignature gas is one produced by life. Life metabolizes and generates metabolic byproducts. Some metabolic byproducts dissipate into the atmosphere and can accumulate as biosignature gases. In exoplanets, then, we focus on a "top down" approach of a biosignature framework. In the top down approach, we do not worry about what life is, just about what life does (i.e. life metabolizes). The "bottom up" approach – the details of the origins and evolution of life – is left to the biologists.

The canonical concept for the search for atmospheric biosignatures is to find an atmosphere severely out of thermochemical redox equilibrium (Lederberg 1965, Lovelock 1965). Redox chemistry adds or removes electrons from an atom or molecule (<u>red</u>uction or <u>ox</u>idation, respectively). Redox chemistry is used by all life on Earth and is thought to enable more flexibility than non-redox chemistry. The idea is that gas byproducts from metabolic redox reactions can accumulate in the atmosphere and would be recognized as biosignatures because abiotic processes are unlikely to create a redox disequilibrium. Indeed Earth's atmosphere has oxygen (a highly oxidized species) and methane (a highly reduced species) several orders of magnitude out of thermochemical redox equilibrium.

In practice it could be difficult to detect redox disequilibrium molecular features. The Earth as an exoplanet, for example (Figure 11.5), has a relatively

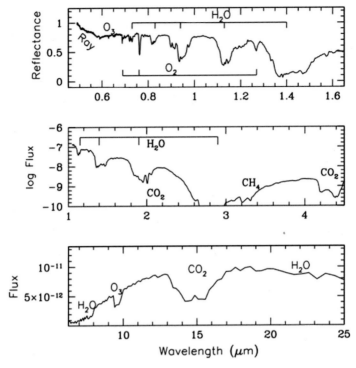

Figure 11.5 Earth as an exoplanet: Earth's hemispherically averaged spectrum. Top: Earth's visible wavelength spectrum from Earthshine measurements plotted as normalized reflectance (Turnbull *et al.* 2006). Middle: Near-infrared spectrum from NASA's EPOXI mission with flux in units of $W\,m^{-2}\,\mu m^{-1}$ (Robinson *et al.* 2011). Bottom: Earth's mid-infrared spectrum as observed by Mars Global Surveyor en route to Mars with flux in units of $W\,m^{-2}\,Hz^{-1}$ (Pearl and Christensen 1997). Major molecular absorption features are noted including Rayleigh scattering. Only strongly absorbing, globally mixed molecules are detectable.

prominent oxygen absorption feature at 0.76 μm, whereas methane at present-day levels of 1.6 ppm has only extremely weak spectral features. During early Earth CH_4 may have been present at much higher levels (1000 ppm or even 1%), as possibly produced by widespread methanogen bacteria (Haqq-Misra *et al.* 2008 and references therein). Such high CH_4 concentrations would be easier to detect, but since the Earth was not oxygenated during early times the O_2-CH_4 redox disequilibrium would not be detectable concurrently (see Des Marais *et al.* 2002).

The more realistic atmospheric biosignature gas is a single gas completely out of chemical equilibrium. Earth's example again is oxygen or ozone, about eight orders of magnitude higher than expected from equilibrium chemistry and with no known abiotic production at such high levels. The challenge with a single biosignature outside of the context of redox chemistry becomes one of false positives. To avoid false positives we must look at the whole atmospheric context. For example, high atmospheric oxygen content might indicate a planet undergoing a runaway greenhouse with evaporating oceans. When water vapor in the atmosphere is being photodissociated with H escaping to space, O_2 will build up in the atmosphere for a short period of time. In this case, O_2 can be associated with a runaway greenhouse via very saturated water vapor features, since the atmosphere would be filled with water vapor at all altitudes. Other O_2 and O_3 false positive scenarios (Selsis *et al.* 2002) are discussed and countered by Segura *et al.* (2007).

Most work to date has focused on mild extensions of exoplanet biosignatures compared to Earth (O_2, O_3, N_2O) or early Earth (possibly CH_4) biosignatures. Research forays into biosignature gases that are negligible on Earth but may play a more dominant role on other planets has started. Pilcher (2003) suggested that organosulfur compounds, particularly methanethiol (CH_3SH, the sulfur analog of methanol), could be produced in high enough abundance by bacteria, possibly creating a biosignature on other planets. Pilcher (2003) emphasized a potential ambiguity in interpreting the 9.6 μm O_3 spectral feature since a CH_3SH feature overlaps with it. Segura *et al.* (2005) showed that the Earth-like biosignature gases CH_4, N_2O, and even CH_3Cl have higher concentrations and therefore stronger spectral features on planets orbiting M stars compared to Earth. The reduced UV radiation on quiet M stars enables longer biosignature gas lifetimes and therefore higher concentrations to accumulate. Seager, Schrenk, and Bains (2012) have reviewed Earth-based metabolism to summarize the range of gases and solids produced by life on Earth. A fruitful new area of research will be from which molecules are there potential biosignatures and which can be identified as such on super-Earth planets different from Earth.

Breaking free from terra-centrism, the NRC Report on *The Limits of Organic Life in Planetary Systems* by Baross *et al.* (2007) proposed that the conservative requirements for life of liquid water and carbon could be replaced by the more general requirements of a liquid environment and an environment that can support covalent bonds (especially between hydrogen, carbon, and other atoms). This potentially opens up a new range for habitable planets, namely those beyond the ice-line in a cryogenic habitable zone where water is frozen but other liquids such as methane and ethane are present. Although no one has yet studied the possibility of exoplanet biosignatures on cold exoplanets, research on what it takes for life to exist in non-water liquids (Bains 2004) and the possibilities for methanogenic life in liquid methane on the surface of Titan (McKay and Smith 2005) are a useful start.

In addition to atmospheric biosignatures, the Earth has one very strong and extremely intriguing biosignature on its surface: vegetation. The reflection spectrum of photosynthetic vegetation has a dramatic sudden rise in albedo around 750 nm by almost an order of magnitude! Vegetation has evolved this strong reflection feature, known as the "red edge," as a cooling mechanism to prevent overheating which would cause chlorophyll to degrade. On Earth, due to clouds, this signature is probably reduced to a few percent (see, for example, Woolf *et al.* 2002, Seager *et al.* 2005, Montanes-Rodriguez *et al.* 2006, and references therein) but such a spectral surface feature could be much stronger on a planet with a lower cloud cover fraction. Recall that any observations of Earth-like extrasolar planets will not be able to spatially resolve the surface. A surface biosignature could be distinguished from an atmospheric signature by time variation; as the continents, or different concentrations of the surface biomarker, rotate in and out of view the spectral signal will change correspondingly.

Water vapor, while not a biosignature gas, is considered a habitability indicator. In the atmosphere of a small rocky planet, water vapor would be photodissociated with the hydrogen escaping to space. Only with a surface liquid-water reservoir could water vapor be present on a rocky planet with moderate temperatures.

Research beyond Earth's conventional and strong biosignature gases is only just beginning to unfold and seems to be fertile ground for new lines of investigation – especially with the considerable interest it draws. Nonetheless we will always be humbled by the warning that in many cases we will not be 100% sure of a biosignature's definite attribution to life. Before we can become too serious about any future detection of biosignatures, however, we must examine how we can detect Earth-like planets that would support life.

Prospects for studying an Earth-like exoplanet atmosphere

The most natural way to think of observing exoplanet atmospheres is by taking an image of the exoplanet. This would be akin to taking a photograph of the stars with a digital camera, although using a very expensive, high-quality detector. This so-called "direct imaging" of planets is currently limited to big, bright, young, or massive planets located far from their stars (see Figure 11.1, Kalas *et al.* 2008, Marois *et al.* 2008). Direct imaging of substellar objects is currently possible with large ground-based telescopes and adaptive optics to cancel the atmosphere's blurring effects. Solar-System-like small planets are not observable via direct imaging with current technology, even though an Earth at 10 pc is brighter than the faintest galaxies observed by the Hubble Space Telescope. The major impediment to direct observations is instead the adjacent host star; the Sun is 10 million to 10 billion times brighter than Earth (for mid-infrared and visible wavelengths, respectively). No existing or planned telescope is capable of

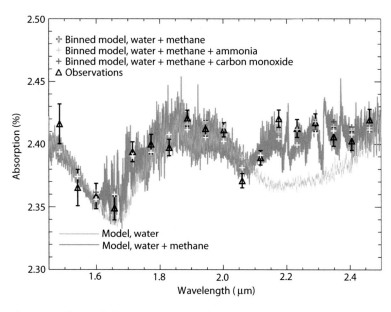

Figure 11.6 Transmission spectrum example: HD 189733. Hubble Space Telescope observations are shown by the black triangles. Two different models highlight the presence of methane in the planetary atmosphere. Reprinted by permission from Macmillan Publishers Ltd: *Nature*, Swain *et al.* (2008), copyright 2008. See also Gibson *et al.* (2011) for analysis of the same data with differing results. A colour version of this illustration appears in the colour plate section between pp. 148–149.

achieving this contrast ratio at 1 AU separations. The high planet–star contrasts are prohibitive. Fortunately much research and technology development is ongoing for space-based direct imaging to enable direct imaging of Earths and Jupiters in solar systems as old as the current Solar System, in the future.

Transiting planets, successfully observed with techniques unique to transiting planets, has been mainly limited to hot giant planet atmospheres (Seager and Deming 2010, and references therein), and is not considered ideal for atmospheric characterization for terrestrial planets for a few reasons. First, transiting Earths orbiting Sun-like stars are rare, and second, their atmospheric signal in transmission will almost certainly be too small (Kaltenegger and Traub 2009).

Transiting super-Earth atmospheres are, however, expected to be observable if the planet is orbiting a small (M) star. We anticipate the discovery of a handful of rare but highly valuable transiting super-Earths in the habitable zones of the brightest low-mass stars. With such prize targets, astronomers will strive to observe the transiting super-Earth atmospheres in the same way we are currently observing transiting hot Jupiters orbiting Sun-like stars (Figure 11.6). Fortunately, the nearest example of a habitable transiting super-Earth is expected to lie closer than

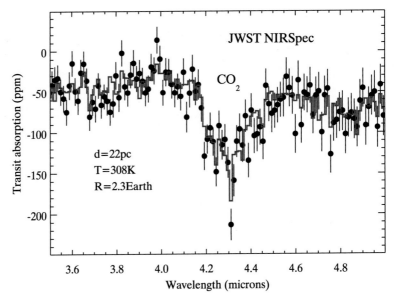

Figure 11.7 Simulated spectrum of a super-Earth as would be observed over many
transits by the James Webb Space Telescope near-infrared spectrometer instrument.
The points and error bars are the simulated spectrum of carbon dioxide absorption in
a habitable super-Earth having $T = 308$ K and $R = 2.3$ $R_\oplus$, with the model spectrum
overlaid (blue line). The aggregate SNR for this detection is SNR $= 28$ for 85 hours of
in-transit observing, and the distance to this planetary system is $d = 22$ parsecs.
Redrawn from Deming *et al.* (2009). A colour version of this illustration appears in the
colour plate section between pp. 148–149.

35 parsecs (Deming *et al.* 2009). If such a system can be found using an all-sky survey
like TESS (Ricker *et al.* 2010), NASA's James Webb Space Telescope (JWST) will be
capable of observing the absorption signatures of major molecules like water and
carbon dioxide. Such observations will require monitoring of multiple transits,
often amounting to ∼100 hours of JWST observation. For example, in our simulations (Deming *et al.* 2009; Figure 11.7), one habitable super-Earth lying at 22 parsecs
distance required 85 hours of JWST/NIRSpec observations to measure carbon dioxide absorption during transit to SNR $= 28$. The atmospheres of super-Earths with
temperatures above the habitable range will be observable by JWST using only a
few transits and/or eclipses. For a comparison of Earth-like planet atmospheres
around both M stars and Sun-like stars, see Kaltenegger and Traub (2009).

Detection prospects for Earth analogs

Despite any near-term fixation on the easier-to-detect super-Earths transiting M stars, we, as humans, will always desire to find an Earth twin. Discovery of an

Earth analog is a massive challenge for each of the different exoplanet discovery techniques (Figures 11.1 and 11.4) and direct imaging is no exception. In order to observe a planet under the very low Earth–Sun flux contrast, at visible wavelengths the diffracted light from the star must be suppressed by 10 billion times.[3] Space-based direct imaging is the best option for detection and characterization of Earth analogs.

A collection of related ideas to suppress the starlight is called "coronagraphs," a term that originated with the artificial blocking out of sunlight to observe the Sun's corona (Lyot 1939). Novel-shaped apertures, pupils, and pupil masks to suppress some or all of the diffracted starlight have been developed in the last few years (see, for example, Trauger and Traub 2007, and references therein). At mid-infrared wavelengths, an interferometer would be needed to avoid the exceedingly large aperture to spatially separate an Earth-like planet from its star for moderate distances from Earth (i.e. because of the λ/D scaling; see Cockell *et al.* 2009 and references therein for a description of the Darwin mission). At the present time, there are no plans to build and launch a Terrestrial Planet Finder or Darwin mission. The price tag and complexity are almost prohibitive.

A recent development has given renewed promise for a Terrestrial Planet Finder type of mission. The idea is to use the already planned JWST together with a novel-shaped, external occulter situated tens of thousands of kilometers from the telescope in order to suppress the diffracted starlight (Cash 2006, Soummer *et al.* 2009). Most of the time the JWST would be functioning as planned, and during this time the star-shade would fly across the sky to the next target star. While there are many technical and programmatic concerns for the JWST plus an external occulter idea, none are without solution.

Conclusion

At a few special times in history, astronomy has changed the way we see the universe. Hundreds of years ago, humanity believed that Earth was the center of everything – that the known planets and stars all revolved around Earth. In the late sixteenth century, the Polish astronomer Nicolaus Copernicus presented his revolutionary new view of the universe, where the Sun was the center, and Earth and the other planets all revolved around it. Gradually, science adopted this "Copernican" theory (solidly after Comet Halley's successfully predicted return), but this was only the beginning. In the early twentieth century, astronomers

---

3 This level of suppression is not possible from ground-based telescopes, even the ELT of the future, because no source is bright enough for adaptive optics; a guide star of magnitude $R = -7$ would be needed (Traub and Oppenheimer 2010).

concluded that there are galaxies other than our own Milky Way. Astronomers eventually recognized that our Sun is but one of hundreds of billions of stars in our Galaxy, and that our Galaxy is but one of upwards of hundreds of billions of galaxies. When and if we find that other Earths are common and see that some of them have signs of life, we will at last complete the Copernican Revolution – a final conceptual move of the Earth, and humanity, away from the center of the Universe. This is the promise and hope for exoplanets – the detection and characterization of habitable worlds.

References

Bains, W. (2004). "Many chemistries could be used to build living systems," *Astrobiology*, Vol. 4, pp. 137–167.

Baross, J. A. *et al.* (2007). *The Limits to Organic Life in Planetary Systems*. Washington, DC: National Academies Press.

Batalha, N. M. *et al.* (2010). "Selection, prioritization, and characteristics of Kepler target stars," *Astrophysical Journal*, Vol. 713, pp. L109–L114.

Batalha, N. M., Borucki, W. J., Bryson, S. T. *et al.* (2011). "Kepler's first rocky planet: Kepler-10b," *Astrophysical Journal*, Vol. 729, 27.

Bean, J. L., Seifahrt, A., Hartman, H. *et al.* (2010). "The proposed giant planet orbiting VB 10 does not exist," *Astrophysical Journal Letters*, Vol. 711, pp. L19–L23.

Borucki, W. J. *et al.* (2010). "*Kepler* planet detection mission: introduction and first results," *Science*, Vol. 327, p. 977.

Borucki, W. J. *et al.* (2011). "Characteristics of Kepler planetary candidates based on the first data set," *Astrophysical Journal*, Vol. 728, 117.

Boss, A. (2009). "The Carnegie astrometric planet search program," *Publications of the Astronomical Society of the Pacific*, Vol. 121, pp. 1218–1231.

Butler, R. P. *et al.* (2006). "Catalog of nearby exoplanets," *Astrophysical Journal*, Vol. 646, pp. 505–222.

Cash, W. (2006). "Detection of Earth-like planets around nearby stars using a petal-shaped occulter," *Nature*, Vol. 442, pp. 51–53.

Cockell, C. S., Léger, A., Fridlund, M. *et al.* (2009). "Darwin – a mission to detect and search for life on extrasolar planets," *Astrobiology*, Vol. 9, pp. 1–22.

Deming, D. *et al.* (2009). "Discovery and characterization of transiting super-Earths using an all-sky transit survey and follow-up by the James Webb Space Telescope," *Publications of the Astronomical Society of the Pacific*, Vol. 121, pp. 952–967.

Des Marais, D. J. *et al.* (2002). "Remote sensing of planetary properties and biosignatures on extrasolar terrestrial planets," *Astrobiology*, Vol. 2, pp. 153–181.

Gaudi *et al.* (2008). "Discovery of a Jupiter/Saturn analog with gravitational microlensing," *Science*, Vol. 319, pp. 927–930.

Gibson, N. P., Pont, F., and Aigrain, S. (2011). "A new look at NICMOS transmission spectroscopy of HD 189733, GJ-436 and XO-1: no conclusive evidence for molecular features," *Monthly Notices of the Royal Astronomical Society*, Vol. 411, pp. 2199–2213.

Haqq-Misra, J. D. *et al.* (2008). "A revised, hazy methane greenhouse for the Archean Earth," *Astrobiology*, Vol. 8, pp. 1127–1137.

Howard, A. W. *et al.* (2010). "The occurrence and mass distribution of close-in super-Earths, Neptunes, and Jupiters," *Science*, Vol. 330, p. 653.

Howard, A. W. *et al.* (2012). "Planet occurrence within 0.25 AU of solar-type stars from Kepler," *Astrophysical Journal*, Vol. 201, p. 15.

Jayawardhana, R. (2011). *Strange New Worlds*. Princeton, NJ: Princeton University Press.

Joshi, M. M., Haberle, R. M., and Reynolds, R. T. (1997). "Simulations of the atmospheres of synchronously rotating terrestrial planets orbiting M dwarfs: conditions for atmospheric collapse and the implications for habitability," *Icarus*, Vol. 129, pp. 450–465.

Kalas, P. *et al.* (2008). "Optical images of an exosolar planet 25 light-years from Earth," *Science*, Vol. 322, pp. 1345–1348.

Kaltenegger, L. and Traub, W. (2009). "Transits of Earth-like planets," *Astrophysical Journal*, Vol. 698, pp. 519–527.

Knutson, H. A. *et al.* (2007). "A map of the day–night contrast of the extrasolar planet HD 189733b," *Nature*, Vol. 447, pp. 183–186.

Koch, D. G., Borucki, W. J., Basri, G. *et al.* (2010). "Kepler mission design, realized photometric performance, and early science," *Astrophysical Journal Letters*, Vol. 713, pp. L79–L86.

Lammer, H. *et al.* (2007). "Coronal mass ejection (CME) activity of low mass M stars as an important factor for the habitability of terrestrial exoplanets, II. CME-induced ion pick up of Earth-like exoplanets in close-in habitable zones," *Astrobiology*, Vol. 7, pp. 185–207.

Lederberg, J. (1965). "Signs of life: criterion-system of exobiology," *Nature*, Vol. 207, pp. 9–13.

Li, C.-H. *et al.* (2008). "A laser frequency comb that enables radial velocity measurements with a precision of 1 cm s^{-1}," *Nature*, Vol. 452, pp. 610–612.

Lissauer, J. J. (2007). "Planets formed in habitable zones of M dwarf stars probably are deficient in volatiles," *Astrophysical Journal*, Vol. 660, pp. 149–152.

Lovelock, J. E. (1965). "A physical basis for life detection experiments," *Nature*, Vol. 207, pp. 568–570.

Lyot, B. (1939). "The study of the solar corona and prominences without eclipses (George Darwin Lecture, 1939)," *Monthly Notices of the Royal Astronomical Society*, Vol. 99, p. 580.

Marois, C., Macintosh, B., Barman, T. *et al.* (2008). "Direct imaging of multiple planets orbiting the star HR 8799," *Science*, Vol. 322, pp. 1348–1352.

McKay, C. P. and Smith, H. D. (2005). "Possibilities for methanogenic life in liquid methane on the surface of Titan," *Icarus*, Vol. 178, pp. 274–276.

Montanes-Rodriguez, P., Palle, E., Goode, P. R., and Martin-Torres, F. J. (2006). "Vegetation signature in the observed globally integrated spectrum of Earth considering simultaneous cloud data: applications for extrasolar planets," *Astrophysical Journal*, Vol. 651, pp. 544–552.

Nutzman, P. and Charbonneau, D. (2008). "Design considerations for a ground-based transit search for habitable planets orbiting M dwarfs," *Publications of the Astronomical Society of the Pacific*, Vol. 120, pp. 317–327.

Pearl, J. C. and Christensen, P. R. (1997). "Initial data from the Mars Global Surveyor thermal emission spectrometer experiment: observations of the Earth," *Journal of Geophysical Research*, Vol. 102, pp. 10875–10880.

Pilcher, C. B. (2003). "Biosignatures of early Earths," *Astrobiology*, Vol. 3, pp. 471–486.

Ricker, G. R., Latham, D. W., Vanderspek, R. K. *et al.* (2010). "Transiting exoplanet survey satellite (TESS)," *Bulletin of the American Astronomical Society*, Vol. 42, p. 459.

Robinson, T. D. *et al.* (2011). "Earth as an extrasolar planet: Earth model validation using EPOXI Earth observations," *Astrobiology*, Vol. 11, pp. 393–408.

Scalo, J. *et al.* (2007). "M stars as targets for terrestrial exoplanet searches and biosignature detection," *Astrobiology*, Vol. 7, pp. 85–166.

Seager, S. and Deming, D. (2010). "Exoplanet atmospheres," *Annual Reviews of Astronomy and Astrophysics*, Vol. 48, pp. 631–672.

Seager, S. and Lissauer, J. J. (2010). "Introduction," in *Introduction to Exoplanets*, S. Seager (ed.). Tucson: University of Arizona Press, pp. 3–13.

Seager, S., Schrenk, M. and Bains, W. (2012). "An astrophysical veiw of Earth-based metabolic biosignature gases," *Astrobiology*, Vol. 12, pp. 61–82.

Seager, S., Turner, E. L., Schafer, J., and Ford, E. B. (2005). "Vegetation's red edge: a possible spectroscopic biosignature of extraterrestrial planets," *Astrobiology*, Vol. 5, pp. 372–390.

Segura, A. *et al.* (2005). "Biosignatures from Earth-like planets around M dwarfs," *Astrobiology*, Vol. 5, pp. 706–725.

Segura, A., Meadows, V. S., Kasting, J. F. *et al.* (2007). "Abiotic formation of O_2 and O_3 in high-CO_2 terrestrial atmospheres," *Astronomy and Astrophysics*, Vol. 472, pp. 665–679.

Selsis, F., Despois, D., and Parisot, J. P. (2002). "Signature of life on exoplanets: can Darwin produce false positive detections?," *Astronomy and Astrophysics*, Vol. 388, pp. 985–1003.

Silvotti, R. *et al.* (2007). "A giant planet orbiting the 'extreme horizontal branch' Star V 391 Pegasi," *Nature*, Vol. 449, pp. 189–191.

Soummer, R., Cash, W., Brown, R. A. *et al.* (2009). "A star-shade for JWST: science goals and optimization," *SPIE*, Vol. 7440, p. 1.

Swain, M. R., Vasisht, G. and Tinetti, G. (2008). "The presence of methane in the atmosphere of an extrasolar planet," *Nature*, Vol. 452, pp. 329–331.

Tarter, J. C. *et al.* (2007). "A reappraisal of the habitability of planets around M dwarf stars," *Astrobiology*, Vol. 7, pp. 30–65.

Traub, W. and Oppenheimer, B. (2010). "Direct imaging of exoplanets," in *Exoplanets*, S. Seager (ed.). Tucson: University of Arizona Press, pp. 111–136.

Trauger, J. T. and Traub, W. A. (2007). "A laboratory demonstration of the capability to image an Earth-like extrasolar planet," *Nature*, Vol. 446, pp. 771–773.

Turnbull, M. C., Traub, W. A., Jucks, K. W. *et al.* (2006). "Spectrum of a habitable world: earthshine in the near-infrared," *Astrophysical Journal*, Vol. 644, pp. 551–5599.

Udry, S., Fischer, D., and Queloz, D. (2007). "A decade of radial velocity discoveries in the exoplanet domain," in *Protostars and Planets*, Vol. v, B. Reipurth, D. Jewitt, and K. Keil (eds.). Tucson: University of Arizona Press, pp. 685–699.

West, A. A. *et al.* (2008). "Constraining the age-activity relation for cool stars: the Sloan digital sky survey data release 5 low-mass star spectroscopic sample," *Astronomical Journal*, Vol. 135, pp. 785–795.

Wolszczan, A. and Kuchner, M. (2010). "Planets around pulsars and other evolved stars: the fates of planetary systems," in *Exoplanets*, S. Seager (ed.). Tucson: University of Arizona Press, pp. 175–190.

Wolszczan, A. and Frail, D. A. (1992). "A planetary system around the millisecond pulsar PSR1257 + 12," *Nature*, Vol. 355, pp. 145–147.

Woolf, N. J., Smith, P. S., Traub, W. A., and Jucks, K. W. (2002). "The spectrum of earthshine: a pale blue dot observed from the ground," *Astrophysical Journal*, Vol. 574, pp. 430–433.

12

Review of Known Exoplanets

CHRISTOPHE LOVIS AND DANTE MINNITI

Introduction

For centuries and even millennia, mankind has been wondering whether there exist other worlds similar to ours populating the universe. Until about 1600 AD, these questions have remained outside the field of scientific investigation due to a lack of observational means able to address the issue. The situation began to evolve with the invention of optical instruments. It all started with Galileo Galilei, who made the first discoveries of new worlds using the first very modest telescopes. He discovered the four largest satellites of Jupiter, as tiny points of light that circle the giant planet. We are now able to measure the masses and radii of these satellites, compute their mean densities, and conclude that water ice is a major constituent of Europa, Ganymede, and Callisto.

Galileo also found that the Milky Way is made of millions of stars. We now know that our Galaxy contains a few hundred billion stars, but critical questions remain: How many of them have planets? Just a few, or most of them? What if every star has planets like our Sun does? Would these planets be similar to the ones we know around the Sun? We are lucky enough to live in an era of large telescopes and powerful instruments, giving us for the first time the opportunity to try to answer these important questions. This chapter deals with the search and study of these *extrasolar* planets, worlds orbiting other stars beyond our Solar System.

We may start by defining what a planet actually is, and what the difference is between planets and stars. Most stars shine and produce light because they possess a powerful internal energy source: they burn H into He in their cores in a process called nuclear fusion. Fusion occurs because of the high densities

Frontiers of Astrobiology, ed. Chris Impey, Jonathan Lunine and José Funes.
Published by Cambridge University Press. © Cambridge University Press 2012.

and temperatures in stellar nuclei. To reach the threshold to ignite hydrogen in the core, a total mass of at least $\sim$0.08 of a solar mass ($M_\odot$) is required. Thus, stars are celestial bodies that have masses reaching at least this value. Below this threshold one finds some intermediate objects called brown dwarfs, which are not massive enough for standard fusion to ignite, but can still burn deuterium (heavy hydrogen) in their interiors.

At even lower masses begins the realm of planets. Below about 13 M_{Jup} (0.014 $M_\odot$), no nuclear reactions can occur and objects are not hot enough to shine by themselves (that is, in the visible/near-infrared domains). Such objects are called planets, and are usually formed as a byproduct of star formation itself. The canonical scenario has them emerging from a proto-planetary disk of gas and dust that surrounds stars during their formation. The details of this process are a major focus of today's astrophysical research, and many open questions remain regarding the general properties of planet populations (mass, size, orbital distance, multiplicity, etc.). However, the past 20 years have witnessed an explosion of discoveries in this field that has revolutionized our vision of planetary systems. As it often happens, extending our knowledge from a single example (the Solar System) to many of them has revealed a much more diverse reality than anticipated, and has forced us to reconsider our theories. In this chapter we will try to highlight these new findings and show how they put our Solar System into a broader perspective.

The detection of extrasolar worlds

There are different ways to search for extrasolar planets, and astronomers are constantly looking for better ways of doing this. The main technical challenges arise from the very high flux contrast between an exoplanet and its host star, coupled to their very small angular separation on the sky due to the large distance of stars from the Sun. Imaging an exoplanet requires an optical system able to separate its light from that of the host star, an extremely challenging requirement given a typical 10^{-7}–10^{-10} flux ratio and sub-arcsecond separation between planet and star. Because of the unavoidable atmospheric turbulence, ground-based observations have to cope with blurred images that offer no details below an angular scale of at best 0.5–1.0 arcsecond, depending on the observing site, weather conditions, and wavelength of observations.

These difficulties have triggered astronomers to explore more indirect methods to search for and study exoplanets. Since the light from the planets themselves is so difficult to isolate, they turned to the stellar light and tried to guess the presence of planets through their effects on their host star. This approach has provided us with the two most successful techniques for detecting exoplanets to date: high-precision radial velocities and transit photometry, which are described in the next two sections.

The radial-velocity technique

As seen by an external observer, a planet does not, strictly speaking, orbit its host star; both planet and star actually orbit around the center of mass of the two-body system. Because the star is much more massive than the planet, the latter describes a larger orbit and moves faster, while the star barely moves around the center of mass. Stellar motion imprints a signature on the emitted light: the Doppler effect. For circular orbits, the periodic movement of the star around the center of mass induces a sinusoidal wavelength shift in the stellar light, which is directly proportional to the radial velocity of the star along the line of sight to the observer. Stellar radial-velocity (RV) variations thus reveal the presence of an orbiting planet. The amplitude of the signal is related to the planet's mass and orbital distance, giving us information on the planet's orbital and physical properties. More precisely, the RV signal is inversely proportional to the square-root of the semi-major axis, and directly proportional to the planet's so-called minimum mass, i.e. its mass times the sine of the inclination angle of its orbit. The main difficulty in using this technique is that we need to measure velocity signals that are very small, of the order of meters per second. In a stellar spectrum, this represents a wavelength shift of order 10^{-8} of the wavelength, or about 1/1000 of the intrinsic width of a stellar spectral line. To achieve this, several stringent instrumental requirements must be met: the use of a high-dispersion spectrograph, a close monitoring of environmental conditions, a very stable wavelength calibration, an equally stable or closely controlled light path within the spectrograph, and obviously also a high signal-to-noise ratio to obtain enough precision on the radial-velocity measurement.

This has been made possible only in the last two decades, starting with the 1995 discovery of the planet around 51 Peg, a solar-type star 48 light-years away, by the Swiss astronomers Michel Mayor and Didier Queloz (Mayor and Queloz 1995). The discovery was the culmination of spectacular instrumental advances that had taken place in the years before, with several groups managing to improve radial-velocity precision from hundreds of meters per second down to less than 10 meters per second. In the years that followed the 51 Peg discovery, many more planets were detected with the radial-velocity technique, and these discoveries changed our vision of the universe. We now know for sure that there are other planets in our Milky Way, in fact billions of them by extrapolating what has been found around stars close to the Sun.

As of 2011, the technique has become so sensitive that it is now able to measure radial-velocity signals at the 1 m/s level (see e.g. the HARPS instrument in

Figure 12.1 The HARPS spectrograph, mounted on the European Southern Observatory 3.6-m telescope in La Silla, Chile (Mayor *et al.* 2003). It is presently the most precise radial-velocity instrument in the world, able to detect exoplanets as small as the Earth on short-period orbits thanks to its sub-m/s precision. Courtesy of the Geneva Observatory.

Figure 12.1). New regions of parameter space are thus constantly being explored, in particular the domain of low-mass planets, i.e. those that are not mainly made of hydrogen and helium like Jupiter and Saturn. Ice giants like Uranus and Neptune and massive telluric planets (super-Earths) are now detectable on orbits out to about 1 AU, the habitable zone around solar-type stars. As of today, the radial-velocity technique thus remains an essential tool for astronomers to detect low-mass exoplanets, probe the architecture of multi-planet systems, and measure the mass of these objects.

The transit technique

Luck also helps astronomers who know how to provoke it: in situations of almost perfect geometric alignment between the observer, the planet's orbital plane, and the star, the planet will be seen transiting the disk of its star from the observer's standpoint, blocking a fraction of the incoming stellar light and thus causing a detectable dimming of the star compared to the out-of-transit situation. The transit occurs once every orbital period and the amount of dimming is given by the relative planet-to-star disk area. Thus, the amount of light blocked by the planet depends on its size, and in this way the planet size can be measured. The two most important pieces of information given by the transits are then the orbital period and the size of the planet. The first successful detection of a planetary transit occurred for the hot Jupiter HD 209458b (Charbonneau *et al.* 2000).

The great advantage of transiting planets is that, provided their masses can be measured with the radial-velocity technique, their mean densities can then be computed from their known radius and mass. This opens the way to the physical characterization of these objects in terms of internal structure and chemical composition. If the mean density is much larger than water, it means that we have a planet made of dense material, i.e. it is solid. However, the mean density of most planets detected so far is much lower than the density of water, meaning that these objects are gas giants like Jupiter and Saturn. Amazingly, the combination of two indirect techniques allows us to determine whether distant exoplanets we cannot directly see are solid like Earth, a combination of solids and gas like Neptune, or gaseous like Saturn.

The detection of planetary transits requires a photometric precision that depends on planet size: for Jupiter-like gas giants the dimming is of the order of 1% of the nominal stellar flux, while an Earth-sized object only causes a 10^{-4} photometric variation. Ground-based telescopes have to struggle against atmospheric turbulence and chaotic variability on short time-scales, which effectively makes it almost impossible to detect planetary transits shallower than $\sim 10^{-3}$ from the ground. To access the domain of ice giants and telluric planets, it is thus necessary to go to space, where the perturbing effects of the atmosphere are no longer present. This fact has led to the development of two transit-finding space missions, the French-European CoRoT and the NASA Kepler satellites (see Figure 12.2). As of 2010, CoRoT has indeed found one super-Earth, CoRoT-7b (Léger *et al.* 2009), which has become the first solid object with a measured radius and mass, while Kepler has found hundreds of small-size candidates whose nature is currently being investigated.

Even more insights into the physics of exoplanets can be gained by studying transit events. In particular, transmission spectroscopy of stellar light passing through the planet's atmosphere during transit is able to reveal the thermal and physical structure of the atmosphere, together with its chemical composition. Several hot Jupiter atmospheres have been probed in this way using telescopes on the ground and in space. This topic is further developed in a dedicated chapter.

Direct imaging

Finally, technical progress has also been very fast in the field of direct imaging, i.e. acquiring actual *images* of exoplanets. To overcome the difficulties described at the beginning of this section, advanced adaptive optics systems are being developed. These allow us to correct the distorted wavefronts of incident

Figure 12.2 The NASA Kepler satellite before launch. The highly stable space environment allows Kepler to obtain light curves of unprecedented precision, which have revealed hundreds of small-size planet candidates. Courtesy of Ball Aerospace & Technologies Corp.

stellar light. By partially removing the blurring effects of the atmosphere, adaptive optics effectively increases spatial resolution, up to the diffraction limit of the telescope which is a linear function of its diameter.

Present-generation planetary imagers have already been able to detect massive planets around nearby young stars, orbiting at large semi-major axes $a > 10$–100 AU. These objects are obviously the easiest to detect thanks to the large angular separation from the host star and their relative brightness. Young systems are high-priority targets since planets are then still relatively hot because of their

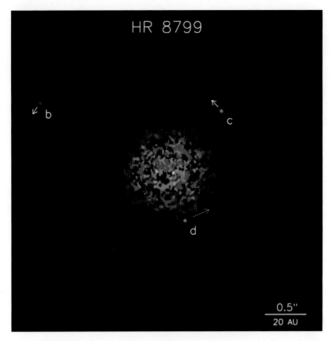

Figure 12.3 The planetary system surrounding the young star HR 8799 (from Marois *et al.* 2008). Three planets more massive than Jupiter are seen on distant orbits. Image courtesy of the National Research Council Canada, C. Marois & Keck Observatory. A colour version of this illustration appears in the colour plate section between pp. 148–149.

recent formation, and thus strongly emit in the near-infrared. Older planets like Jupiter have had ample time to cool and become much more difficult to detect. A spectacular illustration of the capabilities of direct imaging has been provided in 2008 with the detection of three massive planets on distant orbits around the young star HR 8799 (Marois *et al.* 2008, see Figure 12.3). Future advances in instrumentation and techniques like coronographs and interferometry at the next generation of telescopes on the ground and in space will open the way to substantial progress in direct imaging of exoplanets, by probing shorter orbital distances and lower planetary masses.

The diversity of the exoplanetary zoo

There has been a constant trend in the fast-evolving field of exoplanets since the first discovery in 1995: *a priori* expectations and theoretical predictions have been consistently falsified and have been unable to foresee the great diversity in exoplanet properties that has emerged over the years. We summarize below several key aspects of the presently known exoplanet population.

Hot Jupiters and distant massive planets

The first discovery – 51 Peg b in 1995 – also brought the first surprise to the astronomical community: the existence of Jupiter-like planets on incredibly close orbits (a few 0.01 AU, corresponding to orbital periods of a few days). Based on our understanding of the formation of the Solar System, gas giant planets were expected to reside on much more distant orbits (5–20 AU), where the density of solid material is sufficiently high in the proto-planetary disk to aggregate what would become a giant planet embryo. Forming a hot Jupiter *in situ* seems impossible because there is just not enough material to build the planet there, and because high temperatures would prevent the condensation of the required solid material. It took some time to find a plausible explanation to this puzzle, and as a result a new theory was developed that included orbital migration of the forming exoplanets.

Apparently some planets can migrate inwards or possibly outwards within the proto-planetary disk, and there is evidence that this also occurred in our Solar System, to a limited extent. The proposed explanation then is that the hot Jupiters formed far away from their parent stars, and then migrated inwards. This explains their present situation, but immediately raised other interesting questions, such as: What caused migration to stop? Or why did the planets stop migrating inwards and not fall entirely onto their stars? Orbital migration is still far from being completely understood today, and remains an active topic of research. It is probably only a part of the global picture, since some hot Jupiters have recently been found in misaligned orbits with respect to the stellar spin axis, or even on retrograde orbits. These new findings point to a much more complex and violent formation history, involving planet–planet scattering and/or three-body secular interactions.

Because they are relatively easy to find in transit, hot Jupiters have been discovered in large numbers by dedicated transit surveys, and represent the most-studied category of exoplanets. Another mystery remains there: the observed diversity in their mean densities. Many of them seem to have inflated radii, well above the expected radii predicted by internal structure models. An adequate explanation, involving an extra energy source in the interior or other, non-standard physics, is still being actively sought.

More recently, extremely distant massive planets have also been found by direct imaging, this time at semi-major axes (20–100 AU) where densities in the proto-planetary disk were likely much too low for the standard core accretion scenario of planet formation to work. Again, astronomers have yet to come up with a convincing mechanism able to produce the observed planets, but here the more star-like scenario of direct gravitational collapse within a limited region of the disk may be an important part of the process.

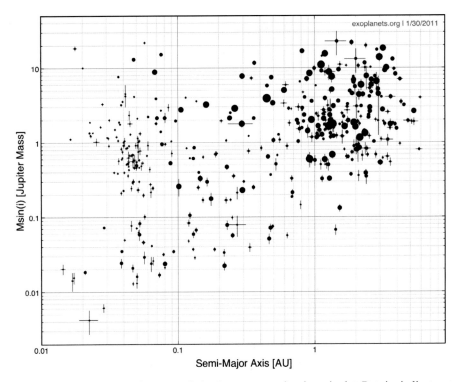

Figure 12.4 Exoplanet population in a mass–semi-major axis plot. Dot size indicates orbital eccentricity. This research has made use of the Exoplanet Orbit Database and the Exoplanet Data Explorer at exoplanets.org.

Distributions of planetary parameters

An easy way of capturing the basic properties of the exoplanet population is a mass–semi-major axis plot as shown in Figure 12.4. The plot contains all known planets discovered with the radial velocity or transit techniques. It reveals a wealth of information, but is at the same time very difficult to interpret correctly due to the numerous observational biases hiding behind it. Indeed, each detection technique has its own sensitivity and stellar search sample, and the data will be biased against planets that are not easily found either by radial-velocity or transit surveys, e.g. low-mass planets on distant orbits. A few interesting characteristics can still be read from this plot and are listed below.

- A clump of data points on short-period orbits indicates the hot Jupiter region. As mentioned above, this is a well-studied domain, but hot Jupiters are actually rare. The high density of points compared to other regions is artificial and is due to the very large stellar samples being searched for hot Jupiters compared to the more limited radial-velocity surveys.

- Another clump of points is seen in the region of "traditional" giant planets, orbiting beyond 1 AU. This reflects a real increase in giant planet occurrence towards the so-called "snow line," the distance to the star where ice condensates and causes an abrupt increase in the solid surface density of the proto-planetary disk. In the conventional view of planet formation, this is the expected region where giant planets can form. Present surveys indicate that about 10% of solar-type stars have at least one giant planet within ~5 AU.

- Another unexpected finding was the rather eccentric nature of many exoplanet orbits (indicated by the dot size in Figure 12.4). It turns out that giant planets in nearly circular orbits as in the Solar System are the minority, since most giant exoplanets have more eccentric orbits. In some cases the orbits are so elongated that they are impossible to explain with the existing theories of planet formation. Today the existence of most of these eccentric orbits is believed to be caused by interactions (e.g. close encounters) with other bodies (giant planets or binary stars).

- Correlations with the properties of the central star have also emerged (not shown in the plot): the higher the abundance in heavy elements in the star, the more likely it is to find giant planets around the star. In other words, high metallicity seems to favor giant-planet formation, probably due to the higher density of solid material in the proto-planetary disk. A similar correlation seems to hold with respect to stellar mass.

- Finally, it is interesting to note that more and more low-mass planets ($m < 0.1\ M_{Jup}$) are being found, despite the significant observational difficulties in detecting them. This indicates a much higher abundance of ice giants and super-Earths with respect to giant planets. See the next section for more details.

Multi-planet systems

Exoplanets are often not found alone, but rather as part of planetary systems. Multi-planet systems represent an outstanding opportunity to better understand the mechanisms of planet formation, through studies of their mass and orbital distribution, dynamical interactions, and comparisons with the Solar System. For observational reasons, the first multi-planet systems found were made of several giant planets, sometimes showing strong mutual gravitational interactions like mean-motion resonances. Many systems of giant planets appear to be on the verge of dynamical instability, and possibly hint at a complicated formation history featuring collisions between proto-planets and ejections of bodies from the system.

At this point, the history of the Solar System seems to have been relatively quiet compared to the dense and complicated architectures that are seen in other giant-planet systems. The orbits of Jupiter, Saturn, Uranus, and Neptune appear remarkably circular and stable in this context. However, one has to bear in mind that we still have a very incomplete picture of multi-planet systems, especially for planets orbiting beyond a few AUs, so that it is impossible to know how common a Solar System-like architecture really is.

The recent discoveries of systems of low-mass planets have opened a new parameter space and are beginning to reveal other dynamical patterns that will be discussed in the next section.

A new era: ice giants and super-Earths

Until 2004, all detected exoplanets had minimum masses above $\sim$0.1 M_{Jup} (or 30 $M_{\oplus}$). The situation suddenly changed that year, thanks to the continuous improvements in sensitivity of radial-velocity surveys. Three Neptune-mass objects were simultaneously found, two around solar-type stars already known to harbor giant planets, and one around an M dwarf, the low-mass star Gliese 436. With masses between 10 and 23 $M_{\oplus}$, these objects are likely to contain a large mass fraction of rocks and ices, albeit with a significant hydrogen envelope. This was confirmed when it was realized that Gliese 436 b actually transits its host star (Gillon *et al.* 2007), revealing a planetary mass and radius very similar to Neptune. Many discoveries of low-mass planets have followed since then, and we briefly describe these below.

A census of low-mass planets

As of 2010, there were 47 known planets with a (minimum) mass below 30 $M_{\oplus}$. The mass distribution of all known planets is shown in Figure 12.5, where a bimodal structure is clearly seen. Again, the distribution is strongly affected by observational biases, in particular at low masses, which makes the low-mass population increase all the more significant. Besides the gas giant and Neptune/super-Earth populations, few objects are found at intermediate masses. This may be explained by the fact that the gas accretion phase is a runaway process occurring on a very short time-scale once it has started. Thus, planets either accumulate a massive hydrogen envelope and become gas giants, or remain in the Neptune-mass range forever because the threshold for rapid accretion could not be reached before disk dissipation.

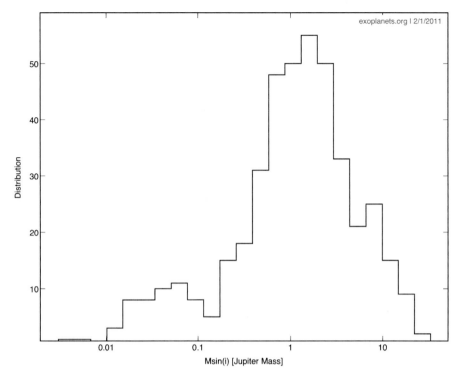

Figure 12.5 Mass distribution of exoplanets as of 2010. An increase in population below ~30 $M_\oplus$ is clearly seen despite strong observational biases towards low masses. The bimodal shape reveals distinct exoplanet populations, gas giants, and Neptunes/ super-Earths, and a relative paucity of intermediate-mass objects. This research has made use of the Exoplanet Orbit Database and the Exoplanet Data Explorer at exoplanets.org.

Interestingly, among the 47 low-mass objects, at least 30 are found in multi-planet systems. The global picture presently emerging is that of a large number of compact systems of low-mass planets. Several such examples have been found in recent years by radial-velocity surveys, starting with the three-Neptune system orbiting HD 69830 (Lovis *et al.* 2006). A system of three super-Earths was found around HD 40307 (Mayor *et al.* 2009b), while HD 10180 revealed up to seven planets, five of which are Neptune-like (Lovis *et al.* 2011).

Ground-based transit surveys have also been able to detect a few transiting low-mass planets despite major observational challenges. The key was to monitor small stars (i.e. M dwarfs), taking advantage of the larger transit signal for a given planet size. Apart from the Neptune-like Gliese 436 b mentioned above, a bloated super-Earth was found around Gliese 1214 (Charbonneau *et al.* 2009). With a mass of 6.5 $M_\oplus$ and radius of 2.7 $R_\oplus$, it likely contains a significant gaseous envelope.

The space missions CoRoT and Kepler have discovered the first two solid planets with a measured mass and radius: CoRoT-7b (Léger *et al.* 2009) and Kepler-10b (Batalha *et al.* 2011). With mean densities between 6 and 9 g cm^{-3}, these are the exoplanets closest to Earth as far as composition is concerned. However, they orbit their stars at a distance of less than 0.02 AU, making them overheated lava worlds rather than hospitable Earth twins. Actually, most of the 47 known low-mass objects have semi-major axes smaller than 0.5 AU, due to observational biases of the radial velocity and transit techniques. Most of these worlds are thus likely too hot to bear any resemblance to Earth. However, radial-velocity surveys are now pushing towards 1 AU, with already a few Neptune-mass objects at these distances, and the Kepler mission is also improving its sensitivity to more distant planets as the total observing time-span increases.

Gliese 581: super-Earths close to the habitable zone?

In an astrobiological context, the detection of temperate planets is obviously of paramount importance. A possible shortcut to the challenging search of planets around 1 AU is the observation of M dwarfs, which have luminosities of only a fraction of those of solar-type stars, and thus have their habitable zones much closer in.

The best example is the multi-planet system orbiting the M-dwarf Gliese 581 (Mayor *et al.* 2009a). Four confirmed low-mass planets have been found to date around this star, at orbital periods of 3.1, 5.4, 13, and 67 days. The third and fourth objects have minimum masses of 5.4 and 7.1 $M_\oplus$, and are thus likely made of rocks and ices (although the absence of transits makes it impossible to exclude a hydrogen envelope). With its effective temperature of only 3500 K and small radius, Gliese 581 has a luminosity of only 1% of the Sun's luminosity. As a consequence, the habitable zone is expected to be located between $\sim$0.08 and 0.2 AU (Selsis *et al.* 2007), just in between the two super-Earths. Gliese 581 c and d are thus arguably the two worlds closest to habitability known today, although they are located at both edges of the temperate zone. Obviously, many other factors enter into the definition of habitability, and it is very premature to draw any firm conclusion about these objects. They however show that exoplanet detection techniques are fast approaching the required sensitivity to explore habitable zones around other stars.

Preliminary statistics

Radial-velocity surveys are able to provide some preliminary statistics on the occurrence of low-mass planets around nearby bright stars. The Keck-HIRES team (Howard *et al.* 2010) surveyed a volume-limited sample of 166 nearby solar-type stars and found that $15^{+5}_{-4}\%$ of stars have at least one planet with a mass between 3 and 30 $M_\oplus$ and an orbital period below 50 days (corresponding to ~0.3 AU). Results from the HARPS survey are even more optimistic (Lovis *et al.* 2009): based on about 300 solar-type stars, it turns out that about 30% of stars harbor at least one planet between 3 and 30 $M_\oplus$ at periods below 100 days (~0.5 AU). This can be considered a good sign for the occurrence rate of telluric planets in the habitable zone.

These numbers at least indicate that low-mass planets are much more numerous than gas giants, contrary to what is shown in Figure 12.5. This is because of the extremely biased radial-velocity and (especially) ground-based transit surveys, which target very large numbers of stars and are not sensitive to low-mass planets. The advent of the Kepler mission will bring a much more balanced view on this issue.

Future prospects

The field of exoplanets is evolving so fast that several future prospects will have become present achievements when this chapter is published. We try here however to anticipate some likely developments, both in the near term and the more distant future.

The Kepler mission

Major results from the Kepler mission are expected to be revealed from 2011 onward. They will likely be of such a large scale that the exoplanet community will need some time to fully exploit them and update our view of planetary systems. Nevertheless, an immediate extraordinary result is the high number of small-size planetary candidates that Kepler has detected in the first months of operations (Borucki *et al.* 2011). As of February 2011, no fewer than 1235 candidates have been found, among which 1051 are in the Earth, super-Earth, or Neptune size category. More than 74% are actually smaller than Neptune. These preliminary numbers fully confirm the findings of radial-velocity surveys, and mark a historical milestone in our knowledge of exoplanets. Kepler is also finding that about 30% of stars have small-size planetary candidates (Neptune-size and below), also confirming the results from the HARPS survey.

Low-mass planets around nearby stars

Although Kepler will revolutionize the statistics of exoplanetary systems, its findings will be very challenging to follow up because of the faint magnitude and large distance of the Kepler target stars. In the meantime, radial-velocity surveys will continue to explore the properties of low-mass planets around nearby bright stars, which are much more amenable to follow-up and, eventually, characterization. Statistically robust results are expected from the HARPS survey for objects down to the super-Earth regime within ~0.5 AU of the parent star. Future Doppler instruments will push these limits further down, like the ESPRESSO instrument to be installed on the European Southern Observatory 8.2-m VLT telescope (Pepe *et al.* 2010). Eventually, we will obtain a complete list of planets accessible to the radial-velocity technique, from a few Earth-masses to gas giants, orbiting solar-type stars and M dwarfs in a well-defined volume around the Sun (e.g. 10–50 pc). In turn, this list will provide the basis for most follow-up studies of exoplanets with future space observatories, like transit searches, transmission spectroscopy, or secondary eclipse detection.

Characterization of low-mass planets

The characterization of low-mass planets, i.e. the study of their atmospheric composition and internal structure, in the habitable zone of nearby stars is obviously the ultimate goal of exoplanetology. Several space missions have been or are being envisaged to achieve this goal, but none has been selected yet because of technological, scientific, and financial limitations. These will certainly be overcome as the field of exoplanets gains in maturity. In the meantime, the James Webb Space Telescope, due to be launched in 2014, will offer some opportunities to characterize planets orbiting nearby low-mass stars. On the ground, the development of 30- to 40-m class telescopes will also open new possibilities. Three projects are currently in an advanced design phase: the Thirty Meter Telescope and Giant Magellan Telescope on the American side, and the European Extremely Large Telescope on the European side. These future facilities will offer unprecedented performance in direct imaging and spectroscopy of exoplanets orbiting nearby stars.

References

Batalha, N. M., Borucki, W. J., Bryson, S. T. *et al.* (2011). "Kepler's first rocky planet: Kepler-10b," *Astrophysical Journal*, Vol. 729, pp. 27–65.

Borucki, W. J., Koch, D. G., Basri, G. *et al.* (2011). "Characteristics of planetary candidates observed by Kepler, II: analysis of the first four months of data," *Astrophysical Journal*, Vol. 736, pp. 19–40.

Charbonneau, D., Brown, T. M., Latham, D. W., and Mayor, M. (2000). "Detection of planetary transits across a Sun-like star," *Astrophysical Journal*, Vol. 529, pp. L45–L48.

Charbonneau, D., Berta, Z. K., Irwin, J. *et al.* (2009). "A Super-Earth transiting a nearby, low mass star," *Nature*, Vol. 462, pp. 891–894.

Gillon, M., Pont, F., Demory, B.-O. *et al.* (2007). "Detection of transits of the nearby hot Neptune GL 436b," *Astronomy and Astrophysics*, Vol. 472, pp. L13–L16.

Howard, A. W., Marcy, G. W., Johnson, J. A. *et al.* (2010). "The occurrence and mass distribution of close-in Super-Earths, Neptunes, and Jupiters," *Science*, Vol. 330, pp. 653–655.

Léger, A., Rouan, D., Schneider, J. *et al.* (2009). "Transiting exoplanets from the CoRoT Space Mission. VIII. CoRoT-7b: the first Super-Earth with measured radius," *Astronomy and Astrophysics*, Vol. 506, pp. 287–302.

Lovis, C., Mayor, M., Pepe, F. *et al.* (2006). "An extrasolar planetary system with three Neptune-mass planets," *Nature*, Vol. 441, pp. 305–309.

Lovis, C., Mayor, M., Bouchy, F. *et al.* (2009). "Towards the characterization of the hot Neptune/Super-Earth population around nearby bright stars," *IAU Symposium*, Vol. 253, pp. 502–505.

Lovis, C., Ségransan, D., Mayor, M. *et al.* (2011). "The HARPS search for southern extrasolar planets. XXXVIII. Up to seven planets orbiting HD 10180: probing the architecture of low-mass planetary systems," *Astronomy and Astrophysics*, Vol. 528, pp. 112–120.

Marois, C., Macintosh, B., Barman, T. *et al.* (2008). "Direct imaging of multiple planets orbiting the star HR 8799," *Science*, Vol. 322, pp. 1348–1352.

Mayor, M. and Queloz, D. (1995). "A Jupiter-mass companion to a solar-type star," *Nature*, Vol. 378, pp. 355–359.

Mayor, M., Pepe, F., Queloz, D. *et al.* (2003). "Setting new standards with HARPS," *The Messenger*, Vol. 114, pp. 20–24.

Mayor, M., Bonfils, X., Forveille, T. *et al.* (2009a). "The HARPS search for southern extrasolar planets. XVIII. An Earth-mass planet in the GJ 581 planetary system," *Astronomy and Astrophysics*, Vol. 507, pp. 487–494.

Mayor, M., Udry, S., Lovis, C. *et al.* (2009b). "The HARPS search for southern extra-solar planets. XIII. A planetary system with 3 Super-Earths (4.2, 6.9, and 9.2 Earth masses)," *Astronomy and Astrophysics*, Vol. 493, pp. 639–644.

Pepe, F., Cristiani, S., Rebolo Lopez, R. *et al.* (2010). "ESPRESSO: the Echelle spectrograph for rocky exoplanets and stable spectroscopic observations," *Proceedings of the SPIE*, Vol. 7735, pp. 14–20.

Selsis, F., Kasting, J. F., Levrard, B. *et al.* (2007). "Habitable planets around the star Gliese 581?," *Astronomy and Astrophysics*, Vol. 476, pp. 1373–1387.

Wright, J. T., Fakhouri, O., Marcy, G. W. *et al.* (2011). "The exoplanet orbit database," *Astronomical Society of the Pacific*, Vol. 123, pp. 412–422.

13

Characterizing Exoplanet Atmospheres

GIOVANNA TINETTI

Introduction

In 1929, the observation of the relative motion among galaxies by Hubble was interpreted as evidence of the cosmological principle, according to which no position (galaxy) in the universe is privileged. The discovery in 1964 of the electromagnetic fossil radiation at close to 3 K provided the missing experimental link between the unique thermodynamic origin of the universe and the present observable stellar era. These are the foundations of the standard cosmological model. The thermodynamic origin and the various stages of evolution provide the framework to understand the process of nucleosynthesis, and we expect that the relative abundances of the chemical elements in the universe are uniform on the large scale.

At this point, we might ask whether it makes sense to talk about the formulation of a "bio-cosmological principle," stating that the probability of finding life in the universe is uniform, with no privilege for our galaxy. But right away we find a difficulty: in the standard cosmological model the observables are well defined by physics. Instead the assumed probability of finding life in the universe refers to something – *life* – that is not defined rigorously. We do not have a definition of life usable everywhere; we are in a pre-Galilean stage.

The acquisition of spectroscopic data of the Earth's atmosphere from artificial satellites has changed the old question of whether the phenomenon of terrestrial life is unique, or not. The observation from a satellite is "objective" while the observation developed historically is "subjective," or anthropomorphic. Seen from outside, the life on Earth appears as one among other possible forms that are,

Frontiers of Astrobiology, ed. Chris Impey, Jonathan Lunine and José Funes.
Published by Cambridge University Press. © Cambridge University Press 2012.

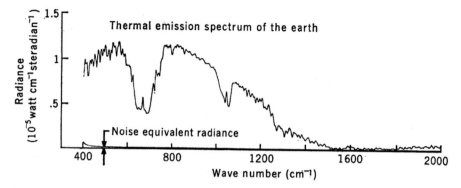

Figure 13.1 Thermal emission spectrum of the Earth obtained with the Infrared Interferometer Spectrometer (IRIS) carried in the Nimbus 3 satellite (Hanel and Conrath 1969). The spectrum was recorded over the equatorial Pacific Ocean on 15 April 1969. From the spectrum the absorptions due to CO_2, water vapour, and ozone are clearly visible.

in principle, detectable by pointing the instrument toward another planet (see Figure 13.1).

The first time a man-made space probe gave us an image of another heavenly body that we could not achieve from Earth was probably Luna 3's image of the far side of the Moon in 1959. Soon after, the Ranger spacecraft gave us close-ups of the lunar surface at better resolution than ground-based telescopes, and the Venera and Mariner spacecraft exploded the myths of the habitability of Mars and Venus with their close-up look at those planets. Voyagers 1 and 2 launched in 1977 were probably the most successful missions in opening our eyes to the diversity and beauty of the Solar System planets. In this way planetary science was born, and a key theme has always been planets as the cradle for life. Venus was eliminated from consideration early on. There is still some hope attached to Mars, though at best residual simple life is the best we could expect. Unexpectedly there also seems to be some chance that satellites of the outer planets could harbor life, thanks to localized energy sources like tidal heating.

In the search for life's signs the theory of what we are looking for has broadened and thrown up a host of questions. It is clear we need to know a lot about a planet to know how suitable it is for life. And no matter how optimistic one is, it seems certain that in the Solar System nowhere else has life in as complex a form as is found on Earth. So our attention turns outwards, boosted by the discovery in 1995 of the first provable planet around a Sun-like star. As the exoplanet count grew, so did the excitement about the prospect of there being other centers where life has arisen. With transit science came the first tangible remote sensing of these planetary bodies and so we could start to tie together what we had learnt from the Solar

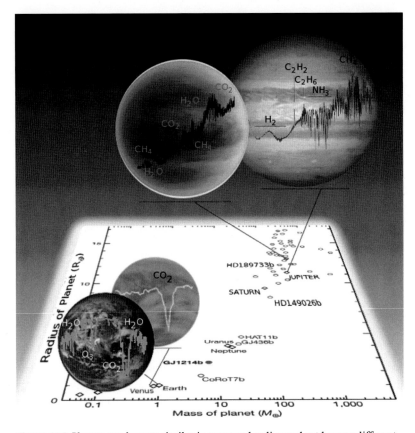

Figure 13.2 Planets can be very similar in mass and radius and yet be very different worlds, as demonstrated by these two pairs of examples. A spectroscopic analysis of the atmospheres is needed to reveal their physical and chemical identities. A colour version of this illustration appears in the colour plate section between pp. 148–149.

System probes to what we might plan to learn about their faraway siblings. Just as our knowledge of planetary science in the Solar System expanded exponentially as they went from dots in ground-based telescopes to clear images and extensive high-resolution remote-sensing data, so will our knowledge of exoplanets depend not on the simple statistics of masses and radii, but on what we can learn from transits and differential photometry and spectroscopy (Figure 13.2).

As we learn more about the atmospheres, surfaces, and near-surfaces of these remote bodies, we will begin to build up a clearer picture of their construction, history, and suitability for life. Even while we have been limited to the larger, hotter, and closer-in bodies we have made significant discoveries. It is in characterizing more bodies in different environments that we will take detailed planetology out of the Solar System into the Galaxy as a whole.

The exoplanet revolution

The "exoplanet revolution" has indeed irreversibly changed our views about planets and stars, and the ways in which they are formed. Before 1995 only nine planets were known. These comprised our Solar System including our own planet, the Earth, and Pluto, subsequently reclassified in 2006 as a "dwarf" planet. Before 1995, planets were divided into giants and terrestrials: the former colder, with reducing atmospheres, and the latter warmer, smaller, denser, and with oxidizing atmospheres. The Solar System model, with planets moving around the Sun on quasi-circular and coplanar orbits, appeared to be the unquestioned paradigm, blessed by Kant's and Laplace's theories of planet formation. But all this was before Mayor and Queloz (1995) found a gas-giant planet orbiting the G-type star 51 Peg with an annual period of only 4 Earth-days.

Over 700 planets later (Schneider 2012) – not including the approximately 1200 planetary candidates detected to date by Kepler (Borucki *et al.* 2011) – our Solar System seems to stand out more as a rarity than the paradigm in our Galaxy. While better statistics and planet detection with multiple discovery methods are needed to eliminate the undoubted bias in the current sample due to the detection techniques adopted, most of the gas-giant planets detected so far orbit very close to their parent star. It seems likely that they migrated close to their star during the formation process, which means that our Jupiter, at 5 AU from the Sun, should have experienced an unusual history (Tsiganis *et al.* 2005) compared to the observed population of hot Jupiters. The European Space Agency Gaia mission, expected to discover thousands of new planets with the astrometric method, will soon provide new insight into the statistics of Jupiters orbiting at larger separation.

Among the exoplanets known today, eccentric planets no longer appear to be oddities. Furthermore "super-Earths," planets up to 10 Earth-masses, are common around other stars (Howard *et al.* 2012, Mayor *et al.* 2011, Cassan *et al.* 2012), but completely absent in our Solar System. These are hircocervi between Neptunic and telluric planets, and about which we know little today. Among those, the trio of Kepler-10b (Batalha *et al.* 2011), CoRoT-7b (Léger *et al.* 2009), and 55 Cnc e (Winn *et al.* 2011) all have high densities and orbit G stars with periods under 1 day. By contrast, GJ 1214b (Charbonneau *et al.* 2009), Kepler-11d, e, and f (Lissauer *et al.* 2011) have much lower densities and are subject to less intense insolation because of their longer period or cooler parent star. Finally, among the super-Earths discovered, we might already have a potential candidate for habitability, GJ 581d (Mayor *et al.* 2009, Wordsworth *et al.* 2011). Interestingly, this planet does not orbit a canonical G-type star like our own Sun, but rather a much dimmer and colder M dwarf, clearly challenging any geocentric concept of habitability. We should expect many more planets like GJ 581d, given that ∼90% of the stars in the solar neighborhood are

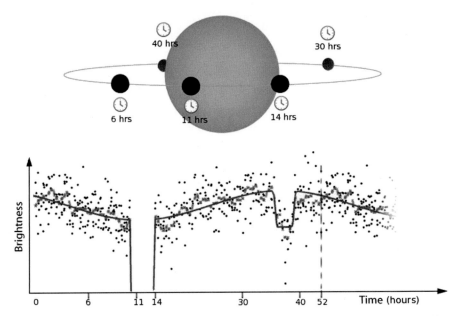

Figure 13.3 Optical phase curve of the planet HAT-P-7b observed by Kepler (Borucki *et al.* 2009), showing primary and secondary transit measurements.

M dwarfs and, from preliminary analysis of Kepler, the occurrence of 2–4 $R_\oplus$ planets in the Kepler field linearly increases with decreasing stellar temperature, making these small planets seven times more abundant around cool stars (Howard *et al.* 2012).

Transit spectroscopy

A critical milestone in our understanding of these faraway objects was the realization that we can probe the atmospheres of a subset of known exoplanets: those which transit their parent star when viewed from Earth (for an example, see Figure 13.3). For those planets, one can effectively use the wavelength-dependent stellar occultation to measure a transmission spectrum of the planetary limb (Charbonneau *et al.* 2002), or use the star as a "natural coronograph" while the planet is passing behind it, to extract a spectrum emitted or reflected by the planet (Charbonneau *et al.* 2005, Deming *et al.* 2005, Rowe *et al.* 2006). Additionally, one can observe the planet at different orbital phases (Harrington *et al.* 2006, Knutson *et al.* 2007a, Snellen *et al.* 2009, Borucki *et al.* 2009, Crossfield *et al.* 2010).

Indeed transit spectroscopy has proved to be very successful in the last 10 years: the use of the transit technique combined with the photometric precision of the Hubble and Spitzer Space Telescopes, has allowed a glimpse of the atmospheric composition (Figure 13.4), albedo, escape processes, and thermal characteristics of a handful of hot Jupiters (see, for example, Knutson *et al.* 2007b, Tinetti *et al.* 2007,

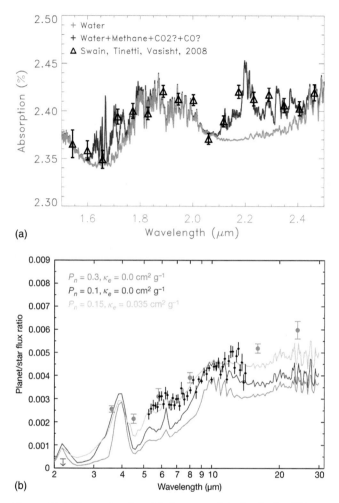

(a)

(b)

Figure 13.4 Observed and modeled spectra of the hot Jupiter HD 189733b.
(a) Transmission spectrum recorded with Hubble-NICMOS during the primary transit
of the planet. The spectrum can be well explained with water vapor and methane as
atmospheric components (Swain *et al.* 2008b). The low spectral resolution does not
allow the detection of CO or CO_2, detected on the "day-side" of the planet. (b) Emission
spectrum recorded with Spitzer-IRS during the secondary eclipse of the planet
indicating the presence of water vapor in the atmosphere (Grillmair *et al.* 2008). A
colour version of this illustration appears in the colour plate section between
pp. 148–149.

2010a, Swain *et al.* 2008b, 2009a, b, Grillmair *et al.* 2008, Vidal-Madjar *et al.* 2003,
Linsky *et al.* 2010) as well as on those of colder and smaller planets, such as the warm
Neptune GJ 436b (Stevenson *et al.* 2010, Beaulieu *et al.* 2011, Knutson *et al.* 2011).

More recently, ground-based telescopes have been combined powerfully with
space-based observations. Ground-based observations have the non-trivial limitation of having the Earth's atmosphere interfering with the measurement,

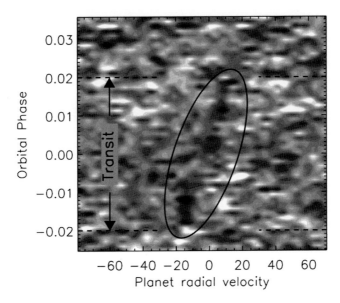

Figure 13.5 Detection of CO in the atmosphere of the hot Jupiter HD 209458b, using the VLT-CRIRES instrument (Snellen *et al.* 2010).

especially in the infrared, where key molecules show stronger spectral features. At the same time, observations from the ground have the great advantage of being more easily repeatable and of offering higher spectral resolution than Hubble and Spitzer in some spectral channels. Among the strategies attempted, the successful ones are based on time-resolved coverage of the transit/eclipse event, providing light curves at multiple wavelengths (low spectral *R*), or time-dependent Doppler shift signatures (high spectral *R*). Each of these techniques involve relative measurements, so they rely less on the absolute correction of the telluric signal.

Using these methods, there has been a rapid escalation in ground-breaking results in the past few years: results include the detection of alkali metals in the optical (Redfield *et al.* 2008, Snellen *et al.* 2008, Sing *et al.* 2011a, Colon *et al.* 2011), and the first detection and atmospheric characterization of the warm super-Earth GJ1214b (Charbonneau *et al.* 2009, Bean *et al.* 2010). In the near-infrared (NIR), a discovery that stands out is the detection of CO (Snellen *et al.* 2010), never conclusively detected from space (Figure 13.5), and the non-local thermodynamic equilibrium (LTE) methane emission (Swain *et al.* 2010), which was completely unexpected.

Waldmann *et al.* (2011) have demonstrated the reproducibility of the methane emission feature on HD 189733b in several observations, and shown that the data analysis method adopted is robust enough to remove the telluric contamination while preserving the exoplanetary signal (Figure 13.6).

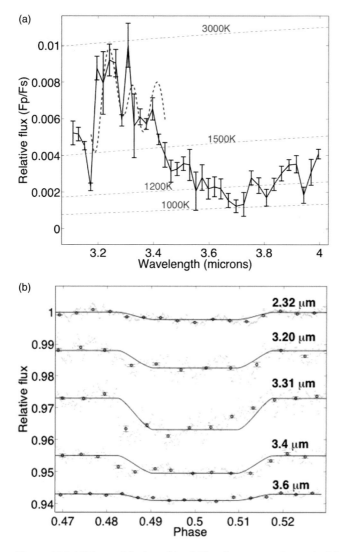

Figure 13.6 (a) Three nights' combined L-band spectrum observed with NASA-IRTF (Waldmann *et al.* 2012). The discontinuous curve shows a comparison of the observations with the non-LTE methane emission model proposed by P. Drossart. Overplotted are blackbody curves at 100, 1500, 2000, and 3000 K. (b) Light curves of the "three-night-combined" analysis for the K and L bands. Light curves are offset vertically for clarity.

Molecular signatures in exoplanet spectra

Combining near-infrared (NIR) with mid-infrared (MIR) spectra from space and ground measurements, it is found that the absorptions due to H_2O, CH_4, CO, and CO_2 explain most of the features present in the observed exoplanet spectra. On the large scale, the spectra of hot Jupiters seem to be dominated by the signature of water vapour (Burrows *et al.* 2007, 2008, 2010, Barman 2007, 2008, Beaulieu *et al.* 2010, Charbonneau *et al.* 2008, Grillmair *et al.* 2008, Knutson *et al.* 2008, Madhusudhan and Seager 2009, Showman *et al.* 2009, Tinetti *et al.* 2007, 2010a, 2011), whereas warm Neptunes are dominated by methane (Beaulieu *et al.* 2011). This trend seems to be confirmed by photochemical models (Moses *et al.* 2011, Line *et al.* 2012), but needs further confirmation on a larger sample of targets.

In the IR the spectral features are more intense and broader than in the visible. Moreover, hazes or clouds seem to affect the transparency of the atmosphere in the visible–NIR spectral range, at wavelengths shorter than 1.5 µm. While this is the case for HD 189733b and HD 209458b (Knutson *et al.* 2007b, Pont *et al.* 2008, Sing *et al.* 2011b), the hot-Jupiter XO-1b spectrum shows distinctive molecular features, among which are H_2O, CH_4, CO_2, and possibly CO, in the 1.2–1.8 µm spectral region (Tinetti *et al.* 2010a). With the loss of NICMOS and Spitzer-IRS/MIPS and partially IRAC, it is technically impossible at present from space to repeat previous observations on other targets. Today, we could potentially observe the atmospheres of about 90 of the more than 200 transiting planets. Most of these targets were discovered by dedicated ground-based radial velocity and transit surveys. Ground-based spectra in the Z, J, H, K, and L bands can be observed to record the exoplanetary signal in the NIR–MIR spectral region (Danielski *et al.* 2012). While water vapor and CO_2 are clearly the most difficult molecules to observe from the ground, repeated observations can be used to break the degeneracy between exoplanet and telluric signals, as was done in the case of methane (Waldmann *et al.* 2011). Also, statistical techniques used in cosmology and communication science to optimize the extraction of a weak signal from a noisy background, find more and more applicability in the analysis of exoplanetary signals (e.g. Carter and Winn 2009, Gregory 2011, Feroz *et al.* 2011) and can be applied effectively to the analysis of ground-based observations of exoplanet spectra (Waldmann 2012).

Within the past few years, efforts to determine the abundances of atmospheric constituents found instead a range of degenerate temperature and composition solutions from the spectra (Swain *et al.* 2009a, Madhusudhan and Seager 2009, Tinetti *et al.* 2010b, Lee *et al.* 2012, Line *et al.* 2012). Using an iterative forward model approach for spectral retrieval, we evaluated a variety of temperatures (T) as a function of pressure (P) together with the molecular absorption effects.

In the case of emission spectra (e.g. Figure 13.7), we obtain a family of plausible solutions for the molecular abundances and detailed temperature profiles illustrated in Figure 13.8. Additional observational constraints on the atmospheric temperature structure and composition require improved wavelength coverage and spectral resolution for the dayside spectrum.

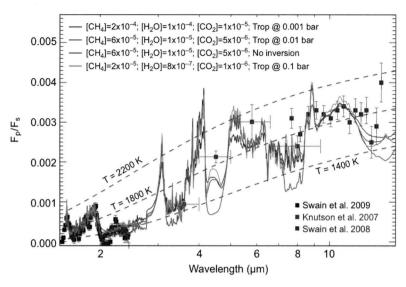

Figure 13.7 Emission photometry and spectroscopy data for HD 209458 b (Swain *et al.* 2009a). The near-infrared and mid-infrared eclipse observations are compared to synthetic spectra for four models that illustrate the range of temperature/composition possibilities consistent with the data. For each model case, the molecular abundance of CH_4, H_2O, and CO_2 and the location of the tropopause is given (see Figure 13.8). These serve to illustrate how the combination of molecular opacities and the temperature structure cause significant departures from a purely single-temperature thermal emission spectrum. A colour version of this illustration appears in the colour plate section between pp. 148–149.

Transmission spectra are almost insensitive to vertical thermal gradients, although the average atmospheric temperature may play an important role in the overall scale-height, hence in the amplitude of the spectral signatures, as well as in the molecular absorption coefficients. By contrast, the derived composition at the terminator depends noticeably on the assumed planetary radius. For example, a ∼1% difference in the estimate of the planetary radius at the ∼1 bar pressure level would result in a variation of the molecular abundances of a factor of 10. Transit data at multiple wavelengths are needed to constrain the molecular abundances, including the UV–VIS part of the spectrum where only Rayleigh scattering absorbs.

We note that some of the temperature profiles/molecular mixing ratios consistent with the observations raise the question of whether the dayside atmosphere is in radiative and thermochemical equilibrium. Although advection of heat and/or photochemistry could support departures from radiative and thermochemical equilibrium, our present lack of knowledge of molecular opacities at high temperatures for species such as CH_4, H_2S, and C_2H_6 limits our ability to determine

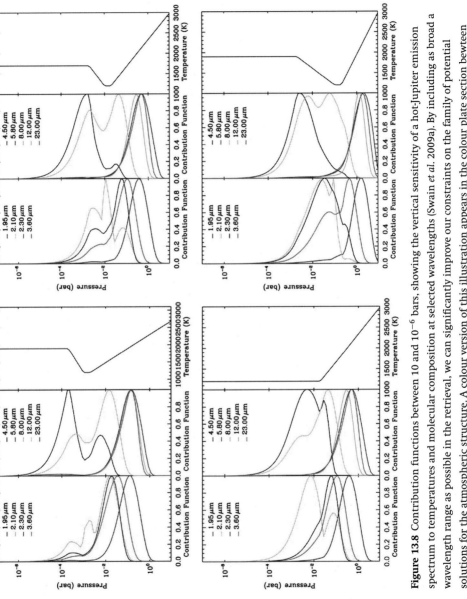

Figure 13.8 Contribution functions between 10 and 10^{-6} bars, showing the vertical sensitivity of a hot-Jupiter emission spectrum to temperatures and molecular composition at selected wavelengths (Swain *et al.* 2009a). By including as broad a wavelength range as possible in the retrieval, we can significantly improve our constraints on the family of potential solutions for the atmospheric structure. A colour version of this illustration appears in the colour plate section bewteen pp. 148–149.

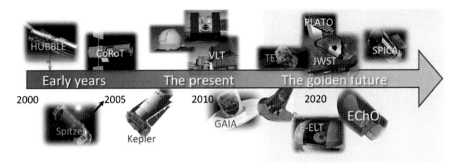

Figure 13.9 In the next decade important new facilities are foreseen with relevance to exoplanetary science. This graphic shows the signature facilities from the past, present, and future of the field.

decisively whether this condition is met or not; thus there is an urgent need for further laboratory studies to obtain molecular databases for determining high-temperature opacities of the most common molecules expected in hot-Jupiter/hot-Neptune atmospheres (e.g. UCL ExoMol and HITEMP, Barber *et al.* 2006, Yurchenko *et al.* 2011, Rothman *et al.* 2010, Freedman *et al.* 2008).

The future: a dedicated mission to probe exoplanet atmospheres?

One could comment that there are many more exoplanets, although they do not transit or they orbit at large separation from their parent star. It is thus impossible to study these with the transit technique. Most recently the first spectrum of a hot giant planet at a projected separation of 38 AU from its host star was observed from the ground with VLT/NACO (Janson *et al.* 2010). Spectroscopy in the shorter wavelength range of the YJHK-band will likely start soon with dedicated integral field units on VLT (SPHERE) and Gemini (GPI). The young exoplanets that these instruments are expected to find and characterize are likely to feature several molecular species.

In the upcoming decade important new facilities are planned to come on line (JWST, E-ELT, TMT, SPICA; see Figure 13.9). But none of these multi-purpose facilities will give us a coherent program of observations of a large sample of objects on long time-scales to test models fully. We need a "Rosetta Stone" for exoplanets, so that we may interpret these alien worlds correctly, and decipher past, present, and future observations beyond any reasonable doubt. Snapshots of this, that, or the other exoplanet, taken with different instruments or by different observers, at different times and wavelength regions, will not allow us to do this. The data are and will be too sparse and inarticulate to provide a coherent view and comprehensive

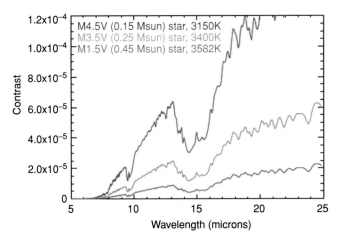

Figure 13.10 Planet–star contrast for a "habitable" super-Earth orbiting different M-types stars. Here the super-Earth is supposed to have an Earth-like atmospheric composition (Tessenyi *et al.* 2012).

classification of the variety of exoplanets we might encounter; this leaves room for misinterpretation.

The way forward is a dedicated mission able to observe a large and diverse sample of planets, spanning a range of masses (from gas giants to super-Earths), stellar companions (F, G, K, M), and temperatures (from hot to habitable). Most importantly the ability to "stare at the planet," simultaneously over a broad wavelength coverage, may allow us to remove effectively the stellar fluctuations (Figure 13.10).

For these reasons, a proposal was made to the European Space Agency (ESA) for a dedicated space mission to observe the atmospheres of tens of known transiting exoplanets over a broad wavelength range: from 0.4 to 16 μm. EChO, the Exoplanet Characterisation Observatory (Figure 13.11), is one of the four M3 mission candidates currently being assessed by ESA, for possible launch in 2022 (http://sci.esa.int/echo; Tinetti *et al.* 2012).

If launched, EChO will provide access to the population of super-Earths orbiting bright stars, on top of GJ 1214b, 55 Cnc e, HD 97658b, etc. In particular, in the quest for habitable worlds outside our Solar System, EChO will be able to observe super-Earths spectroscopically in the temperate zone of late M-dwarfs – e.g. planets similar to GJ 581d. Detecting molecular features in the atmospheres of these planets, which are faint and emit predominantly at wavelengths longer than 5 μm, represents the most challenging case for the mission. However, the current

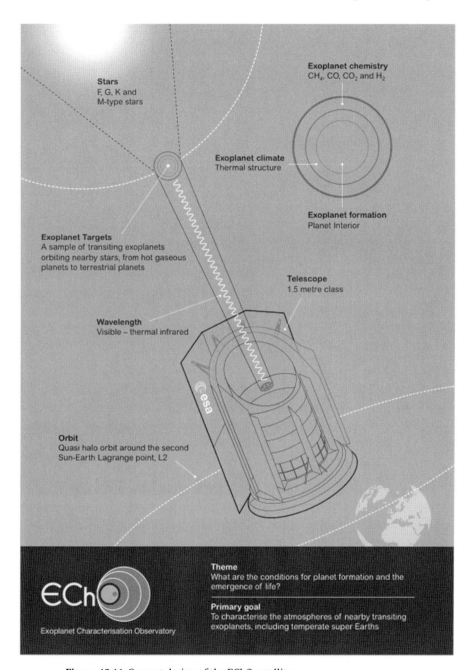

Figure 13.11 Current design of the EChO satellite.

payload design and lifetime for the mission (5 years) will indeed allow these observations to be achieved (Tessenyi *et al.* 2012).

References

Barber, R. J., Tennyson, J., Harris, G. J., and Tolchenov, R. N. (2006). "A high accuracy computed water line list," *Monthly Notices of the Royal Astronomical Society*, Vol. 368, pp. 1087–1094.

Barman, T. (2007). "Identification of absorption features in an extrasolar planet atmosphere," *Astrophysical Journal*, Vol. 661, pp. L191–L194.

Barman, T. S. (2008). "On the presence of water and global circulation in the transiting planet HD 189733b," *Astrophysical Journal*, Vol. 676, pp. L61–L64.

Barnes, J. R., Barman, T. S., Prato, L. *et al.* (2007). "Limits on the 2.2-μm contrast ratio of the close-orbiting planet HD 189733b," *Monthly Notices of the Royal Astronomical Society*, Vol. 382, pp. 473–480.

Barnes, J. R., Barman, T. S., Jones, H. R. A. *et al.* (2008). "HD179949b: a close orbiting extrasolar giant planet with a stratosphere?," *Monthly Notices of the Royal Astronomical Society*, Vol. 390, pp. 1258–1266.

Barnes, J. R., Barman, T. S., Jones, H. R. A. *et al.* (2010). "A search for molecules in the atmosphere of HD 189733b," *Monthly Notices of the Royal Astronomical Society*, Vol. 401, pp. 445–454.

Batalha, N. M., Borucki, W. J., Bryson, S. T. *et al.* (2011). "Kepler's first rocky planet: Kepler-10b," *Astrophysical Journal*, Vol. 729, pp. 27–48.

Bean, J. L., Miller-Ricci Kempton, E., and Homeier, D. (2010). "A ground-based transmission spectrum of the Super-Earth exoplanet GJ 1214b," *Nature*, Vol. 468, pp. 669–672.

Beaulieu, J. P., Carey, S., Ribas, I., and Tinetti, G. (2008). "Primary transit of the planet HD 189733b at 3.6 and 5.8 microns," *Astrophysical Journal*, Vol. 677, pp. 1343–1347.

Beaulieu, J. P., Kipping, D. M., Batista, V. *et al.* (2010). "Water in the atmosphere of HD 209458b from 3.6–8 micron IRAC photometric observations in primary transit," *Monthly Notices of the Royal Astronomical Society*, Vol. 409, pp. 963–974.

Beaulieu, J.-P., Tinetti, G., Kipping, D. M. *et al.* (2011). "Methane in the atmosphere of the transiting hot Neptune GJ436B?," *Astrophysical Journal*, Vol. 731, p. 16–27.

Borucki, W. J., Koch, D., Jenkins, J. *et al.* (2009). "Kepler's optical phase curve of the exoplanet HAT-P-7b," *Science*, Vol. 325, p. 709.

Borucki, W. J., Koch, D. G., Basri, G. *et al.* (2011). "Characteristics of planetary candidates observed by Kepler. II. Analysis of the first four months of data," *Astrophysical Journal*, Vol. 736, pp. 19–40.

Brown, T. M., Charbonneau, D., Gilliland, R. L. *et al.* (2001). "Hubble Space Telescope time-series photometry of the transiting planet of HD 209458," *Astrophysical Journal*, Vol. 552, pp. 699–709.

Burrows, A., Hubeny, I., Budaj, J. *et al.* (2007). "Theoretical spectral models of the planet HD 209458b with a thermal inversion and water emission bands," *Astrophysical Journal*, Vol. 668, pp. L171–L174.

Burrows, A., Budaj, J., and Hubeny, I. (2008). "Theoretical spectra and light curves of close-in extrasolar giant planets and comparison with data," *Astrophysical Journal*, Vol. 678, pp. 1436–1457.

Burrows, A., Rauscher, E., Spiegel, D. S., and Menou, K. (2010). "Photometric and spectral signatures of three-dimensional models of transiting giant exoplanets," *Astrophysical Journal*, Vol. 719, pp. 341–350.

Carter, J. A. and Winn, J. N. (2009). "Parameter estimation from time-series data with correlated errors: a wavelet-based method and its application to transit light curves," *Astrophysical Journal*, Vol. 704, pp. 51–67.

Cassan, A., Kubas, D., Beaulieu, J. P. *et al.* (2011). "One or more bound planets per Milky Way star from microlensing," *Nature*, Vol. 481, pp. 167–169.

Charbonneau, D., Brown, T. M., Noyes, R. W., and Gilliland, R. L. (2002). "Detection of an extrasolar planet atmosphere," *Astrophysical Journal*, Vol. 568, pp. 377–384.

Charbonneau, D., Allen, L. E., Megeath, S. T. *et al.* (2005). "Detection of thermal emission from an extrasolar planet," *Astrophysical Journal*, Vol. 626, pp. 523–529.

Charbonneau, D., Knutson, H. A., Barman, T. *et al.* (2008). "The broadband infrared emission spectrum of the exoplanet HD 189733b," *Astrophysical Journal*, Vol. 686, pp. 1341–1348.

Charbonneau, D., Berta, Z. K., Irwin, J. *et al.* (2009). "A Super-Earth transiting a nearby low-mass star," *Nature*, Vol. 462, pp. 891–894.

Colon, K. D., Ford, E. B., Redfield, S. *et al.* (2011). "Probing potassium in the atmosphere of HD 80606b with tunable filter transit spectrophotometry from the Gran Telescopio Canarias," *Monthly Notices of the Royal Astronomical Society*, Vol. 419, pp. 2233–2250.

Crossfield, I. J. M., Hansen, B. M. S., Harrington, J. *et al.* (2010). "A new 24 micron phase curve for υ Andromedae b," *Astrophysical Journal*, Vol. 723, pp. 1436–1446.

Danielski, C., Deroo, P., Waldmann, I. *et al.* (2012). "1.4–2.4 microns ground-based transmission spectra of the Hot Jupiter HD-189733b," *Astrophysical Journal*, submitted.

Deming, D., Seager, S., Richardson, L. J., and Harrington, J. (2005). "Infrared radiation from an extrasolar planet," *Nature*, Vol. 434, pp. 740–743.

Feroz, F., Balan, S. T., and Hobson, M. P. (2011). "Bayesian evidence for two companions orbiting HIP 5158," *Monthly Notices of the Royal Astronomical Society*, Vol. 416, pp. L104–L108.

Freedman, R. S., Marley, M. S., and Lodders, K. (2008). "Line and mean opacities for ultracool dwarfs and extrasolar planets," *Astrophysical Journal Supplement*, Vol. 174, p. 504–513.

Gamache, R. R., Lynch, R., and Neshyba, S. P. (1998). "New developments in the theory of pressure-broadening and pressure-shifting of spectral lines of H_2O: the complex Robert–Bonamy formalism," *Journal of Quantitative Spectroscopy and Radiative Transfer*, Vol. 59, pp. 319–335.

Gregory, P. C. (2011). "Bayesian exoplanet tests of a new method for MCMC sampling in highly correlated model parameter spaces," *Monthly Notices of the Royal Astronomical Society*, Vol. 410, pp. 94–110.

Grillmair, C. J., Charbonneau, D., Burrows, A. *et al.* (2007). "A Spitzer spectrum of the exoplanet HD 189733b," *Astrophysical Journal*, Vol. 658, pp. L115–L118.

Grillmair, C. J., Burrows, A., Charbonneau, D. *et al.* (2008). "Strong water absorption in the dayside emission spectrum of the planet HD189733b," *Nature*, Vol. 456, pp. 767–769.

Hanel, R. and Conrath, B. (1969). Interferometer experiment on Nimbus 3: preliminary results," *Science*, Vol. 165, pp. 1258–1260.

Harrington, J., Hansen, B. M., Luszcz, S. H. *et al.* (2006). "The phase-dependent infrared brightness of the extrasolar planet Upsilon Andromedae b," *Science*, Vol. 314, pp. 623–626.

Howard, A. W., Marcy, G. W., Bryson, S. T. *et al.* (2012). "Planet occurrence within 0.25 AU of solar-type stars from Kepler," *Astrophysical Journal*, submitted.

Janson, M., Bergfors, C., Goto, M. *et al.* (2010). "Spatially resolved spectroscopy of the exoplanet HR 8799 c," *Astrophysical Journal*, Vol. 710, pp. L35–L38.

Knutson, H. A., Charbonneau, D., Allen, L. E. *et al.* (2007a). "A map of the day–night contrast of the extrasolar planet HD 189733b," *Nature*, Vol. 447, pp. 183–186.

Knutson, H. A., Charbonneau, D., Noyes, R. W. *et al.* (2007b). "Using stellar limb-darkening to refine the properties of HD 209458b," *Astrophysical Journal*, Vol. 655, pp. 564–575.

Knutson, H. A., Charbonneau, D., Allen, L. E. *et al.* (2008). "The 3.6–8.0 μm broadband emission spectrum of HD 209458b: evidence for an atmospheric temperature inversion," *Astrophysical Journal*, Vol. 673, pp. 526–531.

Knutson, H. A., Madhusudhan, N., Cowan, N. B. *et al.* (2011). "A Spitzer Transmission Spectrum for the exoplanet GJ 436b, evidence for stellar variability, and constraints on dayside flux variations," *Astrophysical Journal*, Vol. 735, pp. 70–50.

Lee, J. M., Fletcher, L. N., and Irwin, P. G. J. (2012). "Optimal estimation retrievals of the atmospheric structure and composition of HD 189733b from secondary eclipse spectroscopy," *Monthly Notices of the Royal Astronomical Society*, Vol. 420, pp. 170–182.

Lepine, S. and Gaidos, E. (2011). "An all-sky catalog of bright M dwarfs," *Astronomical Journal*, Vol. 142, p. 138.

Leger, A. Rouan, D., Schneider, J. *et al.* (2009). "The CoRoT space mission: early results," *Astronomy and Astrophysics*, Vol. 506, pp. 287–302.

Line, M. R., Zhang, X., Vasisht, G. *et al.* (2012). "Information content of exoplanetary transit spectra: an initial look," *Astrophysical Journal*, Vol. 749, pp. 99–103.

Linsky, J. L., Yang, H., France, K. *et al.* (2010). "Observations of mass loss from the transiting exoplanet HD 209458b," *Astrophysical Journal*, Vol. 717, pp. 1291–1299.

Lissauer, J. J., Fabrycky, D. C., Ford, E. B. *et al.* (2011). "A closely-packed system of low-mass, low-density planets transiting Kepler-11," *Nature*, Vol. 470, pp. 53–58.

Madhusudhan, N. and Seager, S. (2009). "A temperature and abundance retrieval method for exoplanet atmospheres," *Astrophysical Journal*, Vol. 707, pp. 24–39.

Mandell, A. M., Drake Deming, L., Blake, G. A. *et al.* (2011). "Non-detection of L-band line emission from the exoplanet HD189733b," *Astrophysical Journal*, Vol. 728, p. 18.

Mayor, M. and Queloz, D. (1995). "A Jupiter-mass companion to a solar-type star," *Nature*, Vol. 378, pp. 355–359.

Mayor, M., Bonfis, X., Forveille, T. *et al.* (2009). "The HARPS search for sourthern extra-solar planets," *Astronomy and Astrophysics*, Vol. 507, pp. 487–494.

Mayor, M., Marmier, M., Lovis, C. *et al.* (2011). "The HARPS search for southern extra-solar planets XXXIV: occurrence, mass distribution and orbital properties of Super-Earths and Neptune-mass planets," arXiv e-prints.

Moses, J. I., Visscher, C., Fortney, J. J. *et al.* (2011). "Disequilibrium carbon, oxygen, and nitrogen chemistry in the atmospheres of HD 189733b and HD 209458b," *Astrophysical Journal*, Vol. 737, pp. 15–52.

Partridge, H. and Schwenke, D. W. (1997). "The determination of an accurate isotope dependent potential energy surface for water from extensive *ab initio* calculations and experimental data," *Journal of Chemical Physics*, Vol. 106, pp. 4618–4639.

Pont, F., Knutson, H., Gilliland, R. L. *et al.* (2008). "Detection of atmospheric haze on an extrasolar planet: the 0.55–1.05 micron transmission spectrum of HD 189733b with the Hubble Space Telescope," *Monthly Notices of the Royal Astronomical Society*, Vol. 385, pp. 109–118.

Redfield, S., Endl, M., Cochran, W. D., and Koesterke, L. (2008). "Sodium absorption from the exoplanetary atmosphere of HD 189733b detected in the optical transmission spectrum," *Astrophysical Journal*, Vol. 673, pp. L87–L90.

Richardson, L. J., Deming, D., and Seager, S. (2003). "Infrared observations during the secondary eclipse of HD 209458b. II. Strong limits on the infrared spectrum near 2.2 μm," *Astrophysical Journal*, Vol. 597, pp. 581–589.

Richardson, L. J., Deming, D., Horning, K. *et al.* (2007). "A spectrum of an extrasolar planet," *Nature*, Vol. 445, pp. 892–895.

Rothman, L., Gordon, I., Barbe, A. *et al.* (2009). "The HITRAN 2008 molecular spectroscopic database," *Journal of Quantitative Spectroscopy and Radiative Transfer*, Vol. 110, pp. 533–572.

Rothman, L. S., Gordon, I. E., Barber, R. J. *et al.* (2010). "HITEMP, the high-temperature molecular spectroscopic database," *Journal of Quantitative Spectroscopy and Radiative Transfer*, Vol. 111, pp. 2139–2150.

Rowe, J. F., Matthews, J. M., Seager, S. *et al.* (2006). "An upper limit on the albedo of HD 209458b: direct imaging photometry with the MOST satellite," *Astrophysical Journal*, Vol. 646, pp. 1241–1251.

Schneider, J. (2012). *The Extrasolar Planets Encyclopaedia*, available at: http://exoplanet.eu/.

Seager, S. and Sasselov, D. D. (2000). "Theoretical transmission spectra during extra-solar giant planet transits," *Astrophysical Journal*, Vol. 537, pp. 916–921.

Showman, A. P., Fortney, J. J., Lian, Y. *et al.* (2009). "Atmospheric circulation of hot Jupiters: coupled radiative-dynamical general circulation model simulations of HD 189733b and HD 209458b," *Astrophysical Journal*, Vol. 699, pp. 564–584.

Sing, D. K., Desert, J.-M., Fortney, J. J. *et al.* (2011a). "Gran Telescopio Canarias OSIRIS transiting exoplanet atmospheric survey: detection of potassium in XO-2b from narrowband spectrophotometry," *Astronomy and Astrophysics*, Vol. 527, p. A73.

Sing, D. K., Pont, F., Aigrain, S. *et al.* (2011b). "Hubble Space Telescope transmission spectroscopy of the exoplanet HD 189733b: high-altitude atmospheric haze in the optical and near-ultraviolet with STIS," *Monthly Notices of the Royal Astronomical Society*, Vol. 416, pp. 1443–1455.

Snellen, I. A. G., Albrecht, S., de Mooij, E. J. W., and Le Poole, R. S. (2008). "Ground-based detection of sodium in the transmission spectrum of exoplanet HD 209458b," *Astronomy and Astrophysics*, Vol. 487, pp. 357–362.

Snellen, I. A. G., de Mooij, E. J. W., and Albrecht, S. (2009). "The changing phases of extrasolar planet CoRoT-1b," *Nature*, Vol. 459, pp. 543–545.

Snellen, I. A. G., de Kok, R. J., de Mooij, E. J. W., and Albrecht, S. (2010). "The orbital motion, absolute mass and high-altitude winds of exoplanet HD209458b," *Nature*, Vol. 465, pp. 1049–1051.

Stevenson, K. B., Harrington, J., Nymeyer, S. *et al.* (2010). "Possible thermochemical disequilibrium in the atmosphere of the exoplanet GJ 436b," *Nature*, Vol. 464, pp. 1161–1164.

Swain, M. R., Bouwman, J., Akeson, R. L. *et al.* (2008a). "The mid-infrared spectrum of the transiting exoplanet HD 209458b," *Astrophysical Journal*, Vol. 674, pp. 482–497.

Swain, M. R., Vasisht, G., and Tinetti, G. (2008b). "The presence of methane in the atmosphere of an extrasolar planet," *Nature*, Vol. 452, pp. 329–331.

Swain, M. R., Tinetti, G., Vasisht, G. *et al.* (2009a). "Water, methane, and carbon dioxide present in the dayside spectrum of the exoplanet HD 209458b," *Astrophysical Journal*, Vol. 704, pp. 1616–1621.

Swain, M. R., Vasisht, G., Tinetti, G. *et al.* (2009b). "Molecular signatures in the near-infrared dayside spectrum of HD 189733b," *Astrophysical Journal*, Vol. 690, pp. L114–L117.

Swain, M. R., Deroo, P., Griffith, C. A. *et al.* (2010). "A ground-based near-infrared emission spectrum of the exoplanet HD189733b," *Nature*, Vol. 463, pp. 637–639.

Swain, M., Deroo, P., and Vasisht, G. (2011). "NICMOS spectroscopy of HD 189733b," in *Proceedings of the International Astronomical Union*, IAU 276 symposium, Alessandro Sozzetti, Mario G. Lattanzi, and Alan P. Boss (eds.), pp. 148–153.

Tessenyi, M., Ollivier, M., Tinetti, G. *et al.* (2012) "Characterizing the atmosphere of transiting planets with a dedicated Space Telescope." *Astrophysical Journal*, Vol. 746, pp. 45–69.

Tinetti, G., Vidal-Madjar, A., Liang, M.-C. *et al.* (2007). "Water vapour in the atmosphere of a transiting extrasolar planet," *Nature*, Vol. 448, pp. 169–171.

Tinetti, G., Deroo, P., Swain, M. R. *et al.* (2010a). "Probing the terminator region atmosphere of the hot-Jupiter XO-1b with transmission spectroscopy," *Astrophysical Journal*, Vol. 712, pp. L139–L142.

Tinetti, G., Griffith, C. A., Swain, M. R. *et al.* (2010b). "Exploring extrasolar worlds: from gas giants to terrestrial habitable planets," *Faraday Discussions*, Vol. 147, pp. 369–377.

Tinetti, G. *et al.* (2011). "Exoplanet characterisation observatory," *Experimental Astronomy*, special issue, in press.

Tsiganis, K., Gomes, R., Morbidelli, A., and Levison, H. F. (2005). "Origin of the orbital architecture of the giant planets of the Solar System," *Nature*, Vol. 435, pp. 459–461.

Vidal-Madjar, A., Lecavelier des Etangs, A., Desert, J.-M. *et al.* (2003). "An extended upper atmosphere around the extrasolar planet HD209458b," *Nature*, Vol. 422, pp. 143–146.

Waldmann, I. P. (2012). "Of 'cocktail parties' and exoplanets," *Astrophysical Journal*, Vol. 747, pp. 12–29.

Waldmann, I. P., Tinetti, G., Drossart, P. *et al.* (2012). "Ground-based NIR emission spectroscopy of HD189733b," *Astrophysical Journal*, Vol. 744, pp. 35–45.

Winn, J. N., Matthews, J. M., Dawson, R. I. *et al.* (2011). "A super-Earth transiting a naked-eye star," *Astrophysical Journal Letters*, Vol. 737, pp. L18–L24.

Wordsworth, R. D., Forget, F., Selsis, F. *et al.* (2011). "Gliese 581D is the first discovered terrestrial-mass exoplanet in the habitable zone," *Astrophysical Journal Letters*, Vol. 737, pp. L48–L53.

Yurchenko, S. N., Barber, R. J., and Tennyson, J. (2011). "A variationally computed line-list for hot NH_3," *Monthly Notices of the Royal Astronomical Society*, Vol. 413, pp. 1828–1834.

14

If You Want to Talk to ET, You Must First Find ET

JILL TARTER AND CHRIS IMPEY

Introduction

In 2010, the modern scientific search for extraterrestrial intelligence was 50 years old. Its birth was marked by Frank Drake's *Project Ozma* using the 85-foot diameter Tatel radio telescope at the National Radio Astronomy Observatory in Green Bank, West Virginia (Drake 1961). *Project Ozma* observed two nearby, Sun-like stars (Tau Ceti, 12 light-years away, and Epsilon Eridani, 10.5 light-years away) for a few hundred hours and concentrated on the precise frequency at which neutral hydrogen atoms emit radiation. Since hydrogen is the most abundant element in the galaxy, Drake assumed it would be an obvious choice for any distant engineer to use for deliberate transmission (plus he could assure the observatory's management that the equipment he built, and the money he spent, could be repurposed for traditional scientific research). After 50 years of searches at optical and radio wavelengths, the results are the same as those from *Project Ozma*; no signals of unambiguous extraterrestrial origin have been found, though many signals from our own technologies have confused us.

Should we give up? Conclude that we are alone? Not yet. Though 50 years seems like a long time, it's the blink of an eye to a universe that is 13.7 billion years old, or to our Milky Way Galaxy that is 10 billion years old, or even to our star the Sun that is 4.5 billion years old. The cosmos is not only old, it is vast – we've hardly begun our search. Although not everyone agrees on just how large the task of searching for ET actually is, a pretty good numerical analogy comes from trying to answer a parallel question "Are there any fish in the oceans?" by scooping an 8-ounce glass of water from some ocean and examining it to see if any fish were collected. You

Frontiers of Astrobiology, ed. Chris Impey, Jonathan Lunine and José Funes.
Published by Cambridge University Press. © Cambridge University Press 2012.

might have caught a fish; that experiment could have worked. But if it didn't, you're not likely to be tempted to conclude that there are no fish in the oceans. Exactly the same thing holds for SETI.[1] More searching is required; as noted in the last sentence of the first paper on SETI published in the journal *Nature* in 1959, authored by Guiseppe Cocconi and Philip Morrison: "The probability of success is difficult to estimate, but if we never search, the chance of success is zero."

Earlier chapters in this book have focused on the recent and rapid growth in our knowledge of extremophiles and exoplanets. What we are learning makes the universe appear to be more biofriendly every day. However, appearances can be deceiving, and science is about what is, not what we'd like it to be. This chapter considers whether ET is likely to be out there, and if so, how should we search and what are our chances?

Drake equation

The classical formulation

Soon after *Project Ozma*, in 1961, the National Academy of Sciences convened a meeting at Green Bank to discuss the search for extraterrestrial intelligence. Frank Drake recalls (2003):

> As I planned the meeting, I realized a few day[s] ahead of time we needed an agenda. And so I wrote down all the things you needed to know to predict how hard it's going to be to detect extraterrestrial life. And looking at them it became pretty evident that if you multiplied all these together, you got a number, N, which is the number of detectable civilizations in our galaxy. This, of course, was aimed at the radio search, and not to search for primordial or primitive life forms.

This meeting established SETI as a scientific discipline.

Drake has been clear about the fact that his equation represents a "container" for ignorance rather than a rigorous formalism. But it has achieved iconic status in the scientific community and beyond, and it still serves to focus attention on the most important issues of SETI. The classical formulation is:

$$N = R^* \times f_p \times n_e \times f_l \times f_i \times f_c \times L.$$

In this simple equation, R^* is the average rate of star formation in our galaxy, f_p is the fraction of those stars that have planets, n_e is the mean number of habitable planets for stars that have planets, f_l is the fraction of those that develop life at some point, f_i is the fraction of those that develop intelligent species, f_c is the fraction of intelligent

---

[1] If you are skeptical (and you should be), the numbers to support this analogy can be found in the appendix to this chapter.

species that develop a civilization with the technological means (and the inclination) to communicate across interstellar distances, and L is the duration of such a civilization with that communication capability. One variation of the equation uses the connection between the number of stars in the galaxy now and the star-formation rate, $N^* = \int R^*(t)\, dt$, integrated over the age of the galaxy T_G, and assuming a constant star-formation rate, to get:

$$N = N^* \times f_p \times n_e \times f_l \times f_i \times f_c \times L/T_G.$$

The Drake equation incorporates factors that include astronomy, planetary science, biology, and sociology; it's one of the most interdisciplinary frameworks imaginable. Conceptually, it divides into the first three factors, which are subject to astronomical observation and increasing amounts of data, and the last four factors, which have no data to constrain them at the moment.

Complications and assumptions

The Drake equation can be critiqued because the result is indeterminate, but it has fuelled a lot of creative thinking. As SETI practitioners acknowledge, the current search is for technology, not intelligence, since intelligent species may not have the ability or the inclination to actively communicate beyond their planet. Beyond that, it's a search for a particular set of technologies, which we possess now, but have not in the past and which we may not use in the future. Astrobiology is moving towards the search for biomarkers in the spectra of habitable planets, and since the global alteration of a planetary atmosphere might connote an intelligent civilization, the distance between SETI and conventional astrobiology is shrinking. However, there are distinctions.

Astrobiology in general is concerned with habitability: the natural cosmic abundance, and detectability of sites that might host biology of any kind. It's implicit in the research, based on our knowledge of the Earth, that microbial biospheres are larger and more numerous than biospheres that could host intelligent civilizations. Searches for habitable extrasolar planets are mostly confined to distances of tens to several hundred light-years. SETI, on the other hand, acknowledges the potential scarcity of pen-pals, and aims for techniques that can span a substantial fraction of the Milky Way's diameter, about 100 000 light-years. These attempts at communication are targeted; inadvertent or passive communication is inefficient (Smith 2009). Earth's leakage of radio waves into space reduces below the level of the noise fluctuations in the cosmic microwave background radiation by the edge of the Solar System.

Planets are assumed to be the sites for life, but in the Solar System several plausible biological sites are moons of gas-giant planets. The internal heat and tidal heating of large moons means bodies of liquid water can be created and protected under layers of rock and ice. These "cryogenic biospheres" may include more sites for life on average than the conventional habitable zone. Extrasolar planet searches and SETI have generally concentrated on Sun-like stars. Stars more than about 1.5 solar masses are ruled out based on a requirement for a lifetime of at least a billion years to develop complex life, although that's an untested assumption. Meanwhile, the steepness of the mass function ensures that there's formally more habitable real estate around low-mass dwarfs than around Sun-like stars.

Many of the other adjustments to the Drake equation, and complications in its use, relate to the temporal aspect. Star formation in the Galaxy hasn't been constant over time, and the 11 billion year history has seen a steady build-up in the abundance of biogenic elements. The Drake equation is also overlaid by the metallicity gradient in the disk and the lower enrichment of the halo. The sparseness of heavy elements in the periphery, combined with the high supernova rate and stellar densities in the central disk and bulge, lead to the concept of a Galactic habitable zone: an expanding zone in the disk suitable for complex life (Lineweaver *et al.* 2004). However, the utility of this concept has been challenged (Prantzos 2006).

Another issue is the potential for colonization and the propagation of civilizations, each of which has a lifetime at additional sites, leading to a set of three coupled equations for N (Brin 1983). Civilizations might disappear and reappear in the lifetime of a planet and this would be a new factor in the equation. Also, planets might not be continuously habitable. Finally, Alexander Zaitsev has argued that the inclination to actually communicate, relative to the capability to communication, is a distinct and new factor.

Probabilities and predictions

The limiting cases of the estimate for N are especially intriguing. Absence of contact is consistent with humanity as the only currently communicating civilization. The nature of potential communication incorporates an interplay between the abundance of civilizations and light travel time. For example, in the case where the other factors are such that $N = L$, no "live" communication between extant civilizations is possible unless $L > 3000$ years. More generally, space may contain the ruins of long-extinct civilizations, and N is probably governed not by the mean or median L but by the tail of long-duration civilizations. If we are alone in the Galaxy, it would be a stark challenge to the otherwise successful Copernican Principle. On the other hand, if the later terms in the Drake equation are high, and technological civilizations are abundant, Fermi's question "Where are they?" is well-motivated. But it's difficult to answer since absence of evidence isn't evidence of absence and there are many viable hypotheses to explain what has often been called the "Great Silence" (Webb 2002). Regardless of the scarcity of ETs in the Milky Way, there are approximately 100 million galaxies in the observable universe, so true cosmic loneliness seems extraordinarily unlikely.

At the Green Bank meeting in 1961, Frank Drake and his colleagues settled on $R^* = 10$, $f_p = 0.5$, $n_e = 2$, $f_l = 1$, $f_i = 0.01$, $f_c = 0.01$, and $L = 10\,000$, with the result $N = 10$. He subsequently became more optimistic and nailed his colors to the mast (or at least flattened metal to his car) with a license plate reading $N = L$. Astronomical factors are subject to increasingly good constraints based on observations. Supernova decay products give an indirect handle on massive star formation, indicating $R^* = 10$ (Diehl *et al.* 2006), but the more direct census of young stellar objects gives $R^* = 1$ (Robitaille and Whitney 2010).

Estimates of the incidence of Earth-like planets have taken an enormous leap forward with the first release of data from NASA's Kepler mission. Kepler detects planets by staring at ~160 000 stars and looking for brightness dips due to transits. In early 2011, results from just over four months of data were released, including 68 Earth-like candidates and 288 "super-Earth" candidates, up to twice the Earth's size. The reliability of candidate planets based on follow-up measurements so far is 90%. Kepler excludes giants and supergiants, nearly uniformly samples main-sequence stars with effective temperature from 6000 to 9500 K, and underrepresents late F, G, and K stars from 4000 to 6000 K, although they form the bulk of the sample. Less than 2% of the monitored stars are M dwarfs cooler than 4000 K; the flux-limited sample vastly underrepresents this abundant population. Correcting for sensitivity and various other biases, the Kepler team finds no variation in the incidence rate of Earths and super-Earths with stellar temperature or mass, and uniform incidence for orbital radii of 0.15 AU for Earths and 0.35 AU for super-Earths (Borucki *et al.* 2011). Down to the level of super-Earths, the incidence rates agree with results from radial-velocity surveys (Howard *et al.* 2010). The bottom line is that 6% of the candidate stars have orbiting Earths and an additional 7% have super-Earths. These fractions are likely to increase towards the end of the mission as Kepler more fully samples the longer orbital periods and the incidence of multiple planets.

Formally, these early data imply the product $f_p \times n_e = 0.13$, lower than Drake's 1961 guess, but not by much, and the Kepler number is a hard lower bound that can only increase as the mission reaches its design goal of identifying Earth-like planets in Earth-like orbits. Habitable moons are completely missing from these estimates. It's important to remember that Earth-mass or Earth-size doesn't imply truly Earth-like in terms of habitability. Kepler data is still too sparse for a reliable estimate of the number of Earth-like planets in the habitable zones of their parent stars (where this is a traditional, and conservative, definition in terms of surface temperature in the range where liquid water can exist). The early 2011 data release included 54 planets in habitable zones, five of which are less than twice the Earth's size.

Radial-velocity surveys, which have not yet reached Earth-mass sensitivities, find about 10% of the planets to be in habitable zones. Accepting that additional factor, and extrapolating Kepler's sample to the full census of similar stars in the Milky Way – raw material of about 50 billion stars – implies about 500 million potential biological experiments (Kasting 2010). The cosmic number is a mind-boggling 10 billion billion, or 10^{19}. The temporal dimension is relevant too. Heavy element abundances built up quickly enough for there to be habitable planets where the motor of life turned over 6–7 billion years before Earth formed.

Everything therefore depends on the product $f_l \times f_i \times f_c \times L$. Unfortunately, there's no data to bring to bear on any of these factors in the Drake equation. Various commentators have argued for high or low values, plausibly and with great ingenuity. For example, "rare Earth" proponents claim that the emergence of complex or metazoan life on Earth required an improbable set of astronomical and geological circumstances (Ward and Brownlee 2000). However, countervailing views have been presented (Boss 2009). The relatively rapid emergence of life on Earth and the wide adaptation of extremophiles have been used to argue that biology on Earth was inevitable, and analogous arguments have been made for the evolutionary inevitability of brains and intelligence. On the other hand, it's possible that there are "hard steps" in the evolution towards technology and civilization, such that a "Great Filter" operates to make such an outcome very unlikely. All these arguments – whether they conclude that ET is common or rare – are tinged by anthropic bias because they depend on the history of the only living planet we know and its only space-faring species.

At the dawn of the age of science, Plato and Aristotle offered quite different ways of thinking about the universe. Plato posited ideas as the ultimate reality and argued that the universe could be understood by thought alone. Aristotle believed in an empirical approach, where understanding stemmed from observation and active inquiry. Modern science is based on Aristotle's approach. The emerging census of habitable worlds provides ample motivation for SETI. Cocconi and Morrison's words should ring in our ears – speculation and prediction is not enough, we have to look.

Strategies and search specifics

Twenty-first century technology – not magic

Any technological civilization containing ET will probably have to be older and presumably technologically more capable than we are today. That's because we are such a young technology in such an old galaxy. We've had technologies that can be used for interstellar communication for less than a hundred years. The Milky Way Galaxy is 10 billion years old and most of the stars in our corner of the Galaxy are billions of years older than the Sun (Lineweaver *et al.* 2004). So we are the new kids on the block, an emerging technology. Any civilization younger than we are cannot yet play the communication game. The probability of others being at precisely our technological level cannot be any higher than $\sim$100/10 billion. There may be no other technological civilizations out there, but any that we can detect will be older, much older. Arthur C. Clarke (1973) has suggested in his third law that "Any sufficiently advanced technology will be indistinguishable from magic." This argues that signals that we might actually detect and recognize as the hallmark of a technological civilization will be those that are deliberately broadcast for the benefit of emerging technologies. ET might have an "Institute for Ancient Instruments," tended and maintained for just this purpose. Another possibility arises from continuing to study our universe with astronomical tools of every description. We might just stumble over some unimaginable manifestation of that magical advanced technology; and in time we might conclude that what we have observed is the result of astro-engineering rather than astrophysics.

So should we sit back and just let the astronomers do their job, while we await a future in which we ourselves develop the requisite technology that will allow us finally to discover the "leakage" from routine, internal pursuits of advanced technologies? Or maybe we need to wait until we finally invent that technology with which all other advanced civilizations are conducting their interstellar commerce and communication? No, we should use the tools and technologies of the twenty-first century to explore systematically for signals intended for us (and other emerging technologies). Seth Shostak at the SETI Institute remarks "Columbus didn't wait for a 747 to cross the Atlantic." Columbus used the tools that he had, and they turned out to be adequate, barely.

What tools should we use, where should we look, and for what? Taking the last question first, the answer is signals. Now we must ask, what kind of signals? *The Cyclops Report* (Oliver and Billingham 1971) suggested these properties for the ideal information bearing-signal carriers:

1.   The energy per quantum should be minimized, other things being equal.
2.   The velocity should be as high as possible.
3.   The particles should be easy to generate, launch, and capture.
4.   The particles should not be appreciably absorbed or deflected by the interstellar medium.

Any particle with charge will follow the Galactic magnetic field lines rather than going towards a desired target. Particles with very small capture cross-sections are excluded by the third criterion. Photons traveling at the speed of light fit the bill, and the vast majority of all SETI programs have involved searching for radio and optical electromagnetic signals.

However, there are lots of photons received by astronomers studying the cosmos from observatories on Earth and in orbit. How would a SETI signal be recognized as such? If the signal is deliberately broadcast, it might have characteristics that are impossible for nature to produce and therefore be recognizable as something that was obviously engineered. Radio searches have focused on looking for frequency compression (narrowband continuous or pulsed signals – a sort of cosmic dial tone or busy signal) while optical searches depend on time compression (fast, nanosecond bursts) to distinguish optical SETI (OSETI) signals from all the other things that shine in the night sky.

It might also be possible that a deliberately broadcast signal might be constructed to look "almost" like an astrophysical emission, so that when astronomers build tools to survey the universe for different kinds of objects, this signal will be captured by an astronomical tool. It might take a long time, but eventually a graduate student or postdoc will examine the survey database and realize that there's something "odd" about one of the objects within it: a pulsar that regularly changes its period between two values, for example, or a Cepheid variable star whose precise period is advanced in order to produce a Morse-code-like message from regular-period and shortened-period variations, or a very large, non-circular object orbited around a star, that could be detected from the fine details of a light curve during a search for exoplanets that transit and periodically block the light from their host star.

The possibilities are numerous, and beyond our current technical capabilities, but perhaps not beyond theirs. We need to remind everyone who has access to observational survey data about the natural world to be thoughtful and curious about any anomalies they may uncover there – they may teach us something unexpected, about the astrophysical or technological universe.

Special places, special times, and some magic

Until the past decade, SETI has used observational tools belonging to the astronomers: radio and optical telescopes (see Tarter 2001, and the online archive at http://observations.seti.org). In most cases the observations had to compete for telescope time with other scientific programs engaged in more traditional research. The amount of time actually looking at the sky was minimal. The series of SERENDIP projects, and their augmentation by SETI@home more recently, are exceptions to the rule (Amir 2009). Because radio astronomers measure the electric field at the aperture of the telescope (they register both phase and intensity), it is possible to create multiple copies of the incoming radio signals and distribute them to different detectors on the telescope. Amplification can restore the power after splitting, so that no observing program suffers from this redistribution of incoming data.

In the mid-1970s astronomers at the University of California at Berkeley decided to use the university's 85-foot radio telescope for this sort of piggy-back observing (the proper term is commensal observations). They argued that the astronomer who had been assigned the telescope time would be choosing the direction to point on the sky and the frequency on which to observe, but they might be as reasonable choices as any other and represented a way to have nearly continuous access to the sky. A surplus 100-channel spectrometer was programmed to look for narrowband signals and SERENDIP began its program. It was supplanted by SERENDIP II, III, IV, and today version V.v operating on the seven-beam ALFA feed array at the world's largest telescope in Arecibo, Puerto Rico. The SERENDIP signal processing hardware has evolved to handle many hundreds of millions of spectral channels simultaneously, but the detection algorithm used remains quite simple and insensitive to signals that pulse on and off or happen to change their frequency rapidly in time (the result of acceleration between the transmitter and receiver, as for example the rotation of the Earth). More sophisticated, matched-filter algorithms are possible for narrowband signals, but they require significantly more computational muscle.

In 1999, the Berkeley SETI team decided to record a small portion of the frequency spectrum being observed by SERENDIP and then create small packets of data that could be downloaded by volunteers who were enlisted to donate the unused CPU cycles on their computers to analyze the data more thoroughly. SETI@home became an instant, global success, putting distributed computing into the mainstream. Today there is almost limitless computing power available on the desktops of individuals, and a large, diverse collection of scientific @home projects for interested volunteers. Yet recording technologies are still the bottleneck that restricts the frequency coverage to only a few percent of that analyzed by SERENDIP V.v on site. Radio observatories are remote by design (to protect them from the interference generated by human and commercial neighbors) and the Internet bandwidth available to most observatory sites, and Arecibo in particular, does not offer any better solution than recording and shipping the data.

One of the first optical SETI (OSETI) projects was also commensal and became possible when photodiodes began to count photons fast enough at an affordable cost. OSETI was suggested very early in the SETI game by Schwartz and Townes (1961), but had to wait for the appropriate technology until the beginning of the twenty-first century. Since optical detectors measure only intensity, or count the arrival rate of photons, splitting the data stream reduces the number of photons going into any backend detector and hence reduces its sensitivity. However, the

Harvard OSETI group realized that about 25% of the photons entering the Wyeth telescope at the Oak Ridge Observatory did not get into the slit of the Digital Speedometer exoplanet detector system, and thus were going to waste. They built an OSETI detector to capture those wasted photons. It consisted of two fast photon counters working in coincidence (both counters had to simultaneously register the arrival of photons in order to rule out false positives from high-energy particles interacting with the glass and other apparatus on the telescope).

This system observed the stars being explored for exoplanets for more than five years. Along the way it was partnered with the refurbished Fitz–Randolph telescope on the campus of Princeton University and a duplicate detector. By using GPS clocks to synchronize the observations, and requiring a detection in both photon counters at both observatories with a time differential of about 1.6 milliseconds (the light travel time between Harvard and Princeton) the false positive rate was reduced to a negligible level (Howard *et al.* 2004). OSETI programs looking at thousands of stars are also being (or have been) conducted at UC Berkeley's Leuschner Observatory, UC Santa Cruz's Lick Observatory, and the Campbelltown Rotary Observatory in New South Wales. To speed up the radio and optical searches, two new survey instruments have been purpose-built recently for SETI and OSETI.

The Allen Telescope Array has been built as a partnership between the SETI Institute and the UC Berkeley Radio Astronomy Laboratory in Northern California at the Hat Creek Radio Observatory (Welch *et al.* 2009). Funding for the technology development and first phase of construction (42 antennas out of an eventual 350 antennas) came from Paul Allen, Nathan Myhrvold, other private and corporate donations, plus the US Naval Observatory. This is the first LNSD radio array (a large number of small dishes) and is serving as a prototype demonstration of how to build future large radio telescopes (see Figure 14.1). Because an array of small dishes observes a large area of the sky at any moment, it has been possible for the ATA to extend the idea of commensal searching. Four independent data streams at selectable frequencies from 0.5 to 10 GHz are simultaneously available for observations. As long as astronomical objects of interest and SETI stellar targets can be found in the same piece of sky, multiple SETI and traditional radio astronomical research programs can be conducted simultaneously.

SETI researchers have a catalog of ~250 000 stars selected to be potentially suitable hosts for habitable planets and technologies (at least given our current state of knowledge – perhaps bias is a better term – about planet formation and habitability). This catalog should increase dramatically with the successful conclusion of the proposed ESA Gaia mission to measure the precise distances to a billion or more nearby stars. SETI's goal is to observe these target stars several times over the frequency range from 1 to 10 GHz, where the natural noise from all the astrophysical sources of emission is minimal, and to do so with frequency

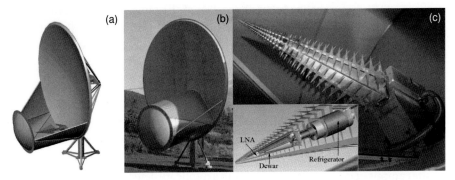

Figure 14.1 The ATA uses a clear-aperture, offset Gregorian design. (a) Schematic of the 6.1-meter offset parabolic primary, the 2.4-meter hyperbolic secondary, the metallic shroud, and the pyramidal feed/receiver. (b) Picture of ATA antenna enclosed by radome cover. (c) The cryogenic feed/receiver system with cutaway showing vacuum dewar and refrigerator (courtesy Seth Shostak).

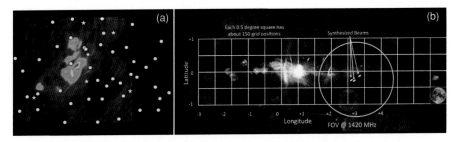

Figure 14.2 ATA Observing. (a) Fifty stellar SETI targets selected from catalogs prepared by Turnbull and Tarter (2003) are observed pair-wise using beam formers and 100 million-channel spectrometers at the same time that the neutral hydrogen gas in the M81 and M82 system is mapped using imaging correlators. (b) Twenty square degrees surrounding the galactic center are explored using a grid of 3518 pointing centers for SETI signal processors at the same time that correlation spectrometers search for transient astronomical signals.

channels as narrow as 1 Hz. (That's nine billion channels on the radio dial per star!) Beyond the catalog, there are more than 100 billion stars in the Milky Way; what do we do about those? One approach is to survey the area of the sky that contains the largest number of stars in the smallest number of square degrees, near the center of the Galaxy. Although these stars are too far away to catalog individually, a systematic search of somewhere between 4 and 10 billion stars can be accomplished by surveying a 20 square degree strip along the Galactic plane, surrounding the Galactic center (see Figure 14.2). Not only are the individual stars unknown, but any transmitter near one of them would need to be about 20 000 times stronger than Earth's strongest radio transmitter in order to be detectable, but that really isn't a very magical stretch.

Daily SETI observations at the ATA have been ongoing since the beginning of 2009 (Tarter *et al.* 2010, 2011). The observations are conducted in a pipeline process where data are collected in one stage and analyzed during the next stage – when candidate signals are detected (and thousands are detected every day) and are compared against known sources of interference. They are then checked to see if they are so constant in frequency that they must be locked up to the observatory frequency standard and therefore being generated internally by some of our computers; we also check to see if they are being detected in other beams that are pointing at other stars. Candidates that pass all these tests cause the pipeline to be interrupted and a sequence of follow-up observations to start. Only if the candidate survives all these follow-up tests do humans look at the data.

All this takes time and requires that the signals be persistent – it is why we aren't sensitive to transient, single pings. The best way to find such transients would be to build a telescope that can look at all directions on the sky at once. There are ways to do that, but currently they require far more computational speed than is available or affordable. Moore's law or a new kind of crowd-sourcing project should make this possible in the future. In the meantime, we'll continue to look for signals with a high duty cycle, i.e. when we look in the right place in the right way, we expect that they will be turned on.

When the fast detectors being used by the Harvard OSETI group began to be manufactured as pixilated arrays of photomultiplier tubes, the group designed and built a new telescope for surveying the sky utilizing this new technology. This dedicated OSETI observing facility on the site of the Oak Ridge Observatory in Massachusetts was constructed by students and faculty at Harvard University to cover the 60% of the sky visible from there. The telescope is housed in a building featuring a roll-back roof and a removable section in the south-facing wall that accommodates drift scans with only a single axis of rotation. Because the system does not require image quality optics, the 72-inch primary and 36-inch secondary mirrors have been manufactured inexpensively by fusing glass over a spherical form and then polishing. The detection system is based on eight pairs of 64-pixel Hamamatsu fast photomultipliers, and custom electronics to permit real-time, pixel by pixel comparison for coincidence detection to eliminate false positives (see Figure 14.3).

This new telescope searches for powerful laser transmitters by conducting meridian transit scans of the sky in $1.6° \times 0.2°$ strips (the telescope is parked at a particular declination and the sky turns overhead, with a dwell time on the detector, due to the Earth's rotation, of about one minute). The sky visible from that site can be scanned in approximately 150 clear nights. The first scan of the

Figure 14.3 Harvard All-Sky Optical SETI Observatory. (a) The roll back roof and removable panel in the south wall. (b) The 72-inch primary and 36-inch secondary mirrors, two of the young builders, and the photomultiplier detector array. (c) Electronics for the pixellated detector array (courtesy Paul Horowitz).

sky has been completed and 12 areas of the sky with interesting results have been followed up using the Whipple Telescope (built for detecting air showers spawned by high-energy cosmic rays) without any further confirmation (Howard *et al.* 2007). Another survey is now underway with upgraded electronics. The survey sensitivity should be adequate to detect laser pulses from the analog of a Helios-class laser being transmitted through a 10-meter telescope up to a distance of 1000 light-years. Although the system can detect a single event, multiple pulses and signals with large duty cycles are still favored in this observing mode. This observatory cannot track a particular sky position; the group is more likely to be able to make a credible case for getting other telescopes to follow up, staring in the right direction until the next time the Earth is illuminated by the transmitter, if the initial detection consisted of multiple events that cannot be explained as anything but engineering. So if we are one of many targets on a list of worlds that ET is hailing sequentially, we are more likely to discover those cosmic pings if they come in rapid bursts and don't take too long before they return.

Get the whole world involved

For the past 50 years, the number of scientists working in the SETI field has remained very small. It's a long-shot, with no guaranteed payoff, scientific papers that cite only negative results to date, and a frustrating confusion in the minds of many that equate SETI and UFOs or other forms of pseudo-science. Yet the "Are we alone?" question is such an important one that the field remains active, and huge technological progress continues to be made. In particular the existence of dedicated, and piggy-back observing facilities, plus the move away from full custom hardware detectors into the realm of software systems now offers an opportunity to let the world help with the searching.

SETI@home (Anderson *et al.* 2000) pioneered distributed, service computing over a decade ago and the world responded with great enthusiasm, leaving their computers turned on so that spare CPU cycles could explore data recorded from Arecibo Observatory. The computers have kept at the job, but the humans didn't really get continuously engaged. We'd like to change that by giving humans challenging and creative tasks to do. Through their engagement, we hope to improve our signal detection capabilities, but even if no signals are found, we also hope to change their point of view to one that embraces a more cosmic perspective capable of understanding the connectedness and sameness of all humans on planet Earth, when compared to any ETs elsewhere. "setiQuest" is an open community (Tarter *et al.* 2010, 2011) being formed as the result of Jill Tarter's 2009 TED Prize. This community will give trained and talented individuals the opportunity to improve the signal detection software used on the ATA, making it faster and capable of better recognizing a much broader class of information-bearing, noise-like signals that lack the narrowband artifacts currently being sought. For those without the specialized training, setiQuest hopes to use the phenomenal pattern recognition capabilities of their eyes and brains to find signals for which no algorithmic detectors exist, or to work through regions of the spectrum so crowded with our own technologies that the automated systems get overwhelmed and confused. We think we can make it fun, so the humans will stay involved. This is another opportunity to engage the world in doing real science. Keep an eye on setiQuest to see how we implement our goals and by all means join the quest.

The archeology of the future

In addition to recommending active SETI searches even in the face of unknown probabilities for success, the physicist Philip Morrison had a lovely way of characterizing the SETI enterprise. He used to refer to it as the archeology of the future. The archeology piece of that is easy to understand; even traveling at the speed of light any signal we detect will tell us about the transmitter's past, not their present. The future that Morrison refers to is humankind's future, and the fact that the detection of a signal (even one without any apparent information content) tells us that it is possible for that future to be long, measured in cosmic time. Signal detection will remain impossible unless technological civilizations are close enough together, in four-dimensional space and time! If, on average, technological civilizations arise, flourish, and then the transmitting technologies cease after only a short time, then in the 10 billion year history of the Milky Way Galaxy, it will be very unlikely that two or more would be co-temporal – exist close enough in time to detect each other's signals. The converse implies that any successful SETI detection informs us of the possibility to continue to survive long into the future. Morrison didn't mean to imply that a message would provide the answers to all our unsolved problems, but that problems of survival are, in fact, solvable.

What if? Planning for success

Champagne is on ice

What will we do if some day we detect a signal from ET? Most SETI researchers will do roughly the same thing. Take a deep breath to calm their excitement, then start trying to do everything possible with the tools at their observatory to try to disprove their discovery. They will have found signals before, and will be pretty good at this process. They will check to make sure it isn't some malfunction in their system, or interference caused by the observatory, local transmitters (did someone forget to turn off their cell phone?), overflying transmitters, or orbital transmitters. One way to check out the last is to determine whether the signal is coming from a fixed point on the celestial sky (using e.g. the differential Doppler signature caused by the Earth's rotation, as measured at two separated observatories, or across the extended baseline of an array), and in so doing they impose a bias that might throw away the detection of an interstellar probe that has entered our Solar System and is particularly close. Perhaps the biggest challenge for SETI is correctly discriminating between our own terrestrial and potentially extraterrestrial technologies.

Beware of hoaxes

A signal that continues to pass all the tests at the local observatory level will escalate the effort to attempting to get an independent confirmation from someone else's hardware and software. This is a crucial defense against deliberate hoaxes, and it may not be possible in many circumstances. The only other observing facilities capable of seeing that direction on the sky may not have any equipment that is suitable to the task (the large multi-channel spectrometers, or fast photon counters, or receivers for that frequency, or even adequate sensitivity). Since there are very few SETI programs, it's a good bet that another observatory won't have SETI-type detectors, but once a signal has been discovered, it may be possible to use non-SETI detectors to confirm it. Except in the case of chasing down deliberate hoaxes, this independent confirmation step hasn't yet been undertaken.

Tell the world

An independent confirmation would get things really jumping because it provides so much more credibility to the claim of detecting ET. This also seems like the milestone that will trigger toasting with the bottle of champagne we keep in the refrigerator at the ATA. If independent confirmation isn't possible, or if it fails, a lot more deliberation and work at the discovery site will be required and the champagne will have to wait. This confirmation step also starts a clock ticking down the moments over which the discovery team might have any hope of containing the news. You've told one person (observatory director, telescope operator, or a good buddy with access to the sky) a secret. It will not stay a secret very much longer. Issuing an *IAU Telegram* (a historically named process that has been updated for the Internet) through the Central Bureau of Astronomical Telegrams run by the International Astronomical Union will send an email alert to all observatories around the world that subscribe, so they can begin to use their tools – whatever they might be – to search for other manifestations from this same direction.

There is no guarantee that the signal will continue, and at any particular observatory it is likely to set below the horizon, so global sites are needed for continuous surveillance. There could in fact be many different types of signals associated with a deliberate broadcast, thus many instruments make sense. No doubt the discovery team will want to stake a scientific claim for this detection, so a paper will be submitted very rapidly to some journal that does not demand an embargo until the date of publication; such an embargo would probably not be possible to enforce. Finally, it will be time to schedule a press conference and tell the world. All data from the discovery team and any other observatories involved must be made available for detailed scrutiny. Care should be taken to ensure that credit is given to everyone involved. If the discovery was the result of philanthropic funding (as almost all SETI programs are today), some discrete "thank you" calls to the visionary sponsors will no doubt precede public announcement.

Your guess is as good as mine

After the announcement happens, how will the world react? In part the reaction will depend on the nature of the discovery and information available about the signal. Your guess is as good as anyone's. At least for a short while, it is likely that this news will capture the world's attention. The world's media representatives will not all be able to contact the discovery team individually, and this is another reason for sending out the *IAU Telegram*. Astronomers are the most likely group of humans to understand the nature and details of the discovery, and they can serve as local resources for their media reps, trying to make the reporting as scientifically accurate as possible. But humans have a limited attention span, and especially if there is no immediate mechanism for extracting information from the signal, something else will replace the detection of ET on the world's home pages. However, over time, the detection will cause a profound paradigm shift in how we see ourselves, completing the Copernican revolution that began by taking us out of the center of the Solar System. We will know ourselves to be one of many intelligent species created by the laws of physics and chemistry in this universe.

One, two, infinity

The discovery team and everyone else with relevant observing capabilities will want to continue collecting information on this signal. And of course, they will want to look for others, because once you have found another technological civilization, you know there will be many. Number two is the crucial number in this exploratory science, the game-changer, after which nothing is ever the same again.

Appendix: Sampling the cosmic ocean

Start by assuming that electromagnetic signals (radio and optical) are indeed the correct manifestation of extraterrestrial technologies for which to search. If we are wrong about that, then all bets are off, and we've done almost no relevant searching. The search volume for signals is at least nine-dimensional (three spatial dimensions, time, two polarizations, frequency, modulation, required sensitivity). There are $\sim 10^{11}$ stars in the Milky Way, and not all of them will have planetary systems, but some that do probably will have more than one habitable planet, so let's take 10^{11} stars as a good measure of the three spatial dimensions to be searched (see below). Radio searches have typically covered both polarizations, so that isn't a factor. In the radio spectrum there are nine billion different 1-Hz wide channels within the 1–10 GHz terrestrial microwave window, so roughly $10^{11+10} = 10^{21}$ combinations of star–Hz to search.

How many different ways of embedding information into signals (modulation schemes) are there? Modern radio searches have mostly looked for circularly polarized, narrowband signals because these propagate nearly undistorted over the long distances that must be traversed through the interstellar medium. But such signals are information-poor and can be masked by local RFI. Communication theory suggests that with adequate computer power it would be possible to search through a large number of modulation schemes for more complex signals with a large number of degrees of freedom, primarily very broadband signals. It might also be necessary to correct for the interstellar distortion (mainly dispersion and scattering) that broadband signals suffer. As a rough guess, the requirement is perhaps a search of 100 different modulation schemes and 100 different dispersion measures, or scattering coherence times, for a total of 10^4 trials.

That brings us to something like 10^{25} different star–Hz–modulations for seven of the nine dimensions (time and sensitivity remain). A 10-year search program run by the SETI Institute and the first two years of observing with the purpose-built Allen Telescope Array (ATA) in Northern California have examined a total of 2×10^{12} star–Hz–modulations or about 2×10^{-13} of the whole job.

What about required sensitivity and time? We'd like to be able to detect the equivalent of our own twenty-first century technology anywhere in the Milky Way Galaxy. If ET is using technologies that we haven't yet invented, then these numbers are meaningless, and we'll just have to survive as a technological species long enough to invent those technologies ourselves. The current Galactic center survey being conducted with the ATA requires a transmitter 20 000 times as strong as our current most powerful transmitter (the Arecibo planetary radar with an effective isotropic radiated power of 2×10^{13} watts), so we could use another factor of $\sim 10^4$ in sensitivity, meaning that our searches to date are only 10^{-17} of the eight-dimensional portion of the nine-dimensional volume, and we haven't yet factored in time.

The duty cycle of any signal (the fraction of the time the signal is turned on and aimed in our direction) is unknown, but if it is a deliberate signal it should be fairly high because the sender wishes it to be received. To date all our searches have required that this duty cycle be unity so that it is detectable with one or a few looks and persists long enough for confirmation. That is to say we have had little or no sensitivity to transient signals, or to signals that rapidly undergo strong fading due to scattering in the interstellar medium. However, this is a small factor compared to the temporal issue of longevity; arbitrarily we set the duty cycle to 1/2. During the 10 billion year history of the Milky Way, the technology has to be transmitting at an epoch when we are searching. The probability of this is crudely $L/10^{10}$ years, where L is the longevity of the transmitter.

Although terrestrial history indicates that technologies outlast the civilizations of the technologists who invented them, L is often thought of as the longevity of the technological civilization, and we know nothing about it. We use our own example to set a lower limit of 100 years. However, in our Galactic neighborhood, most stars are older than the Sun with an average age of ~ 7 billion years. If elsewhere the evolutionary time-scale to reach technological capability is comparable to our own four billion years, then L could be as long as 3×10^9 years. Using these two limits means that the fraction of the nine-dimensional volume searched to date lies between 5×10^{-26} and 1.5×10^{-18}, with a geometric mean of 2.5×10^{-22}. The Earth's oceans hold 1.4×10^{18} m^3 of water, or 6×10^{21} eight-ounce glasses of water. So our search of the nine-dimensional cosmic ocean to date is equivalent to sampling about 1.6 glasses of water from the Earth's oceans. Not much.

References

Anderson, D., Werthimer, D., Cobb, J. et al. (2000). "SETI@home: internet distributed computing for SETI," in *Bioastronomy 99: A New Era in the Search for Life*, Proceedings of a Conference held in Hawaii, ASP Conference Series, Vol. 213, G. A. Lemarchand and K. J. Meech (eds.).

Amir, A. (2009). "SERENDIP takes a great leap forward," Planetary Society website, available at: www.planetary.org/programs/projects/setiathome/setiathome_20090526.html.

Borucki, W. J. *et al.* (2011). "Characteristics of planetary candidates observed by Kepler, II: analysis of the first four months of data," *Astrophysical Journal*, Vol. 736, pp. 19–40.

Boss, A. P. (2009). *The Crowded Universe: The Search for Living Planets*. New York: Basic Books.

Brin, G. D. (1983). "The great silence – the controversy concerning extraterrestrial intelligent life," *Quarterly Journal of the Royal Astronomical Society*, Vol. 24, pp. 283–309.

Clarke, A. C. (1973). *Profiles of the Future: An Inquiry into the Limits of the Possible*. New York: Henry Holt.

Cocconi, G. and Morrison, P. (1959). "Searching for interstellar communications," *Nature*, Vol. 184, pp. 844–846.

Diehl, R. *et al.* (2006). "Radioactive ^{26}Al from massive stars in the galaxy," *Nature*, Vol. 439, pp. 45–47.

Drake, F. D. (1961). "Project Ozma," *Physics Today*, Vol. 14, p. 140–143.

Drake, F. D. (2003). "The Drake equation revisited. Part I," *Astrobiology Magazine* website, available at: www.astrobio.net.

Howard, A. W., Horowitz, P., Wilkinson, D. T. *et al.* (2004). "Search for nanosecond optical pulses from nearby solar-type stars," *Astrophysical Journal*, Vol. 613, pp. 1270–1284.

Howard, A. W., Horowitz, P., Mead, C. *et al.* (2007). "Initial results from Harvard all-sky optical SETI," *Acta Astronautica*, Vol. 61, pp. 78–87.

Howard, A. W. *et al.* (2010). "The occurrence and mass distribution of close-in Super-Earths, Neptunes, and Jupiters," *Science*, Vol. 330, pp. 653–655.

Kasting, J. (2010). *How to Find a Habitable Planet*. Princeton, NJ: Princeton University Press.

Lineweaver, C. H., Fenner, Y., and Gibson, B. K. (2004). "The galactic habitable zone and the age distribution of complex life in the Milky Way," *Science*, Vol. 303, pp. 59–62.

Oliver, B. M. and Billingham, J. B. (1971). "Project Cyclops: a design study of a system for detecting extraterrestrial intelligent life," NASA Technical Report CR-1144445.

Prantzos, N. (2006). "On the galactic habitable zone," invited talk at "Strategies for Life Detection," ISSI Bern, astro-ph/0612316.

Robitaille, T. P. and Whitney, R. A. (2010). "The present-day star formation rate of the Milky Way determined from Spitzer-detected young stellar objects," *Astrophysical Journal Letters*, Vol. 710, pp. 11–15.

Schwartz, R. N. and Townes, C. H. (1961). "Interstellar and interplanetary communications by optical masers," *Nature,* Vol. 190, pp. 205–208.

Smith, R. D. (2009). "Broadcasting but not receiving: density dependence considerations for SETI signals," *International Journal of Astrobiology*, Vol. 8, pp. 101–105.

Tarter, J. C. (2001). "The search for extraterrestrial intelligence (SETI)," *Annual Reviews of Astronomy and Astrophysics*, Vol. 39, pp. 511–548.

Tarter, J. C., Agrawal, A., Ackermann, R. *et al.* (2010). "SETI turns 50: five decades of progress in the search for extraterrestrial intelligence," *Proceedings of SPIE*, Vol. 7819, 781902.

Tarter, J. C., Ackermann, R., Barott, W. *et al.* (2011). "The first SETI observations with the Allen Telescope Array," *Acta Astronautica*, Vol. 68, pp. 340–346.

Turnbull, M. C. and Tarter, J. C. (2003). "Target selection for SETI. II. Tycho-2 dwarfs, old open clusters, and the nearest 100 stars," *Astrophysical Journal Supplement*, Vol. 149, pp. 423–436.

Ward, P. D. and Brownlee, D. (2000). *Rare Earth: Why Complex Life is Rare in the Universe*. New York: Copernicus Books.

Webb, S. (2002). *If the Universe is Teeming with Aliens . . . Where is Everybody?* New York: Copernicus Books.

Welch, W. J. *et al.* (2009). "The Allen Telescope Array: the first widefield, panchromatic, snapshot radio camera for radio astronomy and SETI," *Proceedings of the IEEE*, Vol. 97, pp. 1438–1447.

Index

Page numbers in *italics* refer to figures and tables.

Acasta Gneiss, 52
accretion heating, 205
acetylene, 189–90
Achaean era, 188
Acidiphilium spp, 160
Acidithiobacillus ferrooxidans,
    160
Acidobacteria, 160
Acinetobacter, 160
Actinobacteria, 160, 164
adenosine triphosphate (ATP),
    49, *50*, 136
Aeromonas, 160
aerosols (tholins), on Titan,
    187, 188, 190, 192,
    194
Africa, 53, 55, 95, 100, *107,*
    119, *119*, 126
Akilia island, 92, 95, 108
albedo, 121*n*, 242, 270
ALFA feed array, 294
algae, 122, 165
    photosynthetic, 160
ALH84001 (Martian
    meteorite), 7, 17, 35–7, *35*
alkalinity, 8
Allan Hills Meteorite
    (ALH84001), 7, 17, 35–7,
    35
Allen, Paul, 295

Allen Telescope Array (ATA),
    18, 295–7, *296*, 299, 301,
    302–3
Alpha Centauri B, 234
altimeters, 182, 183, 185
Altman, S., 62
aluminum, 205
Amazonian (cold and dry)
    Period, 158, 172
    Earth analogs for, 162–5
amino acids, *49*
    as biomarker, 25
    in meteorites, 64
    synthesis of, 12, 49, 59
ammonia, 120, 125, 192–3,
    205, 214, 216–17, 219
ammonia hydrates, 204, 208
anaerobic metabolism, 123–5,
    141
analysis, in life theory
    formation, 28, 31
Andes Mountains, 163, 164
animals:
    foundation for evolution of,
    135, *143*, 144
    macroscopic, 143–5
Antarctica, 162, 173, 179
    extreme environment of,
    32, 48, 161
    Mars meteorite in, 35–7

anthropomorphism, 266, 291
antibiotics, in search for
    non-standard life, 39
anti-greenhouse effect, 188–9
aragonite, 64
Archaea (domain), 12, *29*, *57*,
    94, 122*n*, 123, 133, 160,
    162, 164
    metabolism in, 49, 58–9
Archean Eon, 12, *53*, 54–5, 97,
    118, 135, 159
    evolution of habitability in,
    115–28
    fossils from, 102–3, 108
    late, 134
Arecibo, Puerto Rico,
    observatory at, 294, 299,
    303
Ariel, *203*, 212
Aristotle, 291
Arthrobacter, 164
asteroid belt, 90, 205, 213,
    221
asteroids:
    icy, 203–9
    ingredients for habitability
    of, 204–9
    missions to, 203–4
    position and temperature
    of, *204*

property table for, *203*
water on, 84, 201–9, 213–16
astrobiology:
 aims of, 9, 18
 breakthroughs in, 8–9
 as discipline, 3
 habitability as goal of, 288
 history of, 6–9
 and human culture, 18
 and humility, 18–19
 new synthesis in, 5–19
 and origin of life, 11–13
 questions proposed by, 6, 9
 SETI and, 288
 study week on, 3–4
 use of term, 6, 9
astrometry planet search
 technique, 235
astronomers, astronomy,
 historical, 5, 6, 254–5
Atacama Desert, as Mars
 analog, 163–5, *163*
atmosphere:
 blurring effects of, 254, 271
 as dynamic, 168
 erosion of, 237
 of Europa, 184
 of exoplanets, 242–4,
  266–80
 of Mars, 158, 160–1, 165–73,
  170
 spectroscopic analysis of,
  268
 of Titan, 185, *186*, 187–9,
  192, 194
 of transiting super-Earths,
  243
atmosphere, Earth, 52, 63, 89,
 109, 115, 185, 239–40
 effect of early life on, 123–5
 before GOE, 122–3
 oxygen stabilization and
  regulations in, 136–8
 postbiotic, *124*

prebiotic, *124*
 and redox state, 122–8
 vs. Mars, 168
 weakly reduced, 123
 see also Great Oxidation
  Event
atmospheric erosion, 237
Australia, 55, 92, 95, 100, 115,
 159
 cherts in, *98*, *105*, *106*, 125
autotrophic metabolic
 pathways, 12
autotrophs (self-feeding), 48
Axel Heiberg Island, as Mars
 analog, 161, 162
Azua-Bustos, Armando,
 157–73

Bacillus, 164
bacteria:
 fossil record of, 132
 in Mars analogs, 162–4
 metabolism in, 58–9, 60
 microaerophilic, 137
 in Proterozoic oceans,
  138–9, 141
Bacteria (domain), 12, *57*,
 122*n*
 metabolism in, 49
banded iron formations (BIF),
 55, 56, 121, 135–6
Barberton Greenstone Belt, 53,
 95, 99–100, 101, 104, 105,
 107, 108, 109
Baross, John, 5–19
Beacon Valley, as Mars analog,
 161, 162, 165
Beagle, HMS, 5, 17
Benner, Steven, 25–45
Benner laboratory, 43
Benz, Willy, 73–84
benzene, 187, 189, 190, 191
Berkner, L. V., 7
bilateralism, 145–6

biochemistry, unity of, 11
bio-cosmological principle,
 266
biology, advances in, 9
biomarkers, 17
biosignature gasses, 239–41
biosignatures, 239–42
biosynthesis, 189
Black Cloud (character), 27–8
black shales, biomarkers in,
 126–7
Blanc, Michel, 175–97, 202
"boring billion," 135, 139
bottom-up approach, 11, 12,
 239
brain, 145
Brevibacillus, 164
brown dwarfs, 251
brucite, 64, 214
Buick, R., 135

Calamarians (characters), 27–8
calcium–aluminum
 inclusions, 205
California, University of, at
 Berkeley:
 Radio Astronomy
  Laboratory at, 295
 telescope at, 294–5
Callisto, 176, 178, *178*, 192,
 195, 250
Cambrian Period, 116
Campbelltown Rotary
 Observatory, 295
Canada:
 glaciation in, 119
 Mars analogs in, 159, 161
 old rocks in, 52, 90, 92
canali, 157
Canfield, Donald, 136, 138
carbohydrates, 25
carbon:
 isotopes, 10, 102–3, 104,
  139

carbon (*cont.*)
  as life essential, 10, 13, 89,
    100–1, 214, 241
  in microbial metabolism,
    54–5, 67, 96–7
  organic, 121, 139, 142
  in respiration, 121–2
carbonaceous "snowflakes,"
    120
carbon dioxide, 13, 244, *271,
    272,* 274
  as carbon source, 12
  in Earth's atmosphere, 91,
    120–1, 139–40, *267*
  in Mars atmosphere, 165,
    168–71, 172
carbon fixation, 54, 67
Carboniferous Era, 159
Cassini Composite Infrared
    Spectrometer, 191
Cassini–Huygens mission, 8,
    185, 186, *187,* 189, 191,
    192, 195–6, 212, 220
Cassini Imaging Science
    Subsystem (ISS), 187
Cassini Infra–Red
    Spectrometer, 212
Cassini Ion and Neutral Mass
    Spectrometer (INMS), 187,
    191, 212, 220
Cassini Radio Science
    Subsystem, 193
Cassini/VIMS instrument, 189,
    193
Castillo-Rogez, Julie, 201–22
catalysis, as life essential, 50
catalysts, in metabolism,
    60–7, *65, 66*
caves, as Mars analogs, 164–5
Cech, T. R., 62
cells:
  formation of early, 94,
    100–1
  as foundation of life, 36

Cell Theory of Life, 36
cellular signatures, 10, 100–1
Cellulomonas, 164
Cenozoic epoch, 96
Centaurs, 216
Central Bureau of
    Astronomical Telegrams,
    301
Central Park, 117
Ceres, 218, 219, 220–1
  habitability potential for,
    202, 203, 203, 213, 214,
    215, 216
Charon, 201, 216, 218–19,
    221–2
  habitability potential of,
    202, 203, 218, 219–22
chasmolithic organisms, 103
chemical signatures, 10
chemical systems, as life
    essential, 27
chemistry:
  origins paradox in, 40–1
  pre-biotic, 8, 39–41, 63–5
chemolithotrophs, 96–7, *97,*
    100, 104–5, 109, 161
chemoorganotrophs, 97, 100,
    101, 105, 109
cherts, oxygen isotopes in,
    116–17
"chicken and egg" problem,
    40, 127
China, *142, 144*
Chlorella, 160
CHNOPS (key elements for
    life), 180
Christalline Entity (character),
    27
chromium, 135
Chroococcidiopsis, 164
Chryse Planitia, 162–3
Chyba, C. F., 10, 63
citric acid cycle (TCA), 59
Clark, R. N., 189

Clarke, Arthur C., 18, 292
Class III habitats, 176, *176,*
    180, 196
  Europa as potential, 180–1
Class IV habitats, 176, *176,*
    177, 196
clathrate hydrates, 204–5,
    212, 214, 219, 221
clays, as catalysts in origin of
    life, 13, 64
Cleland, C. E., 10
climate:
  in Archean Eon, 116–22,
    125
  of Mars, 169, 172–3
  in Proterozoic, 139–41
Clinton, Bill, 8
Clostridium, 160
clouds, 242, 274
cnidarians, 145
coal beds, 51
Coastal Mountain Range, 163,
    164
cobalt, 13
Cocconi, Giuseppe, 287
collisions:
  with Earth, 90
  of landmasses, 52
  in planet formation, 74–6,
    83
Colour Peak, 161
comets, 90
  in asteroid belt, 213
  as building blocks for life,
    11
commensal observations,
    294–5
common ancestry, 31, 32, 39
complementarity, 41, 43–4, *42*
conduction, 207–8
contamination, interstellar, 7
continents, 95
convection, 207–9, *215*
  mobile-lid, 212

Copernican Principle,
    Copernican revolution,
    16, 245–6, 289, 301
Copernicus, Nicolaus, 245
Copley, Shelley, 48–67
copper, 13
corals, 145
coronographs, 245, 256, 270
corotation (zeroth order)
    resonance, 78–9
CoRoT (CNES) missions, 197,
    254, 262
CoRoT-7b (planet), 254, 262,
    269
Cosmic Vision Plan, 195
Coustenis, Athena, 175–97,
    202
crater-counting, 183
crater relaxation, 210
cratons, 52
    formation of, 133–4
Crick, F. H. C., 41, *42*, 43
crust, Earth's, 52, 91–2
cryogenic biospheres, 289
cryovolcanism, 190–1, 193,
    209, 214, 217, 219
crystals, oldest, 92
Cyanidium, 160, 164
cyanobacteria, 125, 126, 127,
    133, 134, 141, *142*, 162,
    164
Cyclops Report, The, 292

Darwin, Charles:
    voyage of, 5–6, 17, 18
    see also evolution,
        Darwinian; natural
        selection
Darwin mission, 245
Davies, Paul, 25–45
Dawn Misson, 220–1
Death Star, 76
Debengda Formation, *142*
Decadal Survey, 194

De Duve, C., 12
deep hot biosphere, 38
Deinococcus radiodurans, 99
deoxyribonucleic acid (DNA),
    11, 62, 122
    complementarity in, 41,
        43–4
    GACTZP, 44–5
    in LUCA, 58–9
    as molecular genetic
        system, 26, 27, 29
    polymerases, 44
    and protein production, 40
    sequencing, 30
    stability of, 38
    synthetic, 41, 43–4
Derry, L. A., 142
Desch, S. J., 217, 219
deserts:
    life in, 164, 172
    on Mars, 172
    oldest, 163
Desulfosporosinus, 160
deuterium, 251
diagenesis, 116
diamictite, 117
differentiation, 108
diffusion-limited rate, 123
Digital Speedometer
    exoplanet detector
    system, 295
dinitrogen, 185, 186
Dione, *202, 203*
direct imaging, 16, 235,
    242–3, 245, 254–6, 264,
    267
diseases, 31
dissolved inorganic carbon
    (DIC), 54
Doppler method, 16, 182, 197,
    233–4, 252, 264, 272, 300
double helix, 41, 43–4
Doushantuo Formation, *142,*
    144

Drake, Frank, 286, 287, 290
Drake equation, 287–91
    complications and
        assumptions of, 288–9
    formula for, 287–8
    probabilities and
        predictions of, 289–91
drop stones, 117
Drossart, P., *273*
Dunaliella, 160, 165
Dune (Herbert), 172
dust, as building blocks for
    life, 11
dust analyzer, 183
dwarf planets, 213, 216,
    218–19, 269

Earth:
    as appropriate setting for
        life, 89–110, *216,* 232, 291
    biological "Renaissance" of,
        141–6
    climate history of, 116–22
    composition of, 73
    concept of habitability on,
        92, 94
    early environment of,
        89–110, *93, 97*
    evolution of habitability on,
        115–28
    as evolving planet, 132–47
    expanding the definition of,
        235–7
    extraterrestrial material
        imported to, 63–4, 67, 90
    first billion years of, 48–67
    geological "Dark Ages"
        period of, 133–4
    geological "Middle Ages"
        period of, 134–41, 147
    geological time scale of, *118*
    life on, *see* terrestrial life
    Mars analogs on, 157–65
    -Mars system, 37

Earth (*cont.*)
  metabolism on, 48–67
  microbial habitat of, 94–9,
    98
  perceived as exclusive
    domain of life, 5
  spectra of, *240, 267*
  Titan compared to, 15,
    185–6, 188–9, 190–1,
    193–4, 196
  topography of, 51–2
  uniqueness of, 52
  Venus vs., 232
Earth-like planets, 8, 9, 231–7,
    262
  candidates for, 238–9
  definition of, 233, *233*
  detection techniques for,
    233–5
  finding biosignatures on,
    239–45
  incidence of, 238–9, 290
  nomenclature associated
    with, 232
Earth twins (Earth analogs),
    233, 244–5, 262
eclipses, 16, *234*
Eemian Ice Drilling Project,
    161, 162
Eemian Period, 162
EJSM mission (Europa Jupiter
    system mission)
    (proposed), 194–5
electricity, in origin of life, 63
electromagnetic fossil
    radiation, 266
elongation factors (EFs), 31
embryos, fossilized, 143–5
embryos (planetary), 74–6
  growth of planets from, 76,
    83
Enceladus, 179–80, 193, 196,
    201, 203, *203,* 210, *213,*
    220

geysers on, 8
South Polar Terrains (SPTs)
    of, 209, 212
endolithic organisma, 103
energy sources:
  for building biomass, 48–9
  chemical, 6, 10, *181*
  on Earth, 6
  as life essential, 10, 15, 89,
    94, 101, 175, 184, 205–8
  light as, 6, 10, 48
  for microbial metabolism,
    54–5
  oxidation as, 48
environment:
  of early Earth, 89–99
  as setting for early Earth
    life, 12, 13, 99–100
  see also extreme
    environments
enzymes, in progenote, 61–2
Eocene Era, 161
epilithic organisms, 103
EPOXI mission (NASA), *240*
Epsilon Eridani, 286
Equinox Mission, 220
Eris, *203, 217*
erosion, 102, 133
  of Mars atmosphere, 171–2
error catastrophe, 61
ESA (European Space Agency),
    186, 194–6, 269, 278,
    295
ESPRESSO instrument, 264
etching, 102–3
ethane, 15, 241
eubacteria, 31
Euglena, 160
Eukarya, eukaryotes, 12, *57,*
    122, 132, 139, 141–3, *142*
  cell formation in, 134
  differentiation in, 146
  in Mars analogs, 162, 164
  metabolism in, 49, 58–9

Europa, *177,* 218, 250
  age of, 183–4
  atmosphere of, 184
  characterizing
    environments of, 185
  habitability potential of,
    176–7, *178,* 179–85, *181,*
    201, 202, 215, *216, 217*
  as large, icy satellite, 202
  mapping seafloor of, 182–3
  as potential Class III habitat,
    180–1
  proposed future mission to,
    8, 181–5, 194–5
  searching for biosignatures
    on, 185
  subsurface ocean on, 8,
    181–5
  surface composition and
    chemistry, 183
  surface/exosphere/
    magnetosphere
    interactions of, 184–5
  surface morphology and
    dynamics of, 183–4
  Titan compared to, 194
European Extremely Large
    Telescope, 264
European Southern
    Observatory, *253,* 264
European Space Agency (ESA),
    186, 194–6, 269, 278, 295
euxinia, 136
evolution:
  cosmic, 9
  Darwinian, 10, 12, 17, 26–8,
    32, 39–40, 41, 43–4
  human, 291
  non-Darwinian chemistry
    in, 39–41
  in onset of life, 11, 12
  reproduction vs., 26–7
  theory of, 6, 30
exobiology, 7

Exoplanet Characterization
    Observatory (EChO), 278,
    279
Exoplanet Data Explorer, *258,*
    261
Exoplanet Orbit Database, *258,*
    261
exoplanet revolution, 268–9
exoplanets (extrasolar
    planets), 81, 82, 256, *258,*
    261
    anticipated developments
        in search for, 263–4,
        277–80
    atmospheres of, 242–4,
        266–80, *273*
    detection of, 8, 15–16, 76,
        250, 251–6, *253*
    distant, massive, 257
    distribution of parameters
        of, 258–9
    diversity of, 256–60
    eclipses on, *234*
    impediments in discovery
        of, 254, 258, 259, 260,
        263–4
    low mass, 260–4
    molecular signatures in
        spectra of, 273–7
    most Earth-like, 262
    multi-planet systems of,
        260–2
    number of, 15, 16, 252
    past, present, and future
        facilities in science of, *277*
    preliminary statistics for,
        262–3
    review of known, 250–64
    searches for habitable,
        231–46, 250–64
    and terrestrial planet
        formation, 84
    thermal and physical
        nature of, 254

exopolysaccharides (EPS), 160
Expedition Fiord, 161
exploration:
    and example of Earth, *29,*
        37–9
    and example of Mars, 32–7
    in life theory formation, 28,
        30, 32–9
external occulter, 245
extrasolar planets, *see*
    exoplanets
extraterrestrial intelligence,
    search for, 3, 17–18
extraterrestrial life (ET):
    communication with, 6
    definitive signs of, 7, 16–18
    extinct, 289
    hypothetical non-water
        based, 189, 190
    intelligent, 291
    potential sites for, 14–16,
        122, 288–9
    primitive, 25–45
    as radically different from
        terrestrial life, 6, 7, 9–10,
        14–15, 19, 175, 189–90,
        266–8
    signals from, 292–3, 300–3
    terrestrial life as model for,
        10, 14, 19, 30–1, 146–7,
        233
extraterrestrial life (ET),
    search for:
    astrobiological basis of, 9
    attempted communication
        in, 16, 286–303
    challenges in, 17
    compared to Darwin's
        voyage, 5–6
    Drake equation in, *see* Drake
        Equation
    funding of, 8
    null results of, 18, 286,
        289–90

origins of terrestrial life in,
    5, 9, 14
philosophical implications
    of, 6, 18–19
potential for colonization
    in, 289
SETI in, *see* Search for
    Extraterrestrial
    Intelligence (SETI)
social implications of, 6, 18,
    301–2
strategies and search
    specifics in, 291–302
tools for, 292–3
extreme environments, 6, 9,
    32, 94
    desert, 164, 172
    exploration approaches for,
        38–9
    of microbes, 48
    on moons, 193
    non-standard life in, 38
    terrestrial adaptation to,
        146
extremophiles, 287
    limits of, 38
    terrestrial, 6, 9, 14, 38, 291
    water and, 15

Faint Young Sun, 119–22,
    119–22, 158, *158,* 169, *170*
Fairén, A. G., 158, 217
Fennel, K., 127
fermentation, 125
Fermi, Enrico, 289
55 Cnc e (planet), 269
51 Peg (star), 15, 252, 257, 269
Firmicutes, 160, 164
Fischer Tropsch, 96
Fitz-Randolph telescope, 295
Flagstaff, Ariz., astrobiology
    symposium at, 7
"follow the energy" principle,
    10

fossils, 30, *105*, *142*
  animal, 51
  biosignatures in, 103–4,
    127, 141, 143–5
  chemical components of,
    100
  eukarotic, 142–3
  evidence of metabolism in,
    100
  extraction and
    identification of, 102–3
  forests, 161
  fuel, 121*n*
  history recorded in, 132–3,
    159
  living, 39
  microbial, 53, 57–9, 67,
    99–104
  microorganic, 7, 35–6
  mineral preservation
    process of, 101–2, 132
  molecular, 57–9
  molluscan, 116
  physical structure of, 100–1
  trace, 132
fractals, 120
fractionation, 10
  "mass-dependent" vs.
    "mass-independent," 135
Frank, E. A., 204
frequency comb, 234
frequency compression, 293
freshwater environments, 141
Funes, José Gabriel, 4
fungal hyphae, *104*

GACTZP synthetic genetic
  system, 44–5
Gaia mission (ESA), 269, 295
Gaidos, Eric, 132–47
galaxies, 231, 246
  number of, 290
  see also Milky Way galaxy
Galilei, Galileo, 250

Galileo spacecraft, 8, 176, 180,
  182, 183
Gammaproteobacteria, 160,
  161
Gamma Ray and Neutron
  Diffraction instrument,
  221
Ganymede, 176, 178, *178*, 179,
  192, 194–5, 250
Gas Chromatograph and Mass
  Spectrometer (GCMS), 34,
  35, 188, 191
gas giant planets, 16, 254, 257,
  261, 269
  formation of, 74, 76, *77*,
    79–83
  habitability on moons of,
    175–97, *176*
  low-mass planets vs., 263
  moons of, 288–9
  in rocky planet formation,
    84
  see also hot Jupiters; *specific*
    planets
Gaucher, E. A., 31
genes:
  ancestral, 31
  of LUCA, 59–60
  in progenote, 60–1
  sequencing, 31
"genes first" hypothesis, 64
genetic code:
  in DNA, 12
  proto-ribosomes in, 13
  in RNA world, 11
genomes:
  microbial metabolism
    recorded in, 51–67
  replication errors in, 61
geological record, 30, 51–6,
  90–1, 92, 94, 96, 100, 102,
  109, 115–19, 132–3, 135
"gestaltian" issue, 10
Giant Magellan Telescope, 264

Gilmour, I., 219
GJ 436b (planet), 270
GJ 581d (planet), as potentially
  habitable, 269, 278
GJ 1214b (super-Earth), 269,
  272
glacial pavements, 117, 119
glacial till, 117
glaciation, *118*
  on Mars, 205
  Paleoproterozoic, 117, 119,
    125
  Precambrian, 127–8
  Proterozoic, 139, 140, 145
glaciers, rock record found in,
  117, 119
Gliese 436b, 260, 261
Gliese 581, 262
Gliese 1214, 261
Global Positioning Systems
  (GPS), 295
glycolysis, 49
gneisses, 91, 92, 95, 108
Goldblatt, C., 123*n*
granites, 52
gravitational microlensing,
  235
gravity:
  in planet formation, 75–6,
    79, *80*, 84
  in planet migration, 78–80
Great Oxydation Event (GOE),
  55–6
  atmospheric composition
    before, 122–8
  competing hypotheses for,
    126–8, 134
  consequencers for life of,
    134–6
"great silence," 18, 290
Green Bank, W.Va., 286, 287,
  290
greenhouse effect, on Mars,
  165–7, 168–70, 173

greenhouse environment, 117, 120, 121*n*, 140, 160–1

greenhouse gases, 120–1, 123*n*, 125–6, 139–40, *170*, 241

in GOE, 126

Greenland:
ice drilling in, 161, 162
old rocks in, 52, 90, 92, 95, 100, 101, 103, *104*

Ground Penetrating Radar (GPR), 182

gypsum, 164

Gypsum Hill, 161

habitability, 209–19
criteria for, 175, 262, 269
of Earth, 92, *93*, 94
of small bodies, 201–22
use of term, 92

habitable planets:
definition of, 232, 233
formation of, 83–4
SETI's catalog of, 295–7

habitable zones (HZ), 8, 175, 233, 235–7, *236*, 253, 289, 290
definition of, 233
super-Earths close to, 262

Hadean Eon, 9, 52, *53*, 102, 108, *118*

Haldane, J. B. S., 63

Halley's Comet, 245

Hamamatsu fast photomultipliers, 297

Hand, K. P., 180

HARPS (ESO) survey, 197, 252, *253*, 263–4

Harvard All-Sky Optical SETI Observatory, *298*

Harvard University, OSETI at, 295, 297

Hat Creek Radio Observatory, 295

HAT-P-7b (planet), *270*

haze:
organic, 120, 125
on Titan, 188–9

HD 10180 (star), 261

HD 40307 (star), 261

HD 69830 (star), 261

HD 97658b (planet), 278

HD 189733b (planet), *238, 243,* 271, 272–4

HD 209458b (hot Jupiter), 254, *272, 274, 275*

heat retention and transfer, 207–8

heavy metals, 159–60

helium:
-hydrogen fusion, 120, 250–1
in planet formation, 74, 253

hematite, 160

Herbert, Frank, 172

Hesperian (volcanic) Period, 158, 167, 172
Earth analogs for, 160–2

heterotrophs ("other"-feeding), 48, 164

high-resolution cameras, 184

Hill sphere, 79

Himalayas, 52

HNOPS (essential elements), 89, 94

Hooke, Robert, 36

hopanes, 101

horseshoe orbits, 78

hot Jupiters:
atmosphere of, *276*
density of, 257, 274, 277
detection of, 243–4, 254, 257, 258–9, 269, 271, *272,* 274
formation of, 81–2, 257, 269
orbits of, 81–3

Hoyle, Fred, 27

HR 8799 (star), 256, *256*

Hsieh, H., 213

Hubble Space Telescope, 8, 74, 184, 221, 242, *243*, 270, *271, 272*

human beings:
connectedness of, 299
drive toward knowledge of, 5, 17
see also terrestrial life

Huronian glacial interval, 119, 127

Huygens Aerosol Collector and Pyrolyzer (ACP), 188

Huygens Atmospheric Structure Instrument (HASI), 192

Huygens Descent Imager-Spectral Radiometer (DISR), 193

hydoxyl free radical, 127

hydrocarbons, 8, 189

hydrogen, 8, 13–14, 122, 176, 209, 286
atmospheric, 122–5, 260
in ET transmission, 286
-helium fusion, 120, 250–1
in planet formation, 74, 253
in respiration, 121–2

hydrogen bonding complementarity, 41, 43–4, *43*

hydrogen cyanide, 189

hydrogen peroxide, 180

hydrogen sulfide, 13

hydrothermal systems, 13–14, 38, 48, 54, 61, *93, 97, 106,* 117, 139, *215*
on asteroids, 205–6, 208–9
on moons, 180–1, 193, 214
in origin of life, 63–5, 67, 92, 94, 96, 99, 159

hydroxyl radicals, 122
Hymenobacter, 164

IAU Telegrams, 301
Iberian Pyritic Belt, 159–60
ice, 8
   on asteroids, 203–9
   on larger planets, 76, 79
   as Mars analog, 162, 165
   on moons, 176–7, *176, 177,
     178*
   oldest known, 162
ice ages, *see* glaciation
icebergs, 117
ice cores, 162
ice giants, *77*
   advances in detection of,
     260–3
ice house environment, 117,
   140
icy satellites, habitability of,
   176, 177, 178, 209–13
impact erosion, 171
Impey, Chris, 5–19, 286–303
Infrared Interferometer
   Spectrometer (IRIS), *267*
infrared light spectra, 180,
   238, 240, 272–4, *275*
Infrared Space Observatory,
   191
INMS (Cassini Ion and Neutral
   Mass Spectrometer), 187,
   191, 212, 220
intelligence, extraterrestrial,
   3, 17–18, 25
intelligence, technological,
   145, 146
intelligent design, 40
interferometers, 245, 256, *267*
International Astronomical
   Union, 301
Internet, 301
invertebrates, marine, 137
ion mass spectrometers, 103

IRIS (Infrared Interferometer
   Spectrometer), *267*
iron, 13, 89, 125, 139, 141–2,
   145, 159–60, 205
iron silicates, on Mars, 8
isotopic fractionation, 133,
   171, 191
ISS (Cassini Imaging Science
   Subsystem), 187
Isua Greenstone Belt, 92, 95,
   104, 108, 126
Itsaq Gneiss Complex, 52

James Webb Space Telescope
   (JWST; NASA), 17, 244–5,
   244, 264, 277
jarosite, 160
jellyfish, 145
Jewitt, D., 213
Johnson Space Center (NASA),
   35
JUICE mission (proposed), 195
Jupiter, 15, 74, 84, 90, 176,
   178, 205, 234, 250, 253,
   254, 256, *256,* 269
   magnetic field of, 184
   mass of, 81
   moons of, *specific moons*
   orbit of, 260
Jupiter Europa Orbiter (JEO),
   195

Kalahari Manganese field, 126
Kant, Immanuel, 269
Kasting, James, 115–28
Keck–HIRES team, 262
Kepler 10b (planet), 262, 269
Kepler 11d, e, and f (planets),
   269
Kepler mission (NASA), 8, 16,
   84, 197, 233*n*, 238–9, 254,
   255, 262, 263, 269, *270,*
   290
Kepler's law, 75

Kepler Space Telescope (NASA),
   234, 238
kerogen, 54, 67
Kharecha, P., 124
Kirschvink, Joseph, 115–28
Kitty's Gap Chert, *98,* 104, *105,
   106*
Klingons (characters), 28
Knoll, Andrew, 132–47
Kocuria, 164
komatiites, 96
Koshland, Daniel, 25–6, 28, 36
Kuiper Belt, 216, 221, 222

Lafleur, L. J., 6
Lajolo, Giovanni, 3–4
Lakhanda group, *142*
Laplace, Pierre Simon, 269
laser altimeter, 183, 184
Last Universal Common
   Ancestor (LUCA), 12, 14,
   145
   gene content of, 59–60
   metabolism and, 49, 51
   in microbial metabolism,
     65–7
   microbial metabolism in,
     57–60
   progenitor of, 60–2
Later Heavy Bombardment,
   108, 109
"late veneer," 90
Lederberg, Joshua, 7
Leptospirillum ferrooxidans, 160
Leuschner Observatory, 295
Levin, Gilbert, 34
Lick Observatory, 295
life:
   being alive vs., 26
   building blocks for, 11
   character of, *29*
   definition of, 9–10, 25–30,
     36, 266
   differentiated, 108, 146

diversification of, 11, 133, 135, 143–5
early evolution of, 9, 11–13, 63–4, 99–109, 135
essentials for, 10, 13–14, 15, 26–7, 33–4, 89, 90, 94, 95, 100, 241
expanded view of requirements for, 241
future of, 9
historical context of, 30–2
key elements for (CHNOPS), 180
signatures of, 10, 16, 25
survival strategies for, 94
as system, 26
terracentric view of, 28, 33, 241
theory-based definitions of, 9–10, 25–45
three domains of, 58–9
vs. thriving, 94
see also animals; extraterrestrial life; human beings; terrestrial life
life, origin of, 11–13
bottom up vs. top down approach to, 11–12
Earth as appropriate setting for, 89–110, 232, 291
effect on atmosphere of, 123–5
electricity in, 63
extraterrestrial material in, 63–4, 67, 90
mechanisms for, 64–5
metabolism-first vs. replicator-first hypothesis in, 12–13
mineral catalysts in, 12–13
in oceans, 63–4, 138–9, 141–6
settings for, 13–14, 63–4

timing of, 100, 108
two competing theories of, 63–4
light:
curves, *273*
as energy source, 6, 10, 48
infrared, 17
"Limits of Organic Life in Planetary Systems, The" (Baross et al.), 241
Lindblad resonances, 78
lipids, 101
biomarkers in, 133
lipid synthesis, 58
liquid water:
on Earth, 67, 89, 90, 92, 94, 115, 116
on Enceladus, 8, 212
on Europa, 182–3
in Hadean Eon, 52
on surface of Mars, 8
on Titan, 15
lithification, 102
lithopanspermia (transpermia), 37
lithosphere, 210
Lomagundi excursion, 139
Lost Hammer Spring, 161–2
Love number, 209
Lovis, Christophe, 250–64
Lowell, Percival, 157
Luna 3 probe, 267
Lunine, Jonathan, 201–22

McCord, T. B., 208, 213
McKay, David S., 7, 8, 35–6
McKinnon, W. B., 218
mafic rocks, 96
magic, technology compared to, 292
magma, 89–90
magnesite, 214
magnesium silicates, on Mars, 8

magnesium sulfate, 160
magnetic field:
Earth's, 96
Martian, 160–1
magnetite, 64, 121
magnetometers, 183
magnetospheric plasma, 184
Makemake, *203*
Makganyene glaciation, 119, *119*, 126, 128
manganese, 13
Manhattan, Dr. (character), 25
mantle, Earth's, 89, 90, 91
Mariner spacecraft, 267
Mars, 74
atmosphere of, 158, 160–1, 165–73, *170*, 191
attempts to detect life on, 7–8, 14, 17, 32–5, 110
Cauldron theory of, 172
conflicting views of, 158–9
crust of, 52
debate over life-detection on, 33–5, 172–3
determining habitability of, 157–73
early speculations about, 157
Earth analogs for, 162–5
–Earth system, 37
as example of life theory formation, 32–7, 38
extreme environment of, 32, 165–72
geological ages of, 157–65
human habitation on, 18
lack of plate tectonics on, 91
liquid water on, 8, 158–9, 160, 166–8, 169–71, 172–3
low gravity of, 171
meteorites from, 35–7
missions to, 267

Mars (*cont.*)
  temperature on, 165–7, *166*,
    170
  see also Viking Mars mission
"Mars as the Abode of Life"
  (Lowell), 157
Mars environmental
  simulator, 162
Mars Exploration rovers, 8
Mars Global Surveyor orbiter,
  8, 161, *166*, *240*
Mars Odyssey orbiter, 8
Mars Science Laboratory, 14
mass fraction, 209, 210, 218
mass–semi-major axis plot,
  258–9, *258*
Mayor, Michel, 252, 269
melting, 6
melt layers, on Mars, 166–7
Meridiani Planum, 160
metabolism:
  anaerobic, 123–5, 141
  extant vs. early, 65–7
  hydrogen-based, 8
  as life essential, 33–5
  proto-, 12–13
  in RNA world, 62–3
  terran, 48–67, *49*, *53*, *65*, *66*
metabolism first hypothesis,
  13, 64
metamorphism, 133, 140
meteorites, 30, 90
  amino acids in, 64
  as building blocks for life,
    11
"meter-sized barrier," 74–6
methane, 13, 15, 239, 241, *243*,
  271, 272–3, *273*, 274
  as biomarker, 17, 161, 219,
    221
  as greenhouse gas, 120, 122,
    125–6, 140
  in hydrogen-based
    metabolism, 8

  on Mars, 8
  in methanogenesis, 123–5
  on Titan, 185–7, 188–90,
    194, 195–6
methane-based life, 189–90
methanethiol, 241
methanogenesis, 123–5, 140,
  189, 240, 241
methanol, 204, 214, 219, 241
microbes:
  as alien organism on Earth,
    38–9
  in Allan Hills Meteorite,
    35–6
  in Axel Heiberg Island, 161
  early Earth environment of,
    94–9, *98*
  fossilized, 53, 57–9, 67,
    99–100
  habitats of, 48, 139
  hydrogen-based, 8
  as majority terrestrial
    species, 38
  survival strategies of, 48–51
microbially induced
  sedimentary structures
  (MISS), 53, 67
microbial mats, 53, 55, 97, *98*,
  99, 104, 105–8, *107*, 109,
  132, *142*
microbial metabolism, 48–67
  molecular fossil record of,
    57–9, 100
  rock record of, 51–6
microgradiometer, 183
microorganisms, 9, *106*
  chemoorganotrophic, 97,
    99–100, 101
  traded between Earth and
    Mars, 37
  and UV radiation, 99
mid-infrared (MIR) spectra,
  240, 273–4, *275*
migration, 78–9, 80–2

Milky Way galaxy:
  age of, 286, 292, 299, 303
  number of stars in, 296
  possible planets in, 8, 231,
    250, 252, 291
  star formation in, 84
  stars in, 73
  universe beyond, 246, 290
Miller, Stanley, 12
Miller–Urey experiment, 63,
  188
Mimas, 211
minerals, catalytic, 13–14
Minniti, Dante, 250–64
Miranda, *203*, 212
mirror life, 39
mirror soup, 39
mitochondria, 134, 141
modulation schemes, 302
Mojzsis, S. J., 204
molecular "clocks," 143
molybdenum, 13, 135, 136,
  139, 143
Moon:
  cratering on, 52
  crust of, 52
  effect on Earth of, 90
  formation of, 83, 90
  missions to, 267
moons:
  exploratory missions to,
    179
  future explorations of,
    194–7
  habitability issues for,
    176–9
  icy, 176–7
  large, habitable, 175–97
  position and temperature
    of, *204*
  property table for, *203*
  as sites for ET life, 6, 288–9
Moore's Law, 297
Morrison, Philip, 287, 299

mountain building, 52

M stars, 260, 269, *278*, 290
 habitable zones of, 236–7,
  244, 262

Mullen, G., 120

multiple striated cobble,
 117

mutations, 26–7, 44

Mycrobacterium, 160

Myhrvold, Nathan, 295

Nama Group, *144*

Namibia, *144*

Nanedi Vallis, 8

nanobacteria (nanobes), 39

NASA Astrobiology Institute
 (NAI), 7, 8, 9

National Academy of Sciences,
 287
 Space Studies Board of, 7

National Aeronautics and
 Space Administration
 (NASA), 7, 194, 196, *220*
 definition of life by, 26–8,
  41
 funding for, 8, 195
 see also specific missions

National Radio Astronomy
 Observatory, 286

natural history, 29

natural selection, 6, 12, 13, 17,
 145

Nature, 287

Naval Observatory, U.S., 295

Navarro-Gonzalez, R., 163

Near Infrared Mapping
 Spectrometer (NIMS),
 183

near-infrared (NIR) spectra,
 240, 272–4, *275*

Neptune, 15, *77,* 79, 216, 219,
 234, 253, 254, 260, 263

neural networks, 145

Newfoundland, fossils in, 145

New Horizons Mission, 218,
 221–2

Nice model, 205

nickel, 13, 89

Nimbus 3 satellite, *267*

Nimmo, F., 219

nitrate, 134

nitrite, 134

nitrogen, 55, 192, 219, 221
 isotopes, 10, 13, 159
 in Mars atmosphere, 168

Nitrospira, 160

nitrous oxide, 17, 140

Noachian (warm and wet)
 Period, 157–8, 161, 167–8,
  171–2
 Earth analogs for, 159–60

North Pole Dome, as Mars
 analog, 159, 162

nuclear fusion, 73

nucleic acids, as catalysts for
 metabolism, 62–3

nucleosynthesis, 266

nucleotides, 43–4, *43,* 49, *49,*
 65

Oak Ridge Observatory, OSETI
 telescope at, 295, 297

obliquity cycle, 167, 169, 172

observation, in life theory
 formation, 28–9, 31

oceans:
 deep, 38, 48, 109–10, 135–6
 diversification of life in,
  133, 134–5
 early, *93, 95, 96*
 early life in, 63–4, 138–9,
  141–6
 magma, 89–90
 mapping of, 182–3
 on moons, 180–5, 186,
  191–3, 194
 oxygen stabilization and
  regulation in, *53,* 136–8

planets with, 176
 Proterozoic, 135–6
 redox state of, 122
 reducing of, 52
 on small bodies, 211–12,
  214, 217–19

Odishaw, H., 7

olivine, 13, 52, 64

Oparin, A. I., 63

Opportunity rover, 160

orbital migration, 257

orbits, planetary:
 eccentric, 81–3, *207, 258,*
  259
 of exoplanets, 257, *258, 259*
 formation of, 78–80
 of gas giants, 81–3
 periods of, 253
 of rocky planets, 83

Orcas, *203*

organic acids, 8

organic carbon burial, 142–3,
 146

organic haze, 120, 125

organic solvents, 8

Origin of Species, The (Darwin),
 6

Orion Nebula, 73

OSETI (Optical Search for
 Extraterrestrial
 Intelligence), 293, 294–5,
 297

other Earths, *see* Earth-like
 planets

oxidants, 122

oxydation:
 of Earth atmosphere, 55–6,
  67, 89, 91, 97
 as energy source, 48
 in redox process, 48, 209,
  239–41
 second event of, 142–5
 see also Great Oxydation
  Event

oxygen, 13
  atmospheric, *53*, 120, 122–8,
    134, 239–41
  as biomarker, 17
  isotopes, 10, 116–17
  as life essential, 89, 107,
    145, 147, 175–6
  in ocean, 138–9, 141–4
  in photosynthesis, *see*
    photosynthesis, oxygenic
  Proterozoic rise in, 134–6,
    140–1
  rise of, 115–16
  stabilization and regulation
    of levels of, 136–8
  see also oxydation
oxygen whiffs, 127
ozone:
  atmospheric, 120, 123, *267*
  as biomarker, 17, 240–1

paleogenomics, 31–2
paleosols, 56, 121
perchlorates, 163–4
permafrost, 94, 161–2, 165
Phaeton, 215
Phanerozoic Eon, *53*, 107, *118*,
  135
Phoebe, 209, 217
Phoenix mission, 161, 163–4,
  165
phosphorus, 13
photons, in ET
  communication, 293,
  294–5, 300
photosynthesis:
  anoxygenic, 14, 97, 99–100,
    105, 106, 109, 125, 162,
    164
  oxygenic, 17, 33, 53, 55, 67,
    94, 97, *98*, 99, *107*, 109,
    122–3, 125, 126, 127–8,
    134, 136–7, 147, 242
Photosystem-II, 127, 147

phyllosilicates, 157*n*, 159
PICERAS definition of life, 26,
  28–9, 36
Pierrehumbert, Raymond,
  157–73, *170*
Pilbara Craton, 53, 54, 95, *98*,
  99, 100, 104, *105*, *106*, 108,
  109, 159
Pilcher, C. B., 241
Pinnularia, 160
placozoans, 145
Planetary Protection officer, 7
planetesimals, *202*, 213, *215*,
  221
  belt, 216–17
  growth of planets from, 76,
    83, 89
  hydrated, 83
  see also asteroids
planet–planet scattering, 82
planets:
  accretion of, 52
  beyond Solar System, *see*
    exoplanets
  definition of, *232*, 250
  dwarf, 213, 216, 218–19,
    269
  Earth-like, 290
  formation of, 8, 73–84, *77*,
    80, 251, 269
  giant vs. terrestrial, 269
  measuring mass of, 236
  measuring of, 253
  migration of, 78–84, 90, 269
  physical and chemical
    identities of, *268*
  search for ET life on, *see*
    extraterrestrial life,
    search for
  stars vs., 250–1
  transiting, 234, 243, 253–4,
    274
  see also specific planets and
    types of planets

plankton, 51, 109
plate tectonics:
  in Earth's topography, 51–2
  as energy source, 15
  origin of, 91
  in origin of life, 14, 91–2
Plato, 291
Pluto, 201, 202, *218*
  habitability potential of,
    203, 219, 221–2
  reclassification of, 269
polymerases, 58
polymers, 100
Pongola glaciation, 119, 127
Pontifical Academy of
  Sciences, 3–4
potentially habitable planet,
  definition of, 232, 233
prebiotic chemistry, *29*
predation, 134, 145
present atmospheric level
  (PAL), of oxygen, 56, 135,
  137
pressure:
  in melting, 6
  temperature as function of,
    274–5
primordial soup, 12, 30
  mirror, 39
Princeton University, 295
progenote:
  definition of, 62
  vs. LUCA, 60–2
Project Ozma, 286, 287
prokaryotes, 122*n*, 134, 160
proteins:
  catalysis, 50
  as life essential, 36–7, 38,
    40, 64
  sequencing, 31
  synthesis, 11, 58, 59, 60
  use of term, 62
Proterozoic Eon, *53*, 107, *118*,
  132–47

climate in, 139–41
life in, 138–9
oxydation in, 134–8, 140–1
Proterozoic era, 119
protocontinents, 95, 108, 109
proton probes, 103
protoplanetary disks, *80*
composition of, 76
cooling of, 78
in gas giant formation, 81, 257, 259
in planet formation, 8, 73–9, 251
in rocky planet formation, 83
proto-ribosomes, vs. replicating RNA, 13
pulsars, 15, 235
pulsar timing, 235
pyrite, 56, 136
as catalyst for organic synthesis, 13
pyroxine, 52

Q (character), 27–8
Quaternary glacial interval, 127
Queloz, Didier, 252, 269

Radial Velocity, 197
radial velocity (RV) technique, *232*, 234, 252–3, *253*, 254, 258, 260, 261, 262–4, 274, 290–1
radioisotopic decay, heat from, 205–6, *206*, 209–11, *211*, 214, 221
radio telescopes, 293
Ranger spacecraft, 267
"Rare Earth" hypothesis, 291
Raymond, Sean N., 73–84
recrystallization, 52
"red edge," 242

redox (oxydation and reduction) processes, 48, 209
searching for biosignatures through, 239–41
reef systems, 51
refractory elements, 76
regolith, 205
religion:
ET life and, 18
and science, 5
replication:
in LUCA, 60
mutations in, 26–7, 44, 61
in progenote, 60–1
replicator-first hypothesis, 13
replicators, 11, 12
reproduction:
as life essential, 26–7
replication vs., 26–7
in synthesis, 43–4
respiration, 122, 146
oxydation and global, 134, 137–8
resting stages, 145
resurfacing, 209, 210
ribonucleic acid (RNA), 31, 41, 43, 57
in LUCA, 58–9, 60
messenger, 40
as metabolic catalyst, 62–3, 66–7
primitive life as dependent on, 32, 37, 40
ribosomal, 56–9
as self-replicating, 11, 12, 13
ribose, 49
ribosomes, 31, *35*, 36–7, 38, 56–9, *57*, 60
Rich, Alex, 40
rifting, 52

Rio Tinto, 170, 172
extreme environment of, 32
as Mars analog, 159–60
"RNA first" hypothesis, 40
RNA world, 11, 12, 32, 39
Martian, 37
metabolism in, 62–3, 66–7
pre-biotic chemistry before, 63–5
Robuchon, G., 219
rock mass fraction, *211*
rocks:
basaltic vs. continental, 52
history recorded in, *see* geological record
isotopic life signatures in, 100–1
Mars analogs in, 164
oldest, 52–3, 91, 92, 100
on Titan, 193
traded between Earth and Mars, 37
volcanic, 95–6
–water interactions, 8
see also fossils; sedimentary rocks
rocky (terrestrial) planets, 16
formation of, 74, 76, *77*, 83–4
plate tectonics and size of, 91
potentially habitability of, 83, 146–7
see also Earth; Mars; Venus
Rodinia, breakup of, 135, 145
Rosing, M. T., 52–3, 95, 121
Ruiz, J., 218
Russian Space Agency (Roskosmos), 195
rybozymes, 62
Rye, R., 121

Sagan, Carl, 63, 120
salts, in melting, 6, 214

satellites, small, icy, 202, *210*, *211*

Saturn, 74, 90, 196, 253, 254, 260

moons of, 8, 15, 120, 176–8, 195–6, *202*, 206, 209, *213*, 216, 217

Schiaparelli, Giovanni, 157

Schleyden, M. J., 36

Schmidt, B. E., 216

Schmidt telescope, 238

Schwann, Theodor, 36

Schwartz, R. N., 294

science fiction, theories of life from, 27–8

Science in Space (Berkner and Odishaw), 7

scientific method:
establishment of, 5
value of errors in, 3

seafloor spreading, 14

Seager, Sara, 231–46

Search for Extraterrestrial Intelligence (SETI), 17–18, 25–6, 287, 288, 291–3
analysis of data from, 297
as archeology of the future, 299–300
catalogs of, *296*
dedicated observation sites of, 293–8
eliminating hoaxes in, 300
established as discipline, 287
funding for, 295, 301
null results of, 18
optical (OSETI), 293, 294–5, 297
projected response to ET contact through, 300–2
signals in, 292–3, 302–3
world involvement in, 298–9

sedimentary rocks:
formation of, 51
history recorded in, 132–3, 135–6, 143
microbial metabolism recorded in, 51–6, 67
oldest preserved, 53, 95
transport and deposition in, 51–2

Sedna, *203*

Segura, T. L., 172

seismometers, 182

self-sustainability, as life essential, 26, 28, 45

Sephton, M. A., 219

SERENDIP projects, 293–4

serpentine, on Mars, 8, 64

serpentinization reactions, 8, 208–9

SETI@home, 293–4, 299

setiQuest, 299

shadow biosphere, 38

Sheldon, N. D., 121

Shock, E. L., 218

Shostak, Seth, 292

Siberia, fossils in, *142*

siderite, 121

silica, 96, 102–4, *105*, *106*, 179, 180, 182–3, 194

silicates, 89, 176, 179, 180, 182–3, 194, *215*, 217

16S ribosomal RNA (rRNA), 57, 59

65 Cybele, *203*, 205, 214

size complementarity, 41, 43–4

small bodies:
future exploration of, 219–22
geophysical evolution of, *210*
habitability of, 201–22
habitable, 209–19

Snowball Earth, 119, 126, 140

snow line, 76, 83, 259

sodium chlorate, replication in, 27

solar extreme ultraviolet flux, 171

Solar System:
as atypical, 81, 259–60
formation of, 77, 205
as unique in Universe, 268–9

solar wind, 91

Sorondo, Sanchez, 4

Sotin, C., 208, 213

Soviet Union, Sputnik launched by, 7

spacecraft:
sterilization of, 7
see also specific vehicles

Spain, 159

spectrographs, 252

spectrometers, 34, 35, 103, 183, 187, 188, 191, 212, 300

Sphingomonas, 164

Spitzer Space Telescope, 270, *271*, 272

sponges, 143, 145

springs, 160, 161

Sputnik launch, 7

sputtering, 183, 184

starlight, 250
suppression of, 245, 254

stars:
convective zones of, 82–3
dead neutron, 235
dwarf, 16
exoplanets and, 250–1, 256, 259, 263–4, *263*, *256*
formation of, 73–4, 77, 289, 290
low-mass, 236–7, 244
number in Milky Way, 296

planets vs., 250–1
Sun-like, 16, 237, 238, 243,
    244, 267, 286, 289
star spots, 237
Star Trek: The Next Generation,
    27–8
Star Wars, 76
Strelley Pool Formation, 53–4,
    108
Streptomyces coelicolor, 60
stromatolites, 53–4, 55, 67,
    100, 107–8, 109, 132, *142,*
    159
strontium, weathering
    measured in, 140
subduction, 52, 133
sublimation, 184, 214
subsiding basins, 52
sulfate, 134, 136, 138, 140,
    157*n*
sulfide, 56, 63–4, 125, 136,
    140–1, 142, 145
    oxidation, 138–9
sulfur, 13, 14, 121*n*, 135, 141,
    168
    isotopes, 10, 143, 159
    in microbial metabolism,
        54–5, 56
sulfur cycle, on Mars,
    169–70
Summons, Roger, 48–67
Sun, 219
    age of, 286
    energy production of, 119
    Faint Young, *170*
    heat from, 206, 207
sunlight:
    as energy source, 107, 125
    increased, 140
    in oxydation, 55, 99
    suppression of, 245
supercontinents, breakup of,
    135, 140

super-Earths (telluric planets),
    15, 91, 158, 196–7, 236–7,
    243, 253, 259, *261, 278*
    close to habitable zone, 262
    as common in other
        systems, 269
    definition of, 234
    detection of, 260–3, 264,
        265, 278, 290
    discovery of, 234
    simulated spectrum of, *244*
supernovas, 73, 290
superoxide free radical, 127
synchrotron radiation, 103
synthesis:
    of life from scratch, 41–5
    in life theory formation, 29
    of new life forms, *29, 30*
synthetic biology, 41–5

talc, 64
Tarter, Jill, 286–303, *296*
    TED prize of, 299
Tatel radio telescope, 286
Tau Ceti, 286
technology:
    advances in, 30, 252–3,
        254–6, 263–4, 267–8,
        277–80
    limits of, 17, 18, 29–30
    in search for ET, 17–18, 288,
        291–8, 301–3
    see also specific technologies
tectonism, 51–2, 133, 136,
    145, 159
    on small bodies, 209, 210,
        212
telescopes, 8, 16, 17, 74, 182,
    184, 202, *220,* 221, 233,
    234, 238, 242–5, 250,
    254–6, 264, 267, 270–1,
    277, 298
    of SETI, 293–8

telluric planets, *see*
    super-Earths
temperature:
    in Archean Eon, 116,
        117–22, 125
    condensation, 76
    effect of greenhouse gases
        on, 120–1
    extreme, 6, 38
    as a function of pressure,
        274–5
    and global respiration, 138
    heat from Sun, 206, 207
    of low-mass stars, 236
    on Mars, 165–7, *166, 170*
    ocean, 96, 138
    plate tectonics and, 91
    in Proterozoic, 139–41
    radioisotopic decay and,
        206, 207
    on small bodies, 205–8, 209,
        212
    for sustaining life, 205–8
    10 Hygeia, *203*
terrestrial life:
    as analog for ET life, 14,
        30–1, 233, 291
    astrobiological questions
        about, 6, 9
    carbon-based, 9
    Earth as appropriate setting
        for, 89–110, *93*
    ET life radically different
        from, 6, 7, 9–10, 14–15,
        19, 175, 189–90, 266–8
    as limited to one form, 32
    metabolism in, 48–67
    molecular level of, 31
    non-standard, 37–9
    origins of, 5, 9, 40–1, 109,
        179
    reconstruction of primitive,
        32

Terrestrial Planet Finder
    mission, 245
TES instrument, *166*
TESS (Transiting Exoplanet
    Survey Satellite)
    space-based survey, 237,
    244
Tghallophyca, *142*
Tharsis, 172
Themis family, 216
thermal emission spectrum,
    275, 276
thioesters, 12
Thiomicrospira, 161
Thirty Meter Telescope, 264
tholins (aerosols) on Titan,
    187, 188, 190, 193, 194
Tibetan Plateau, 52
tidal stress, 207, 211
tides:
    heating from, 206–8, *207,*
    209, *210,* 212, *213,* 218,
    219, 220, 267, 288
    in orbital excentricities, 82,
    207
Tiger Stripes, 212
time perturbations, 235
timing discovery method, 235
Tinetti, Giovanna, 266–80
Titan, 241
    atmosphere of, 120, 185,
    187–9, 192, 194
    compared to Earth, 15,
    185–6, 188–9, 190–1,
    193–4, 196
    extreme environment of, 32
    hydrogen lakes on, 189
    interior models for, 192–4
    lakes on, 191
    as large, icy satellite, 202
    organic chemistry on,
    187–91, 196
    organic factory and habitat
    of, 185–94

organic solvent lakes on, 8
possible sub-surface ocean
    on, 191–3
as potential Class IV habitat,
    185
potential for life on, 17,
    176, 179, 185–94, 202
potential for water on, 15,
    176–7, 185–6, *187,* 190,
    191–3, *192,* 195–6, 205
proposed future missions
    to, 30, 195–6
temperature of, 189–90,
    191
tonalite–trondjemite gneisses
    (TTGs), 92
Toon, O. B., 120
top down approach, 11–12,
    239
top-of-atmosphere erosion,
    171
torques, in planetary orbits,
    78
Townes, C. H., 294
trace fossils, 132
transit photometry, 238,
    251
transit spectroscopy, 270–2
transit technique, 234, 235–7,
    236, 253–4, 257–8, 261,
    267, 270–7, *270*
transmission spectrum, *243*
translation, 60, 62
trans-Neptunian objects
    (TNOs), 201–2, 203, 206,
    209
    habitability potential of,
    216–19, *217*
    position and temperature
    of, *204*
    properties table for, *203*
transpermia
    (lithopanspermia), 37
tree of life, 57–9, *57*

triangle of habitability, 180–1,
    181
tricarboxylic acid (TCA) cycle,
    49
Triton, *203,* 218, 219
tungsten, 13
turbidites, 95
Turnbull, M. C., *296*
24–isopropylcholestane,
    143
24 Themis, 205, 214
2 Pallas, 221
    habitability potential of,
    203, 215
type I migration, 78, 80, 81
type II migration, 80, 81–2

ultramafic rocks, 96
ultraviolet (UV) radiation, 99,
    106, 120, 123, 127, 135,
    171, 180, 184, 188, 194,
    202, 237, 241
Umbriel, *202, 213*
universe:
    age of, 286
    ancient views of, 291
    galaxies in, 231, 246, 266
    humans perceived as center
    of, 5, 245, 301–2
    origin of, 9
    uniform chemical
    components of, 266
University Valley, as Mars
    analog, 163, 165
unknown biosphere, 9
Upper Eleonore Bay Group,
    142
Urey reaction, 140
Uranus, 77, 79, *202, 213,* 216,
    253
    orbit of, 260
    possible habitability of,
    212
Utopia Planitia, 162, 163

Vatican Observatory, 4
vegetation, as biosignature, 242
Venera spacecraft, 267
Venus, 74, 191
  atmosphere of, 168
  Earth vs., 232
  extreme environment of, 32
  missions to, 267
Vesta, 220
Vicuña, Rafael, 157–73
Vienna Pee Dee Belemnite (VPDB) standard, 54
Viking Mars mission (NASA), 7, 14, 30, 162–3
  life-detection experiments of, 33
  in life theory formation, 33–5
viruses, 9
volatiles, in Earth's environment, 89–90
volcanism, 51, *106*, *107*, 109, 121, 210–11
  on Europa, 182, 184
  as habitat for early terrestrial life, *93*, *97*, *98*, 104–5, 161–2
  on Mars, 158, 159, 160–2
  on moons, 180
  outgassing from, 123

rocks in, 95–6, *105*
  see also cryovolcanism
Voyager missions, 267

Waldman, I. P., 272
Walker, J. C. G., 123, 124
warm Neptunes, 270, 274
Warrawoona Group, 53, 54
Watchman, The, 25
water:
  anoxic, 141–2
  on asteroids, 83, 203–9
  fog as source of, 164
  and hot basalt, 13
  as life essential, 10, 15, 83, 89, 90, 94, 95, 115, 122, 175–9, 180, 193, 204–5, 233, 235, 241, 244
  liquid organic solvents vs., 6
  living in vs. living on, 175–6
  on Mars, 8, 158–9, 160, 166–8, 169–71, 172–3
  on moons, 176–9, *177*, *178*, 185–6, 190, 191–3, 195–6
  other liquids in place of, 241
  in photosynthesis, 55, 97, 165
  for potentially habitable planets, 83, 290

-rock interactions, 8, 95, 209
  in sediment transport, 51
  transition, 76
  see also ice; liquid water
water vapor, 120, 242, *267*, *271*, 274
Watson, James D., 41, *42*, 43–4
weakly reduced atmosphere, 123
weathering, 121, 136–7, 140
weird life, 6, 28, 30
  RNA-based, 32
Westall, Frances, 89–110
Whipple Telescope, 298
White Cliffs of Dover, 51
Woese, Carl, 60–1, 62
Wolf, E. T., 120
Wyeth telescope, 295

Xanthe Terra region, 8
XO-1b (hot Jupiter), 274

Zaitsev, Alexander, 289
zeroth order (corotation) resonance, 78–9
zinc, 13
zircon crystals:
  as nature's time capsules, 52, 115
  oldest, 92, 95
Zolotov, M. Yu., 213